A JEWISH PHILOSOPHER OF BAGHDAD

# ISLAMIC PHILOSOPHY THEOLOGY AND SCIENCE

*Texts and Studies*

EDITED BY

H. DAIBER and D. PINGREE

VOLUME LXV

# A JEWISH PHILOSOPHER OF BAGHDAD

## *'Izz al-Dawla Ibn Kammūna (d. 683/1284) and His Writings*

BY

REZA POURJAVADY AND SABINE SCHMIDTKE

BRILL

LEIDEN · BOSTON
2006

This book is printed on acid-free paper

**Library of Congress Cataloging-in-Publication Data**

ISSN 0169-8729
ISBN-13: 978-90-04-15139-0
ISBN-10: 90-04-15139-7



PRINTED IN THE NETHERLANDS

*For*

*Wilferd Madelung*

# CONTENTS

CHAPTER THREE

## EDITIONS OF SELECTED WRITINGS OF IBN KAMMŪNA

## ACKNOWLEDGEMENTS

During the preparation of this study we were supported by numerous institutions and libraries. In Turkey, we thank the Ministry of Culture and the Ministry of Foreign Affairs for allowing us to conduct research; Nevzat Kaya, the Director of the Süleymaniye and satellite libraries; and the staff of the following libraries: Bayezit Devlet, Köprülü, Nuruosmaniye and Süleymaniye. In Iran, we thank the directors and staff of the following libraries: Āstān-i Quds Library, Mashhad; Central Library of Tehran University; Majlis-i Shūrā-yi Islāmī Library, Tehran; Mahdawī Library, Tehran; Āyat Allāh Marʿashī Najafī Library, Qum; the Library of the Khānaqāh-i Aḥmadiyya, Shiraz; Sayyid ʿAlī Muḥammad Vazīrī Library, Yazd. We thank the directors and staff of the following libraries in Europe: Staatsbibliothek Preussischer Kulturbesitz, Berlin; Bayerische Staatsbibliothek, Munich; Chester Beatty Library, Dublin; Uppsala Universitetsbibliotek; the Bodleian Library, Oxford; the British Library, London; Österreichische Nationalbibliothek, Vienna; Bibliotheek der Rijksuniversiteit te Leiden. In Russia, we thank the director and staff of the Rational National Library, St. Petersburg. In Israel, we thank the director and staff of the Institute of Microfilmed Hebrew Manuscripts at the Jewish National and University Library. In the United States of America we wish to thank the director and staff of the following libraries: the Beinicke Library at Yale University; the Library of Congress, Washington. The acquisition of materials and research was supported by a grant from the German-Israeli Foundation (GIF) awarded to Sabine Schmidtke.

We wish to acknowledge our gratitude for valuable comments and suggestions on earlier drafts of this study to Camilla Adang, Bruno Chiesa, Annabel Keeler, Elizabeth Pond, Nasrollah Pourjavady, Ulrich Rudolph, Sarah Stroumsa, and particularly to Wilferd Madelung who also helped us with establishing critical editions of Ibn Kammūna's writings and to whom this volume is dedicated in gratitude. Furthermore, we are grateful for assistance in acquiring indispensible materials (manuscripts and published sources) and for moral support to Shahin Aawani, Fariba

Afkari, Abdurrahman Aliy, Ömer Mahir Alper, Mohammad Ali Amir-Moezzi, Hasan Ansari, Leila Rahimi Bahmani, Encieh Barkhah, Hans Daiber, Heidrun Eichner, Ayala Eliyahu, Frank Griffel, Omar Hamdan, Margaretha T. Heemskerk, Hamid Naji Isfahani, Qasim Kakayi, Morteza Kariminia, Nader Motallebi Kashani, Gudrun Krämer, Rita Kuczynski, Y. Tzvi Langermann, Sayyed Mahmud Marʾashi Najafi, Roxanne D. Marcotte, Mohsen Mohammadi, Ursula Müller, Sayyed Hossein Sayyed Musawi, Osman Gazi Özgüdenli, Amir Hossein Pourjavady, Ahad Faramarz Qaramaleki, Jawad Qasemi, Ahmed Al-Rahim, Ahmad Reza Rahimi-Riseh, Börte Sagaster, Abdurrahman al-Salimi, Emilie Savage-Smith, Gregor Schwarb, Fuat Sezgin, David Sklare, Tony Street, Sara Sviri, Daniel Tsadik, Muhammad Karimi Zanjani Asl, Hossein Ziai. Remaining errors are ours alone. We also thank Hans Daiber for accepting the volume in the series *Islamic Philosophy Theology and Science*, Isabel Miller for carefully editing the manuscript, and Trudy Kamperveen and Tanja Cowall at Brill for their patience in seeing it through the press.

## CHAPTER ONE

# INTRODUCTION

### 1. *Survey of Previous Scholarship*

Interest in Ibn Kammūna in Europe dates back to the 17th century, when Barthélemy d'Herbelot (1624–1695) included some of the information provided by Ḥājjī Khalīfa (or Kātib Čelebi, 1609–1657) on two of his works—his comparison of Judaism, Christianity and Islam, *Tanqīḥ al-abḥāth li-l-milal al-thalāth*, and his commentary on Ibn Sīnā's (d. 428/1037) *al-Ishārāt wa-l-tanbīhāt*—in his *Bibliothèque orientale*, published posthumously in 1697.[1] Research on Ibn Kammūna started near the end of the 19th century, and the point of departure for the first generation of scholars was a codex, held in the Berlin State Library, containing his *Tanqīḥ al-abḥāth*, together with his investigation into the differences between, and mutual accusations of, Rabbanites and Karaites.[2] Given the relevance of these two texts for Jewish studies and given the fact that the entire codex was written in Hebrew characters, it was scholars of Jewish studies who initiated the research on Ibn Kammūna. Though still in doubt about its authenticity, Hartwig Hirschfeld included an edition of the treatise on Rabbanites and Karaites in his *Arabic Chrestomathy in Hebrew Characters* published in 1892,[3] and in 1893 Leo Hirschfeld published an edition in Hebrew characters with a translation of a section

[1] Barthélemy d'Herbelot: *Bibliothèque orientale ou dictionnaire universel: contenant generalement tout ce qui regarde la connaissance des peuples de l'Orient.* Paris 1697, pp. 476, 851.—D'Herbelot was the source for the accounts on Ibn Kammūna in Johann Christoph Wolff: *Bibliotheca Hebraea.* Hamburg 1715–33, vol. 3, p. 643 no. 1337b, and in Carolus Fridericus Kochius: "Observatio nova de vero scriptore libri in 3 impostoribus mohammedano." In *Annales Academiae Juliae ex editis et manuscriptis monumentis compositi.* Julio, Augusto, September 1721, § viii [= p. 144].

[2] Ms. Or. Oct. 256; see infra, Chapter Two, Sections 1.4.1 and 1.4.2.

[3] Hartwig Hirschfeld (ed.): *Arabic Chrestomathy in Hebrew Characters. With a Glossary.* London 1892, pp. 69–103.

of the chapter on Judaism from *Tanqīḥ al-abḥāth*.[4] Little attention was paid at the time to the philosophical works known to have been written by Ibn Kammūna; the principal source was still Ḥājjī Khalīfa, who in addition to the *Tanqīḥ*[5] and the commentary on Ibn Sīnā's *Ishārāt*[6] listed the following three titles: *Tadhkira fī l-kīmiyā*,[7] a commentary on Shihāb al-Dīn al-Suhrawardī's (d. 587/1191) *Talwīḥāt*,[8] and *al-Ḥikma al-jadīda fī l-manṭiq*.[9] Several manuscripts of philosophical writings were, moreover, known to be held in Western libraries: his commentaries on the *Talwīḥāt* and the *Ishārāt*, his excerpts from Naṣīr al-Dīn al-Ṭūsī's (d. 672/1274) *Talkhīṣ al-muḥaṣṣal*, and a treatise on the pre-eternity of the soul.[10] The latter was incidentally the only independent treatise by Ibn Kammūna on a philosophical issue to be analyzed by Ignaz Goldziher in 1896 on the basis of the manuscript in the Carlo Landberg collection, which was believed to be the only extant copy of the text.[11] By contrast, the rich manuscript holdings of writings by Ibn Kammūna housed in public and private libraries in Turkey, Iraq and Iran remained unexploited and in fact completely unknown to the scholars of the time.

---

[4] Leo Hirschfeld: *Sa'd b. Manṣūr Ibn Kammūna und seine polemische Schrift* تنقيح الأبحاث للملل الثلاث. Berlin 1893.

[5] *Kashf al-ẓunūn ʿan asāmī al-kutub wa-l-funūn. Lexicon bibliographicum et encyclopaedicum a Mustafa Ben Abdallah Katib Jelebi dicto et nomine Haji Khalfa celebrato compositum* 1–7. Ed. Gustav Flügel. Leipzig 1835–58 [hereafter cited as *Kashf al-ẓunūn*], vol. 2, p. 443 no. 3672.

[6] *Kashf al-ẓunūn*, vol. 1, p. 303. See also below, Chapter Two, Section 1.1.1.

[7] *Kashf al-ẓunūn*, vol. 2, p. 271 no. 2865. See also below, Chapter Two, Section 2.6.

[8] *Kashf al-ẓunūn*, vol. 2, p. 420. See also below, Chapter Two, Section 1.1.2.

[9] *Kashf al-ẓunūn*, vol. 3, p. 102 no. 4584. See also below, Chapter Two, Section 1.3.1.

[10] Namely his *Maqāla fī anna wujūd al-nafs abadī wa-baqāʾahā sarmadī*. See infra, Chapter Two, Section 1.3.10.—For an overview of what scholars of the first generation knew about Ibn Kammūna's writings, see Leo Hirschfeld: *Ibn Kammūna*, pp. 11–13.—Max Horten pointed out the significance of Ibn Kammūna's philosophical writings for later Iranian philosophers such as Ṣadr al-Dīn al-Shīrāzī ("Mullā Ṣadrā", d. 1050/1640) and Sabzawārī (d. 1295 or 1298/1878 or 1881). See Max Horten: *Das philosophische System von Schirazi*. Straßburg 1913, *passim*; idem: "Die philosophischen und theologischen Ansichten von Lahígi (ca. 1670) nach seinem Werke ‚Die aufgehenden Sterne der Offenbarung', einem Kommentar zur Dogmatik des Ṭūsī (1273 *)." *Der Islam* 3 (1912), p. 129.

[11] Ignaz Goldziher: "Saʿd b. Manṣūr ibn Kammūna's Abhandlung über die Unvergänglichkeit der Seele." In *Festschrift zum achtzigsten Geburtstage Moritz Steinschneider's*. Leipzig 1896, pp. 110–14.

In the first decades of the 20th century, two scholars explored manuscripts of writings by Ibn Kammūna in Eastern libraries: Hellmut Ritter and Āghā Buzurg al-Ṭihrānī. In his catalogue of Istanbul manuscripts related to Shihāb al-Dīn al-Suhrawardī published in 1937 and 1939, Ritter provided information on numerous copies of Ibn Kammūna's works.[12] The holdings of libraries in Iraq (among others the Gharawiyya library in Najaf holding numerous autographs of Ibn Kammūna) and Iran were consulted by Āghā Buzurg, who included information on the writings of Ibn Kammūna in his *Dharīʿa ilā taṣānīf al-shīʿa*, which he had started to compile in 1911.[13]—The next scholar to advance our knowledge of Ibn Kammūna was Dawid Hartwig Baneth with a study published in 1925.[14] Although his point of departure was again the Berlin codex, he managed to correct some of the errors of earlier scholars. He established the authenticity of the treatise on Rabbanites and Karaites and showed that Ibn Kammūna must have lived and worked in Baghdad, and not in Egypt as Moritz Steinschneider had argued earlier.[15] This was later confirmed when the chronicle *al-Ḥawādith al-jāmiʿa wa-l-tajārib al-nāfiʿa*, containing a biographical entry on Ibn Kammūna, was published in 1932.[16] The information provided by the *Ḥawādith* was later supplemented by the entry on Ibn Kammūna in Ibn al-Fuwaṭī's *Talkhīṣ majmaʿ al-ādāb fī muʿjam al-alqāb*.[17]

---

[12] Hellmut Ritter: "Philologika IX. Die vier Suhrawardī. Ihre Werke in Stambuler Handschriften." *Der Islam* 24 (1937), pp. 270–86 [hereafter cited as Ritter I]; idem: "Philologika IX. Die vier Suhrawardī. (Fortsetzung und Schluß.)" *Der Islam* 25 (1939), pp. 35–86 [hereafter cited as Ritter II].

[13] Āghā Buzurg al-Ṭihrānī: *Al-Dharīʿa ilā taṣānīf al-shīʿa* 1–25. Beirut 1403–6/1983–86 [henceforth cited as *Dharīʿa*].

[14] Dawid Hartwig Baneth: "Ibn Kammuna." *Monatsschrift für Geschichte und Wissenschaft des Judentums* 69 (1925), pp. 295–311.

[15] Steinschneider: *Arabische Literatur*, p. 239.

[16] [Anonymous]: *Al-Ḥawādith al-jāmiʿa wa-l-tajārib al-nāfiʿa*. Ed. Muṣṭafā Jawād [as a work by Ibn al-Fuwaṭī]. Baghdad 1351/1932, p. 441; see Walter Fischel: "Arabische Quellen zur Geschichte der babylonischen Judenheit im 13. Jahrhundert." *Monatsschrift für Geschichte und Wissenschaft des Judentums* 79 [N.F. 43] (1935), pp. 302–22, esp. 319–22.

[17] Ibn al-Fuwaṭī: *Talkhīṣ majmaʿ al-ādāb fī muʿjam al-alqāb*. Ed. Muṣṭafā Jawād. Damascus 1962, pp. 159–61 (= *Majmaʿ al-ādāb fī muʿjam al-alqāb*. Ed. Muḥammad Kāẓim. Tehran 1416/1995–96, vol. 1, pp. 190–91 no. 199). See Leon Nemoy: "New Data for the Biography of Saʿd Ibn Kammūnah." *Revue des études juives* 123 (1964), pp. 507–10.

Despite these advances, scholars in the West were still unable to exploit the manuscript materials in Iran, Iraq and Turkey adequately. The focus of their attention remained the treatises located in Western libraries and known of since the 19th century, particularly those relevant to Jewish studies. Moshe Perlmann continued the work of Leo Hirschfeld by preparing a critical edition of *Tanqīḥ al-abḥāth* published in 1967,[18] to be followed in 1971 by an annotated translation into English.[19] In addition to the Berlin codex and the manuscript from the Bodleian Library in Oxford that had already been used by Leo Hirschfeld, Perlmann used a copy of the text in the Biblioteca Angelica in Rome and two manuscripts from Istanbul and Tehran.[20] Nemoy worked on the other treatise preserved in the Berlin codex, dealing with the differences between Rabbanites and Karaites, which was still believed to be the only extant copy, publishing an improved edition of the text and an annotated translation into English.[21] In addition, Nemoy continued the advances of Goldziher and worked on the treatise on the eternity of the soul on the basis of the manuscript in the Landberg collection that, again, he believed was the only extant copy.[22] Although Nemoy was aware that Ibn Kammūna had written more extensively on the

---

[18] Moshe Perlmann (ed.): *Saʿd b. Manṣūr Ibn Kammūna's Examination of the Inquiries into the Three Faiths. A Thirteenth-Century Essay in Comparative Religion.* Berkeley/Los Angeles 1967 [hereafter cited as *Tanqīḥ*].

[19] Moshe Perlmann: *Ibn Kammūna's Examination of the Three Faiths. A Thirteenth-Century Essay in Comparative Study of Religion.* Translated from the Arabic, with an Introduction and Notes. Berkeley / Los Angeles 1971.

[20] For details, see below, Chapter Two, Section 1.4.1.

[21] Leon Nemoy: "Remarks on Ibn Kammuna's Treatise on the Differences between the Rabbanites and the Karaites." [Hebrew] *Tarbiẓ* 24 (1955), pp. 343–53, vi and in *Sefer yovel mugash le-khevod Yisrael Elfenbeyn.* Ed. Yehuda L. Maimon. Jerusalem 1962, pp. 201–8; idem: "Ibn Kammunah's Treatise on the Differences between the Rabbanites and the Karaites." *Proceedings of the American Academy for Jewish Research* 36 (1968), pp. 107–65; idem: "Ibn Kammunah's Treatise on the Differences between the Rabbanites and the Karaites." *Jewish Quarterly Review* 13 (1972–73), pp. 97–135; 222–46.

[22] Leon Nemoy: *The Arabic Treatise on the Immortality of the Soul by Saʿd ibn Manṣūr ibn Kammūna (XIII Century). Facsimile Reproduction of the Only Known Manuscript (Cod. Landberg 510, fol. 58–70) in the Yale University Library. With a Bibliographical Note.* New Haven 1944; idem: "Ibn Kammūna's Treatise on the Immortality of the Soul." In *Ignace Goldziher Memorial Volume.* Eds. Samuel Löwinger, Alexander Scheiber and Joseph Somogyi. Jerusalem 1958, vol. 2, pp. 83–99; idem: "Ibn Kammunah's Treatise on the Immortality of the Soul." *Ha-Rofe ha-ʿIvri. The Hebrew Medical Journal* 1 (1962), pp. 239–213.

issues dealt with in this treatise,[23] he had no access to any of his other philosophical writings and he was unaware of the author's two other extant treatises on the soul.[24]

Only slowly did scholars follow the path taken by Ritter and Āghā Buzurg to provide information on Ibn Kammūna's writings in tracing manuscripts in Eastern collections. Carl Brockelmann took into consideration the findings of Ritter and Āghā Buzurg in the supplementary volumes to his *Geschichte der arabischen Litteratur* published between 1937 and 1942.[25] Henry Corbin consulted Ibn Kammūna's commentary on Suhrawardī's *Talwīḥāt* when preparing a critical edition of the metaphysics section of the *Talwīḥāt*.[26] Fuat Sezgin checked the Turkish libraries for writings by Ibn Kammūna, although his findings have not been published.[27] In 1958 and 1965 Muḥammad Riḍā al-Shabībī published his descriptions of autographs by Ibn Kammūna in the Khizānat al-Mashhad al-ʿAlawī (i.e., al-Khizāna al-Gharawiyya) in Najaf, which he had consulted in 1911.[28] The information provided by Shabībī and Āghā Buzurg was supplemented by the manuscript catalogues of the collection of Ḥusayn ʿAlī Maḥfūẓ who visited the

---

[23] At the beginning of the treatise, Ibn Kammūna refers in a general manner to earlier writings related to the subject (transl. Nemoy, p. 86).

[24] Namely *Maqāla fī l-taṣdīq bi-anna nafs al-insān bāqiya abadan* and *Maqāla fī anna l-nafs laysat bi-mizāj al-badan wa-lā kāʾina ʿan mizāj al-badan*; for these two treatises, see below, Chapter Two, Sections 1.3.8 and 1.3.9.

[25] Carl Brockelmann: *Geschichte der Arabischen Litteratur* 1–2. Leiden 1943–49 [henceforth cited as GAL]. *Supplementbände* 1–3. Leiden 1937–42 [henceforth cited as GALS], vol. 1, pp. 768ff; vol. 3, p. 1232; see also his *Vorwort* in GALS, vol. 1, pp. xi–xvii. Cf. also Ritter I, p. 270 n. 1: "Dieser Aufsatz hat Brockelmann für sein *Supplement* zur *Geschichte der arabischen Litteratur* im Manuskript vorgelegen. Da der betreffende Teil des *Supp.* inzwischen erschienen ist, zitiere ich nun meinerseits Brockelmann's Werk." Of the *Dharīʿa*, Brockelmann had only the first two volumes at his disposal; see Etan Kohlberg: "al- Ḏarīʿa elā Taṣānīf al-Šīʿa." In *Encyclopaedia Iranica*, vol. 7, p. 36.

[26] Particularly ms. Ahmed III 3244; cf. Shihāb al-Dīn al-Suhrawardī: *Opera Metaphysica et Mystica I*. Ed. Henry Corbin. Istanbul 1945, pp. lxivff.

[27] Professor Sezgin has kindly made his notes available to the present writers.

[28] Muḥammad Riḍā al-Shabībī: *Muʾarrikh al-ʿIrāq Ibn al-Fuwaṭī* 1–2. Baghdad 1378/1958, vol. 2, pp. 121–25; idem: *Turāthunā al-falsafī. Ḥājatuhū ilā l-naqd wa-l-tamḥīṣ*. Baghdad 1385/1965, pp. 100–9. Al-Shabībī wrote another work entitled *Falāsifat al-yahūd fī l-islām* in which he also deals with Ibn Kammūna and which perhaps contains additional information on him (cf. *Dharīʿa*, vol. 16, p. 304 no. 1337); Āghā Buzurg does not provide any bibliographical details, and it is not certain whether the book ever appeared in print.

library in 1958 and of Aḥmad al-Ḥusaynī.[29] Apart from these efforts by individual scholars, the activities in Iran and in Turkey to catalogue the holdings of public and private libraries have brought to light further manuscripts of writings by Ibn Kammūna.[30] Two Istanbul manuscripts served as the basis of an edition of his major independent work on philosophy, *al-Jadīd fī l-ḥikma*, published in 1982 in Baghdad.[31] Because of a too narrow textual basis and the careless manner in which it was produced, this publication, which can hardly be traced in any Western library, is of little use to modern studies.

Contemporary Iranian scholars have concentrated on Ibn Kammūna's philosophical writings and edited some of his works in this field. Independently of each other, Sayyid Ḥusayn Sayyid Mūsawī and Hossein Ziai together with Ahmed Alwishah edited Ibn Kammūna's commentary on Suhrawardī's *Talwīḥāt*.[32] Sayyid Mūsawī also edited his

---

[29] Ḥusayn ʿAlī Maḥfūẓ: "Fihris al-khizāna al-Gharawiyya bi-l-Najaf fī mashhad Amīr al-Muʾminīn al-Imām ʿAlī b. Abī Ṭālib ʿalayhi al-salām." *Revue de l'Institut des Manuscrits Arabes* 5 (1378/1959), pp. 23–30; Aḥmad al-Ḥusaynī: *Fihrist makhṭūṭāt al-Rawḍa al-Ḥaydarīya fī l-Najaf al-Ashraf.* Najaf 1391/1971.—Hideo Fujii and Kazumi Oguchi give two codices containing works by Ibn Kammūna as having been looted from the Kirkuk Museum (Makhṭūṭāt Dār Ṣadām) during the war in 1991, namely manuscript no. 14757 (no. 263 in the list of Fujii and Oguchi), a *majmūʿa* containing *Lubāb al-manṭiq wa-khulāṣat al-ḥikma* [by Nakhjawānī], *Talkhīṣ al-kitābayn al-madhkūrayn* [sic], and *al-Maṭālib al-muhimma min ʿilm al-kalām wa-ghayruhā*; and manuscript no. 14780 (no. 285 in the list of Fujii and Oguchi) containing an autograph by Ibn Kammūna of his *Maqāla fī anna nafs al-insān abadiyat al-wujūd*, meaning presumably his *Maqāla fī l-taṣdīq bi-anna nafs al-insān bāqiya abadan.* See Hideo Fujii and Kazumi Oguchi: *Lost Heritage. Antiquities from Iraq's Regional Museums.* Fascicle 3. Tokyo 1996, pp. 26, 27, 40, 41. That Kirkuk was the location for Ibn Kammūna's autograph seems unlikely.

[30] For manuscripts of Ibn Kammūna in Turkish libraries, cf. specifically Ramaḍān Shīshīn (Ramazan Şeşen): *Nawādir al-makhṭūṭāt al-ʿarabiyya fī maktabāt Turkiyya* 1–3. Beirut 1975, vol. 1, pp. 163–67 [henceforth cited as *Nawādir*]; idem: *Mukhtārāt min al-makhṭūṭāt al-ʿarabiyya al-nādira fī maktabāt Turkiyā.* Istanbul 1997, pp. 142–45 [henceforth cited as *Mukhtārāt*]. Both lists contain numerous errors and misidentifications.

[31] Edited with introduction by Ḥamīd Marʿīd al-Kabīsī (Baghdad 1403/1982).

[32] Ibn Kammūna: *Al-Tanqīḥāt fī sharḥ al-Talwīḥāt.* Ed. Sayyid Ḥusayn Sayyid Mūsawī. PhD thesis: Dānishgāh-i Tihrān, 1375–1376/1996–97; Ibn Kammūna: *Al-Tanqīḥāt fī sharḥ al-talwīḥāt. Refinement and commentary on Suhrawardī's Intimations. A Thirteenth Century Text on Natural Philosophy and Psychology.* Critical Edition, with Introduction and Analysis by Hossein Ziai and Ahmed Alwishah. Costa Mesa, California 2003. The edition contains only part II (physics); Hossein Ziai is currently preparing part III (metaphysics) for publication.

concise philosophical treatise, *al-Maṭālib al-muhimma min ʿilm al-ḥikma*,[33] and Insīyah Barkhʷāh did the same for Ibn Kammūna's three treatises on the soul.[34] One of the two exchanges by letter between Naṣīr al-Dīn al-Ṭūsī and Ibn Kammūna was recently published by ʿAbd Allāh Nūrānī.[35] Apart from our own contributions,[36] the most recent studies on Ibn Kammūna are by the Turkish scholar Ömer Mahir Alper on Ibn Kammūna's notion of the soul[37] and by Y. Tzvi Langermann on some aspects of his *al-Jadīd fī l-ḥikma*.[38]

---

[33] In: *Kheradnāme-ye Ṣadrā* 8 xxxii (Tābestān 1382/Rajab 1425/September 2003), pp. 64–86.

[34] Insīyah Barkhʷāh: *Taṣḥīḥ-i Risāla-yi Azaliyyat al-nafs wa-baqāʾihā taʾlīf-i Ibn Kammūna.* MA thesis: Madrasa-yi Murtaḍā Muṭahharī (Sipahsālār), 1379/2000.

[35] Naṣīr al-Dīn al-Ṭūsī: *Ajwibat al-masāʾil al-naṣīriyya mushtamil bar 20 risāla.* Ed. ʿAbd Allāh Nūrānī. Tehran 1383/2005, pp. 25–34.

[36] Sabine Schmidtke: "Studies on Saʿd b. Manṣūr Ibn Kammūna (d. 683/1284). Beginnings, Achievements, and Perspectives." *Persica. Annual of the Dutch-Iranian Society* 29 (2003), pp. 105–21; Reza Pourjavady and Sabine Schmidtke: "Quṭb al-Dīn al-Shīrāzī's (d. 710/1311) *Durrat al-Tāj* and Its Sources. (Studies on Quṭb al-Dīn al-Shīrāzī I)." *Journal Asiatique* 292 i–ii (2004), pp. 309–28.

[37] Ömer Mahir Alper: *Alkın Hazzı. İbn Kemmûne'de Bilgi Teorisi.* Istanbul 2004.

[38] Y. Tzvi Langermann: "Ibn Kammūna and the 'New Wisdom' of the Thirteenth Century." *Arabic Sciences and Philosophy* 15 (2005), pp. 277–327.—See also the various recent entries on Ibn Kammūna in Turkish, Iranian, and Western encyclopaedias: Sayyid Jaʿfar Sajjādī: "Ibn Kammūna." In *Dāʾirat al-maʿārif-i buzurg-i islāmī*, vol. 4, pp. 524–26; Aydin Topaloğlu: "Ibn Kemmûne." In *Türkiye Diyanet Vakfi Islâm Ansiklopedisi*, vol. 20, pp. 127–29. See also Ḥamīd Ḥamīd: "Naẓarī bih mīrāth-i maʿnawī-yi Ibn Kammūna. Zindigī, andīsha u shubahāt." *Īrānshināsī* 11 iii (1999), pp. 632–43. Y. Tzvi Langermann: "Ibn Kammuna (d. 1284)." In *Routledge Encyclopaedia of Philosophy*, vol. 4, pp. 621–23; Lukas Mühlethaler: "Ibn Kammuna." In *Metzler Lexikon jüdischer Philosophen. Philosophisches Denken des Judentums von der Antike bis zur Gegenwart.* Eds. Andreas B. Kilcher and Otfried Fraisse. Stuttgart / Weimar 2003, pp. 68–70.—Earlier encyclopaedic entries on Ibn Kammūna are: M. Zobel: "Ibn Kammuna, Sad ibn Mansur." In *Encyclopaedia Judaica. Das Judentum in Geschichte und Gegenwart*, vol. 8, pp. 354–55; Moshe Perlmann: "Ibn Kammūna, Saʿd b. Manṣūr." In *Encyclopaedia of Islam.* New Edition, vol. 3, p. 815; idem: "Ibn Kammūna, Saʿd ibn Manṣūr." In *Encyclopaedia Judaica*, vol. 8, pp. 1186–87; idem: "Ibn Kammūna." In *Dānishnāma-yi Īrān u Islam*, fasc. 6, pp. 808–9.—According to Muḥsin Kadīwar and Muḥammad Nūrī (*Maʾkhadh shināsī-yi ʿulūm-i ʿaqlī. Manābiʿ chāpī-yi ʿulūm-i ʿaqlī az ibtidāʾ tā 1375* 1–3. Tehran 1378/1999, vol. 1, p. 196 no. 2423) ʿAbd al-ʿAzīz ʿIzzat authored a study entitled *Ibn Kammūna. Falsafa akhlāqiyya wa-maṣādiruhā* (n.p., n.d., 244 pp.). This is not the case. The correct bibliographical details of ʿIzzat's study are: *Ibn Miskawayh. Falsafatuhū al-akhlāqiyya wa-maṣādiruhā* (Cairo 1946). See Hans Daiber: *Bibliography of Islamic Philosophy* 1–2. Leiden 1999, vol. 1, p. 503 no. 4805.

## 2. *Notes on Ibn Kammūna's Biography*

Much of ʿIzz al-Dawla Saʿd b. Manṣūr b. Saʿd b. al-Ḥasan b. Hibat Allāh Ibn Kammūna al-Baghdādī's biography can only hypothetically be reconstructed as little concrete data is available. He presumably originated from Baghdad and spent most, if not all of his life in this city. His ancestors, to whom he refers at the beginning of his writings when he gives his full name,[39] were most likely learned as well, as is suggested by Ibn Kammūna's accurate reference not only to his father (Manṣūr) and his grandfather (Saʿd), but also to his great-grandfather (al-Ḥasan), and his great great-grandfather (Hibat Allāh).

Nothing is known about Ibn Kammūna's early life and education. He was born into a Jewish family[40] and he must have received a thorough education in both Jewish and Islamic letters, as is evident from the erudition of his writings in both fields. His knowledge of Jewish literature is evident from his treatise on the differences between the Rabbanites and Karaites, although he states in the introduction that he does not consider himself an expert on the details of law.[41] The text is

---

[39] See the beginning of texts as quoted below in Chapter Two, Sections 1.1.3, 1.2.1, 1.3.1, 1.3.3, 1.3.4, 1.3.5, 1.4.1, 1.4.2, 1.5.1.

[40] On the basis of documents from the Ben Ezra Genizah in Fustat, Shlomo Dov Goitein draws attention to three further Jewish personalities with the surname "Ibn Kammūna" living in Iraq, some or all of whom may well have been ancestors of ʿIzz al-Dawla Ibn Kammūna. See his *A Mediterranean Society. Volume III: The Family*. Berkeley 1978, p. 13: "The surname Ibn Kammūna, 'Son of the Grain of Cumin [fem.]'. . . is quite another story. It appears first in the Geniza in the year 1121, in a report about messianic troubles in Baghdad, when an Abū Sahl Ibn Kammūna was received by the caliph and intervened with him for the Jewish community. Another Ibn Kammūna, clearly a prominent personality, died in an underground prison in Wāsiṭ, Iraq, in 1204/5, reported by a Muslim historian [i.e., Ibn al-Sāʿī: *al-Jāmiʿ al-mukhtaṣar*. Baghdad 1934, p. 165; R.P. & S.S.] as a noteworthy happening. A third one, bearing the title 'Pride of the Efficient [servants of the government]' [i.e. Fakhr al-kufāt; R.P. & S.S.], is referred to as a benefactor in a letter by the Gaon of Baghdad, Samuel b. Eli, dated 1206. All these as well as the famous oculist and philosopher ʿIzz al-Dawla Ibn Kammūna... might have belonged to the same family. For all of them lived in Iraq and were connected with the government in one way or another." For the first mentioned incident, involving Abū Sahl Ibn Kammūna, see also idem: "A Report on Messianic Troubles in Baghdad in 1120–21." *The Jewish Quarterly Review* 43 (1952–53), pp. 57–76.

[41] See his introduction to this tract (quoted in Chapter Two, Section 1.4.2): . . . *wa-ḥarrartu al-bayānāt wa-l-ajwiba maʿa qillat maʿrifatī bi-juzʾiyyāt al-sharīʿa*. . .

replete with Talmudic and Rabbinic references, while his main source was Judah ha-Levi's (1075–1141) *Kitāb al-Khuzarī*.[42] The numerous Hebrew terms, notes and quotations from the Hebrew Bible indicate that Ibn Kammūna mastered Hebrew. His erudition in Jewish religious literature is also evident from his *Tanqīḥ*, in which he relied not only on the *Khuzarī*, but also on Maimonides' (1138–1204) *Dalālat al-ḥaʾirīn*.[43] Presumably, Ibn Kammūna also studied Islamic literature under at least some scholars, although for the field of philosophy he states in his *Maqāla fī l-taṣdīq bi-anna nafs al-insān bāqiya abadan* that he was self-taught.[44] This is confirmed by his habit of noting down useful points (*fawāʾid*) for himself from works he was studying. Autographs of excerpts from several philosophical and scientific works by Naṣīr al-Dīn al-Ṭūsī, Najm al-Dīn Abū Bakr Muḥammad al-Nakhjawānī (d. after 626/1229), Najm al-Dīn al-Dabīrān al-Kātibī al-Qazwīnī (d. 675/1277), Muʾayyad al-Dīn al-ʿUrḍī (d. 664/1266), and Abū Rayḥān al-Bīrūnī (d. after 442/1050), written between 670/1271–72 to 675/1277, are extant.[45]

The earliest dated evidence of his life is the date of completion of his short work on philosophy, *al-Maṭālib al-muhimma*, which he finished in Shaʿbān 657/July-August 1259.[46] A few months later, in Dhu l-Ḥijja 657/November 1259, Ibn Kammūna completed his concise philosophical treatise *Taqrīb al-maḥajja wa-tahdhīb al-ḥujja*.[47] Apparently these treatises were not his first output, as in the *Maṭālib* he alludes to a treatise he wrote on the immortality of the soul.[48] It is likely that he was referring

---

42 Detailed references to his sources are provided by Nemoy in the second *apparatus criticus* of his edition ("Ibn Kammūnah's Treatise on the Differences between the Rabbanites and the Karaites," pp. 160–65 [54–59]). See also his "Contributions to the Textual Criticism of Judah ha-Levi's Kitāb al-Khazarī." *The Jewish Quarterly Review* 26 (1936), pp. 221–26.

43 Detailed references to the Jewish sources used in the *Tanqīḥ* are provided by Perlmann in his edition and translation of the work. See also Simone Rosenkranz: "Judentum, Christentum und Islam in der Sicht des Ibn Kammūna." *Judaica. Beiträge zum Verstehen des Judentums* 52 (1996), pp. 7–12.

44 Ed. Barkhʷāh, p. 115.

45 See below, Chapter Two, Sections 1.2.1 to 1.2.7.

46 See below, Chapter Two, Section 1.3.2.

47 See below, Chapter Two, Section 1.3.3.

48 At the end of his proof for the eternity of soul (*fī abadiyyat al-nafs*), Ibn Kammūna states: *wa-qad afradtu lahū maqāla dhakartu fīhā mā yumkinu īrāduhū ʿalayhi wa-ʿalā muqaddamātihī fī l-iʿtirāḍāt wa-bayyantu wajh al-jawāb ʿanhā ʿalā ghāyat al-īḍāḥ*; see Ibn Kammūna: "al-Maṭālib al-muhimma," p. 85:3–4.

here either to his *Maqāla fī l-taṣdīq bi-anna nafs al-insān bāqiya abadan,*[49] or to his *Maqāla fī anna wujūd al-nafs abadī wa-baqāʾahā sarmadī.*[50] Cross references in both writings to another work on a different aspect of the soul, *Maqāla fī anna l-nafs laysat bi-mizāj al-badan wa-lā kāʾina ʿan mizāj al-badan,*[51] suggest that the latter treatise precedes the two former *maqāla*s. Moreover, in his *Taqrīb al-maḥajja* he refers to glosses (*taʿālīq*) in which he had elaborated his thoughts on the soul.[52] None of his extant writings fits this description and it is possible that he is referring here to a work hitherto unknown.

Some time before 667/1268, Ibn Kammūna wrote glosses on the critical remarks of Kātibī on Fakhr al-Dīn al-Rāzī's (d. 606/1209) *Kitāb al-Maʿālim fī l-uṣūlayn*. As he indicated in the introduction, he was at that time in Iraq,[53] by which he presumably meant Baghdad. The *terminus ante quem* for the glosses is determined by Ibn Kammūna's commentary on Suhrawardī's *Talwīḥāt*, completed in the beginning of 667/1268, in which he refers to them.[54]

Ibn Kammūna stated in the introduction to his commentary on the *Talwīḥāt* that he had composed this work at the request of some scholars (*jamāʿa min al-sāda al-nubalāʾ wa-l-aṣḥāb al-fuḍalāʾ*).[55] He did not dedicate either the commentary, nor any of his previous works to any individual, indicating that he had no patron at the beginning of his scholarly career. It was apparently through his commentary on the *Talwīḥāt* that he established himself as a philosopher. Thereafter he seems to have been patronized by the family of the *ṣāḥib al-dīwān*, Shams al-Dīn al-Juwaynī (d. 683/1284).[56] On 28 Shaʿbān 671/19 March 1273 Ibn Kammūna completed his commentary on Ibn Sīnā's *Ishārāt* which he had

---

49 For this *maqāla*, see below, Chapter Two, Section 1.3.8.

50 For this *maqāla*, see below, Chapter Two, Section 1.3.10.

51 For this *maqāla*, see below, Chapter Two, Section 1.3.9.

52 *Wa-qad basaṭtu al-kalām fīhā wa-fī ghayrihā mimmā istanbaṭtuhū bi-fikrī fī taʿālīq ghayr hādhā l-mūjaz*. See below, Chapter Three, Text Three, p. 205.

53 Quoted below Chapter Two, Section 1.1.3.

54 *Sharḥ al-Talwīḥāt*. Ed. Mūsawī, pp. 701:10–11, 714:19–20.

55 For the text of his introduction see below, Chapter Two, Section 1.1.2.

56 On Juwaynī and his family, see Bertold Spuler: "Dj̲uwaynī, Shams al-Dīn Muḥammad b. Muḥammad." In *The Encyclopaedia of Islam*. New Edition, vol. 2, p. 607; see also Hend Gilli-Elewy: *Bagdad nach dem Sturz des Kalifats. Die Geschichte einer Provinz unter ilḫānischer Herrschaft (656–735/1258–1335)*. Berlin 2000, *passim*; George Lane: *Early Mongol rule in thirteenth-century Iran. A Persian renaissance*. London 2003, pp. 177–212.

dedicated to Sharaf al-Dīn Hārūn, one of the sons of Juwaynī, who was apparently a governor for part of Anatolia at that time.[57] On 19 Dhu l-Qaʿda 676/13 April 1278 Ibn Kammūna completed his most extensive independent treatise on philosophy, which later circulated under the titles *al-Kāshif* or *al-Jadīd fī l-ḥikma* or other similar titles, and which is dedicated to Dawlatshāh b. Sanjar al-Ṣāḥibī (fl. 724/1324).[58] Dawlatshāh was a government official who probably held a high position in Baghdad. His title, *ṣāḥibī*, suggests that he must have had good relations with the Juwaynī family. Ibn al-Fuwaṭī, who saw him in 679/1280–81 in the Niẓāmiyya *madrasa* of Baghdad, remarked that he was well versed in astronomy, mathematics, philosophy and literature and that he wrote poetry in Arabic and Persian.[59] The wording of Ibn Kammūna's introduction[60] suggests that at the time Dawlatshāh was still a novice in the study of philosophy. Ibn Kammūna also mentioned here that the work was written with the latter's encouragement, his company,

---

[57] See below, Chapter Two, Section 1.1.1.—Kātibī also dedicated a work on logic, *Jāmiʿ al-daqāʾiq fī kashf al-ḥaqāʾiq*, to Sharaf al-Dīn Hārūn.

[58] See below, Chapter Two, Section 1.3.1. The work will henceforth be cited as *al-Jadīd fī l-ḥikma*.

[59] Ibn al-Fuwaṭī has two entries on Dawlatshāh: *Majmaʿ al-ādāb*. Ed. Muḥammad Kāẓim, vol. 1, pp. 182–83 no. 185: ʿIzz al-Dīn Abu l-Faḍl Dawlatshāh b. Sanjar b. ʿAbd Allāh *al-ṣāḥibī al-adīb al-kātib*; vol. 3, pp. 225–26 no. 2521: Fakhr al-Dīn Abu l-Faḍl Hindū b. Sanjar *al-ṣāḥibī al-ḥakīm al-munajjim al-adīb*. Dawlatshāh produced three works, two of which have been published: *Ṣaḥāḥ al-ʿajam* (ed. Ghulām Ḥusayn Baygdilī. Tehran 1361/[1982]), and a work on history, *Tajārib al-salaf dar tawārīkh-i khulafāʾ u wuzarā-yi īshān* (ed. ʿAbbās Iqbāl. Tehran 1313/1934 [reprinted Tehran 1344/1965]; *Tajārib al-salaf dar tārīkh*. Taṣwīr-i nuskha-yi muʾarrakh-i 846 h.q. Ed. Amīr Sayyid Ḥasan Rawḍātī. Isfahan 1361/1982; cf. also GALS, vol. 2, p. 202), and an anthology entitled *Mawārid al-adab*, which is extant in manuscript; cf. GAL, vol. 2, p. 192; GALS, vol. 2, p. 256. In his *Tajārib al-salaf* Dawlatshāh states that he had studied under Shams al-Dīn al-Kīshī (p. 200). Furthermore he is known to have copied Kātibī's *Jāmiʿ al-daqāʾiq fī kashf al-ḥaqāʾiq* (copy dated 683/1284–85; cf. Ibn al-Fuwaṭī: *Muʿjam*, vol. 3, p. 226 n. 1) as well as Quṭb al-Dīn al-Shīrāzī's *Sharḥ Ḥikmat al-ishrāq* (ms. Turhan Valide 207; copy dated 4 Muḥarram 721/2 February 1321); he also wrote poetry praising Quṭb al-Dīn al-Shirāzī; see John Walbridge: *The Science of Mystic Lights. Qutb al-Dīn Shīrāzī and the Illuminationist Tradition in Islamic Philosophy*. Cambridge, Mass. 1992, p. 74. In addition, he copied Kātibī's and Ibn Kammūna's glosses on Fakhr al-Dīn al-Rāzī's *Maʿālim* (copy dated 11 Ramaḍān 721/4 October 1321); see ms. Esad Efendi 1932, f. 126b. Moreover, extant in manuscript form (ms. Esad Efendi 1932, f. 140a) is a note by his son Muḥammad b. Hindūshāh (dated 28–30 Rabīʿ II 743/28–30 September 1342) at the beginning of a copy of Shahrazūrī's *Sharḥ Ḥikmat al-ishrāq* stating that he studied the work with his father and providing the latter's *isnād* for the work.

[60] See infra, Chapter Two, Section 1.3.1.

and his helpful remarks, suggesting that Dawlatshāh studied with Ibn Kammūna.[61]

Sometime before Shaʿbān 678/December 1279 Ibn Kammūna was asked by a government official by the name of Shams al-Dīn Muḥammad b. al-Muʾmin al-Qazwīnī to write a philosophical work for the library of Bahāʾ al-Dīn Muḥammad (d. 678/1279), another son of Shams al-Dīn al-Juwaynī who was governor of Isfahan and the central regions of western Iran (*ʿIrāq-i ʿajam*) from 673/1274–75 until his death in Shaʿbān 678/December 1279.[62] Ibn Kammūna subsequently composed his *Kalimāt wajīza mushtamila ʿalā nukat laṭīfa fī l-ʿilm wa-l-ʿamal* (known also under the titles *al-Lumʿa al-juwayniyya fī l-ḥikma al-ʿilmiyya wa-l-ʿamaliyya* and *Risāla fī l-ʿilm wa-l-ʿamal*), which he dedicated to Bahāʾ al-Dīn.[63] The work includes various pieces of ethical and moral advice to the ruler.[64] In the introduction Ibn Kammūna mentions the fact that it was during his sojourn in Baghdad that Shams al-Dīn asked him to write the work and further that it was composed before he left for Isfahan, taking the book with him.[65]

Ibn Kammūna was in contact with many of the scholars and philosophers of his time. Twice he wrote to Naṣīr al-Dīn al-Ṭūsī seeking answers to philosophical problems. On one occasion, he asked Ṭūsī about a problem related to Kātibī's view on the issue of the createdness

---

[61] Other members of Dawlatshāh's family also held high positions at the time. His father Sayf al-Dīn Sanjar is referred to by Ibn Kammūna as "the great ruler" (*al-amīr al-kabīr*). See infra, Chapter Two, Section 1.3.1. Dawlatshāh's brother Sayf al-Dīn Maḥmūd was ruler of Kāshān, and in 674/1275 Dawlatshāh for a time succeeded his brother in that position before going to Baghdad. See Hindūshāh: *Tajārib al-salaf*, p. 301 and introduction p. v.

[62] See Waṣṣāf [ʿAbd Allāh b. Faḍl Allāh al-Shīrāzī]: *Tārīkh-i Waṣṣāf*. Lith. Bombay 1269/1853, vol. 1, pp. 60–66.

[63] See below, Chapter Two, Section 1.3.4.

[64] See particularly *jumla* two, *bāb* two, Chapters (*faṣl*) Two, Three, Four, and Five. See infra, Chapter Three, Text One, pp. 176–85. Inter alia, Ibn Kammūna advises him not to persecute religious minorities.—Moral and ethical advice to rulers is also included in Ibn Kammūna's *Ithbāt al-mabdaʾ wa-ṣifātihī wa-l-ʿamal al-muqarrib ilā llāh*; see below, Chapter Three, Text Two, pp. 194–95. Both works therefore can be considered as belonging, at least partially, to the advice literature known as "mirrors for princes" in modern scholarship. For this genre, see Patricia Crone: *Medieval Islamic Political Thought*. Edinburgh 2004, pp. 148–64.

[65] See below, Chapter Three, Text One, p. 140.

of the world (*ḥudūth al-ʿālam*), and on another occasion he requested Ṭūsī's views on seven philosophical issues.[66] Of the first exchange, only Ṭūsī's reply is available, in which he addressed Ibn Kammūna as *al-dāʿī al-mukhliṣ*, while the whereabouts of Ibn Kammūna's initial query and the subsequent letter he wrote in response to Ṭūsī's reply are lost. The second exchange is complete. In it, Ṭūsī addressed Ibn Kammūna respectfully as *al-ḥakīm al-fāḍil akyas al-aqrān wa-awḥad al-zamān*. Neither of the two exchanges is dated. The only secure *terminus ante quem* for both is Ṭūsī's death on 18 Dhu l-Ḥijja 672/25 June 1274. Moreover it is likely that the letter in which Ibn Kammūna is addressed simply as *al-dāʿī al-mukhliṣ* is the earlier of the two. Presumably it was written when Ibn Kammūna did not yet have the standing of an acknowledged philosopher. Another scholar Ibn Kammūna had contact with was Kātibī, with whom he had two exchanges of correspondence. In one of them,[67] Kātibī addressed Ibn Kammūna as *Mawlā al-ʿallāma afḍal al-ʿaṣr* and *Qudwat al-ḥukamāʾ al-muḥaqqiqīn*, and it is evident that the correspondents regarded each other as colleagues on equal standing. The correspondence further indicates that Ibn Kammūna had once participated in the *majlis* of Kātibī at which a problem of logic was discussed. Ibn Kammūna took this as a point of departure for the correspondence, consisting of three letters addressed to Kātibī and two replies by the latter, all of which are extant. The second correspondence dealing with the liar paradox is only partly extant in a brief quotation by Jalāl al-Dīn al-Dawānī.[68] Ibn Kammūna also corresponded with Maytham b. Maytham al-Baḥrānī (d. 699/1300),[69] a Twelver Shīʿī scholar teaching in Ḥilla, who also had connections with the Juwaynī family; he dedicated several of his writings to ʿAlāʾ al-Dawla ʿAṭāʾ Malik al-Juwaynī (gov. 657–81/1259–83).[70] With all three scholars it was Ibn Kammūna who took the initiative, addressing an epistle to them.

---

[66] See below, Chapter Two, Sections 1.5.1 and 1.5.2.

[67] See below, Chapter Two, Section 1.5.3 and Chapter Three, Text Six.

[68] See below, Chapter Two, Section 1.5.4.

[69] See below, Chapter Two, Section 1.5.5 and Chapter Three, Text Seven.

[70] For a biographical sketch of Maytham al-Baḥrānī, see Ali Ahmed Al-Oraibi: *Shiʿī Renaissance: A Case Study of the Theosophical School of Bahrain in the 7th/13th Century*. Diss. McGill University 1992, pp. 46ff.

Two additional exchanges, in which Ibn Kammūna is the addressee, are extant, to some degree or other. One correspondent was a certain Fakhr al-Dīn al-Kāshī,[71] about whose identity nothing is known.[72] The second was Ibn al-Fuwaṭī, as he himself reports in his *Majmaʿ al-ādāb*, quoting Ibn Kammūna's reply.[73] He adds that he had approached him in writing in 683/1284–85, i.e., shortly before Ibn Kammūna's death.

With the exception of the exchange with Ibn al-Fuwaṭī, none of the letters mentioned are dated, and with the exception of the exchange with Kātibī none of the extant letters contain any indication that Ibn Kammūna had met his correspondents. They all belonged to Naṣīr al-Dīn al-Ṭūsī's circle at Maragha (with the exception perhaps of Fakhr al-Dīn al-Kāshī). The possibility that Ibn Kammūna paid a visit to Maragha, presumably at an earlier stage of his life, should not be ruled out. Yet there is no conclusive evidence for this. In his report on Ibn Kammūna in *Majmaʿ al-ādāb*, Ibn al-Fuwaṭī, who had lived for nearly twenty years (660–79/1261–62–1280–81) at Maragha serving as a librarian, writes as if he had never met Ibn Kammūna. This suggests that Ibn Kammūna never went to Maragha or, that, if he ever had, it was before 660/1261–62.[74]

During the period in which the Juwaynī family occupied the leading bureaucratic positions in the state, Ibn Kammūna apparently held a high-ranking position in their administration. A reliable indication of this is his title, ʿIzz al-Dawla.[75] This is further supported by the fact that

---

[71] See below, Chapter Two, Section 1.5.6 and Chapter Three, Text Eight.

[72] Fakhr al-Dīn also had a correspondence with Quṭb al-Dīn al-Shīrāzī of which only the latter's reply is extant (ms. Ayasofya 4862, ff. 268a–269b).

[73] See below, Chapter Two, Section 1.5.7.

[74] *Majmaʿ al-ādāb*, vol. 1, pp. 190–91 (transl. Nemoy: "New Data," p. 508): "...He was sought by many people who wished to profit from his instruction. It so happened that I myself was prevented from meeting him personally by an illness which overtook me, but I wrote him requesting to let me have a token of his instruction, so that I might embellish my work with it. He wrote in reply, through my colleague and friend Shams al-Dīn Muḥammad b. Abī al-Rabīʿ, the calculator, known as al-Ḥashaf, in the year 683..."

[75] For the use of this and similar titles during the Ilkhānid era, see Amnon Netzer: "Rashīd al-Dīn and His Jewish Background." *Irano-Judaica III*. Jerusalem 1994, pp. 121–23. See also Fischel: "Arabische Quellen zur Geschichte der babylonischen Judenheit im 13. Jahrhundert," p. 319 n. 95.

in the introduction to his *al-Jadīd fī l-ḥikma* and elsewhere he indicates that he was engaged in some professional activity which was in complete contrast to his scholarship.[76] Moreover, there is no evidence that he was ever a regular teacher at one of the *madrasa*s in Baghdad, and we do not know of any students of his, apart perhaps from the above-mentioned Dawlatshāh. He may have given some informal philosophical lectures, particularly near the end of his life. In some manuscripts of his works he is referred to by the scribes as *mawlānā, sayyidunā*, and *shaykhunā*, indicating that they regarded him as their teacher.[77] Moreover, there is at least one piece of evidence that the copyist of one of his writings read the copy to him,[78] and Ibn al-Fuwaṭī mentions for the year 683/1284 that "he was sought by many people who wished to profit by his instruction."[79]

At the end of his *Kalimāt wajīza*, completed sometime before Shaʿbān 678/December 1279, Ibn Kammūna remarks that weakening eyesight was preventing him from writing any longer.[80] However, his physical

---

[76] Ibn Kammūna: *Al-Jadīd fī l-ḥikma*, p. 146:12–13: *fa-ʿamaltu hādhā l-kitāb fī athnāʾi mā qad uljiʾtu ilayhi min mulābasat al-umūr al-daniyya wa-l-shawāghil al-dunyawiyya*... See similarly his concluding words in his commentary on Ibn Sīnā's *Ishārāt*, completed in 671/1273, quoted below, Chapter Two, Section 1.1.1.

[77] Cf., e.g., the title page of ms. Damat İbrahim Paşa 820 (undated): *al-qism al-awwal min Sharḥ al-Talwīḥāt wa-l-thānī minhu wa-huwa kāmil al-kitāb li-Mawlānā Mullā ʿIzz al-Dawla wa-huwa kāmil sharḥ al-fannayn al-manṭiqiyyāt wa-l-ṭabīʿiyyāt*. See also the beginning of the text of *Taqrīb al-maḥajja* as preserved in all three extant manuscripts, namely mss. Ayasofya 2446 (f. 293b), 2447 (f. 77b), and Esad Efendi 1933 (f. 49b): *qāla mawlānā wa-sayyidunā wa-shaykhunā ʿIzz al-Dawla*... The formula *sayyidunā wa-mawlānā* is also used at the beginning of *Tanqīḥ al-abḥāth* in Hebrew script, ms. Berlin Ms. Or. Oct. 256/1, f. 1b.

[78] See the colophon on f. 253a of ms. Şehid Ali Paşa 1740 which contains a copy of Ibn Kammūna's *Sharḥ al-Talwīḥāt*. The anonymous scribe completed his copy in 686/1287 in the Niẓāmiyya *madrasa* and he states that he collated part II of the commentary with a copy that had been read to Ibn Kammūna. No further details about the latter copy are provided.

[79] *Majmaʿ al-ādāb*, vol. 1, p. 190; transl. Nemoy: "New Data," p. 508.

[80] See below, Chapter Three, Text One, p. 184. Ibn Kammūna's expression *ḍuʿf bukhārī fī l-baṣar*, literally a 'vapourous weakness in vision', occurs in Ibn al-Nafīs' *al-Muhadhdhab fī l-kuḥl al-mujarrab*. Eds. Muḥammad Ẓāfir al-Wafāʾī and Muḥammad Rawwās Qalʿajī. Rabat 1988, pp. 451–52. There it appears to be a weakness of vision occurring in one of the tunics of the eye. It is presumably a weakness of vision due to excessive discharge which was attributed at the time to a vapourous weakness associated with one of the tunics. (We are grateful to Emily Savage-Smith for the reference and the ophthalmological explanation of Ibn Kammūna's phrase).

condition still allowed for substantial writing activity afterwards. In Jumādā II 679/September-October 1280 he completed *Tanqīḥ al-abḥāth*, and between 21–30 Dhu l-Qaʿda 679/14–23 March 1281 he finished a treatise on philosophy and/or theology, possibly entitled *Risāla fī l-kalām*.[81]

In his *Tanqīḥ* Ibn Kammūna discussed the arguments for and against Judaism, Christianity and Islam as known from the earlier polemical literature. Being basically a work on interreligious theology, its contents are strikingly different from the author's other philosophical writings. The only exception is his treatise on the differences between Rabbanites and Karaites, which is written in an equally objective style. In his introduction to the *Tanqīḥ*, Ibn Kammūna states that it was a discussion that had induced him to write the work, i.e., possibly a debate between representatives of the different religions. In 683/1284, four years after he had completed the *Tanqīḥ*, Ibn Kammūna was persecuted for having written it. The anonymous author of *al-Ḥawādith al-jāmiʿa* relates the following incident for that year:

> In this year... it became known in Baghdad that the Jew ʿIzz al-Dawla Ibn Kammūna had written a volume entitled *The Inquiries on the Three Faiths*, in which he displayed impudence in the discussion of the prophecies. God keep us from repeating what he said. The infuriated mob rioted, and massed to attack his house and to kill him. The amīr Tamaskai, prefect of Iraq, Majd-al-din b. al-Aṯīr, and a group of high officials rode forth to the Mustanṣirīya madrasa, and summoned the supreme judge and the [law] teachers to hold a hearing on the affair. They sought Ibn Kammūna but he was in hiding. That day happened to be a Friday. The supreme judge set out for the prayer service but, as the mob blocked him, he returned to the Mustanṣirīya. Ibn al-Aṯīr stepped out to calm the crowds, but these showered abuse upon him and accused him of being on the side of Ibn Kammūna, and of defending him. Then, upon the prefect's order, it was heralded in Baghdad that, early the following morning, outside the city wall, Ibn Kammūna would be burned. The mob subsided, and no further reference to Ibn Kammūna was made. As for Ibn Kammūna, he was put into a leather-covered box and carried to Hilla where his son was serving as an official. There he stayed for a time until he died.[82]

---

[81] See below, Chapter Two, Section 1.3.7.

[82] Translated in Perlmann (transl.): *Examination*, p. ix.

The report does not explain why it was only after four years that the attention of the people of Baghdad was drawn to the work, leading to such a violent reaction, and there is no independent source confirming the report or providing further information. The time difference between the completion of the book and the riot suggests that there were additional circumstances leading to the persecution. Some time before the incident, on 4 Shaʿbān 683/16 October 1284, *ṣāḥib al-dīwān* Shams al-Dīn al-Juwaynī and several of his sons were executed, and it is likely that the immediate cause for Ibn Kammūna's persecution was the execution of his patrons.[83]

Ibn Kammūna's son was serving as an official in Ḥilla, which was possibly the reason why Ibn Kammūna chose to seek refuge there. Another reason may have been his contacts with some of the Shīʿī scholars of Ḥilla. He had corresponded with Maytham b. Maytham al-Baḥrānī, who had taught in that city for a time.[84] Moreover, the fact that Ibn Kammūna's *Maṭālib* were later copied by a Shīʿī scribe also points to some connections with the Shīʿī community. The scribe who refers to Ibn Kammūna with the title *al-muḥaqqiq al-muftī* even claims that he converted to Shīʿism at the end of his life.[85] A further indication for the reception of Ibn Kammuna's writings among the Shīʿīs of Ḥilla are the glosses on his *Sharḥ al-Talwīḥāt* apparently written by Naṣīr al-Dīn ʿAlī b. Muḥammad al-Kāshī (or al-Kāshānī) al-Ḥillī (ca. 675–755/1276–1354),

[83] Cf., by contrast, Perlmann's introduction to his edition of *Tanqīḥ* (p. x), where he suggests that the persecutions may have been the result of the irritation of the Muslim population following the overthrow in 683/1284 of Ahmad Tegüder, who was a convert to Islam, by Arghun the son of Abaqa, who was a Buddhist. Dominique Urvoy (*Les penseurs libres dans l'Islam classique. L'interrogation sur la religion chez les penseurs arabes indépendants*. Paris 1996, p. 203) suggests, as a possible explanation for the difference in time between the composition of the work and the persecution, that initially it may have been written for a closed circle but that it reached the wider public later on when it caused deep resentment.—For an overview of Ilkhanid dynastic history and the issue of conversion among the Ilkhans, see Reuven Amitai: "Il-Khanids. i. Dynastic History." In *Encyclopaedia Iranica*, vol. 12, pp. 645–54; idem: "The Conversion of Tegüder Aḥmad to Islam." *Jerusalem Studies in Arabic and Islam* 25 (2001), pp. 15–43.

[84] See above, p. 13.

[85] Ms. Āstān-i Quds 901, f. 15a: *hādhā l-kitāb taṣnīf al-Shaykh al-muḥaqqiq al-muftī ʿIzz al-Dawla Saʿd b. al-Ḥasan... b. Hibat Allāh b. Kammūna alladhī ṣāra ʿalā madhhab al-Imāmiyya ākhiran wa-māta ʿalayhi...*

a Shīʿī scholar of Ḥilla and Najaf,[86] and a copy of his commentary on the *Talwīḥāt*, completed on 5 Rabīʿ II 762/12 February 1361 by Ḥasan b. Muḥammad b. Ḥasan b. al-Muṭahhar, apparently a grandson of the ʿAllāma al-Ḥillī (d. 726/1325).[87] Again according to the *Ḥawādith*, Ibn Kammūna died in Ḥilla a short while after his arrival. This is confirmed by the biographical entry on Ibn Kammūna contained in Ibn al-Fuwaṭī's *Majmaʿ al-ādāb*, in which 683 is given as the year of his death.[88]

There has been debate in modern scholarship as to whether Ibn Kammūna converted to Islam at some stage of his life.[89] In his late work, the *Tanqīḥ*, completed in 679/1280, i.e., some four years before his death, his loyalty to Judaism is beyond doubt, whereas his critical attitude towards Islam is equally clear. Particularly telling is the following passage towards the end of his chapter on Islam:

> Thus it becomes clear that there is no proof that Muhammad attained perfection and the ability to perfect others as claimed, nor that anything mentioned on this point has been proved at all—for example, the allegation that the world turned from falsehood to truth, from lie to veracity, from darkness to light, and so on. Yet precisely this has been the subject of dispute.
>
> That is why, to this day we never see anyone converting to Islam unless in terror, or in quest of power, or to avoid heavy taxation, or to escape humiliation, or if taken prisoner, or because of infatuation with a Muslim woman, or for some similar reason. Nor do we see a respected, wealthy, and pious non-Muslim well versed in both his faith and that of Islam, going over to the Islamic faith without some of the aforementioned or similar motives.[90]

---

[86] See below, n. 162 and Chapter Two, Section 1.1.2.

[87] Ms. Ayasofya 2423. For a description of this manuscript, see below, Chapter Two, Section 1.1.2.

[88] Ibn al-Fuwaṭī: *Majmaʿ al-ādāb*, vol. 1, p. 191.

[89] For a summary of the discussion before his time, see the introduction by Habib Bacha to the *Ḥawāshī (Notes) d'Ibn al-Maḥrūma sur le "Tanqīḥ" d'Ibn Kammūna. Etude et édition critique*. Rome 1984, pp. xxiv–xxxiv.

[90] Perlmann (transl.): *Examination*, p. 149:1–15; Perlmann (ed.): *Examination*, p. 102: 10–17.—Ibn Kammūna's position, as it appears in his *Tanqīḥ*, was the main point of Leo Hirschfeld's argument (*Ibn Kammûnah und seine polemische Schrift*, p. 9), Baneth ("Ibn Kammuna," pp. 304–5) and Fischel ("Arabische Quellen zur Geschichte der babylonischen Judenheit," pp. 320–21) against Ibn Kammūna's conversion to Islam, and a similar view has been maintained more recently by Rosenkranz ("Judentum,

For a variety of reasons, some scholars have argued nevertheless that Ibn Kammūna converted to Islam even before he composed the *Tanqīḥ*. This was maintained by Moritz Steinschneider, who considered the eulogies for the Prophet Muḥammad in the *Tanqīḥ* to be clear indications that the author had become a Muslim.[91] Another scholar who took Ibn Kammūna's conversion to Islam prior to the composition of the *Tanqīḥ* as a historical fact was Carl Brockelmann. The reason was evidently a confusion between Samawʾal al-Maghribī (d. 570/1175), the Jewish convert to Islam and author of the anti-Jewish polemical treatise *Ifḥām al-yahūd*, and Ibn Kammūna that occurred in the card catalogue of the Süleymanie Library in Istanbul. There, the correspondence between Samawʾal and an anonymous accuser who had doubted the sincerity of the latter's conversion (as preserved in ms. Fâtih 3141/2) is ascribed to Ibn Kammūna.[92] The entry in the Süleymanie Library's card catalogue

---

Christentum und Islam in der Sicht des *Ibn Kammûna*," pp. 20–21) and Barbara Roggema ("Epistemology as Polemics. Ibn Kammuna's Examination of the Apologetics of the Three Faiths." In *The Three Rings. Textual Studies in the Historical Trialogue of Judaism, Christianity, and Islam*. Eds. Barbara Roggema, Marcel Poorthuis and Pim Valkenberg. Leuven 2005, pp. 47–70, esp. 51 n. 15). See also the reviews on Perlmann's edition of the *Tanqīḥ* by Kurt Rudolph (in *Orientalistische Literaturzeitung* 67 ix/x (1972), pp. 462–65) and Rudi Paret (in *Der Islam* 46 (1970), pp. 102–104) who both reject the assumption of conversion on the basis of this text. This is also the view of Sarah Stroumsa ("On Jewish Intellectuals Who Converted in the Early Middle Ages." In *The Jews of Medieval Islam. Community, Society, and Identity*. Ed. Daniel Frank. Leiden 1995, pp. 179–97, esp. 197). Friedrich Niewöhner, in contrast, denies any bias towards Judaism and against Islam on the basis of the *Tanqīḥ*; he considers Ibn Kammūna to have been a religious sceptic beyond religion; see his "*Die Wahrheit ist eine Tochter der Zeit*. Ibn Kammūna's historisch-kritischer Religionsvergleich aus dem Jahre 1280." In *Religionsgespräche im Mittelalter*. Eds. Bernard Lewis and Friedrich Niewöhner. Wiesbaden 1992, pp. 357–69 esp. p. 364: "Ibn Kammūna argumentiert in den Entgegnungen auf das Judentum von einem Standpunkt aus, der weder christlich noch islamisch ist, aber auch nicht jüdisch sein kann, da er sich ja gegen das Judentum richtet. Es ist somit ein Standpunkt *außerhalb aller drei* Religionen. Immer wenn Ibn Kammūna sagt ‚I say', bedeutet das, daß er seine eigene Meinung sagt, die unabhängig von jeder Religion ist. Ibn Kammūna's Standort ist der der Vernunft, der des eigenen Verstandes."

[91] Steinschneider: *Polemische und apologetische Literatur*, pp. 39–40. Moshe Perlmann accepts the validity of the argument, yet suggests that these may have been inserted by later copyists. See the introduction to his edition of *Tanqīḥ*, p. xii.

[92] The card on ms Fâtih 3141/2 is to be found under the title *Asʾila va Acviba İbn Kamūnat al-Isrāʾīlī*. An edition of the correspondence with translation into English is included in Samawʾal al-Maghribī: *Ifḥām al-yahūd. Silencing the Jews*. Ed. and transl. by

was doubtless Brockelmann's source, who ascribes the correspondence as preserved in ms. Fâtih 3141/2 to Ibn Kammūna. Brockelmann also states that Ibn Kammūna composed the *Tanqīḥ* following his conversion to Islam. He was evidently unfamiliar with the text and the studies of Leo Hirschfeld and Baneth.[93] Confusion between Ibn Kammūna and Samawʾal also occurs in ms. Köprülü 1612, containing Ibn Kammūna's *al-Jadīd fī l-ḥikma*, copied in 691/1291–92 by ʿAbd al-ʿAziz b. Ibrāhīm b. al-Khīmī al-Māridānī in Baghdad. The title page of the work reads as follows: *Kitāb al-mulaqqab bi-l-Jadīd li-l-Shaykh al-ajall al-ʿālim al-kāmil ʿIzz al-Dawla Ibn Kammūna... bal aslama qabla mawtihī wa-ṣannafa Kitāb Ifḥām al-yahūd bayyana fīhi faḍāʾiḥahum wa-kidhbahum wa-mukābaratahum fī l-ḥaqq wa-iftirāʾahum ʿalā llāh...*[94]

These two cases of confusion between Ibn Kammūna and Samawʾal notwithstanding, the majority of Muslim scholars and scribes either considered him to have been Jewish, as can be seen from epithets such as "al-Yahūdī" or "al-Isrāʾīlī",[95] or else that the matter was of no concern, as is indicated by the numerous "neutral" title pages and colophons.[96]

---

Moshe Perlmann. New York 1964.—For Samawʾal al-Maghribī, see the introduction to *Samawʾal al-Maghribī's (d. 570/1175) Ifḥām al-yahūd. The Early Recension*. Introduced and edited by Ibrahim Marazka, Reza Pourjavady and Sabine Schmidtke. Wiesbaden 2006, which includes an improved edition of the correspondence.

[93] GAL, vol. 1, p. 431: "11. Beantwortung einer Frage nach dem Grund seines Übertritts. [Fâtih 3141] f. 263a/264b." Cf. also GALS, vol. 1, p. 768: "ʿIzzaddaula... war zum Islām übergetreten und schrieb bald darauf seinen *Tanqīḥ*.... 1. *Tanqīḥ al-abḥāṯ lil-milal aṯ-ṯalāṯ*, Verteidigung des Islams gegen Juden und Christen..." Brockelmann includes bibliographical details of Leo Hirschfeld's partial edition of the text, but gives no reference to Baneth's study on Ibn Kammūna.

[94] For a facsimile of the titlepage, see ed. Ḥamīd Marʿīd al-Kabīsī, pp. 130, 137.

[95] In the colophon to ms. Landberg 510/5 which is a copy of Ibn Kammūna's *Maqāla fī anna wujūd al-nafs abadī wa-baqāʾahā sarmadī* (= Chapter Two, Section 1.3.10), Ibn Kammūna is described as *al-Shaykh al-muʿaẓẓam wa-l-ḥakīm al-muqaddam qudwat al-ḥukamāʾ al-yahūdiyyīn*. See also mss. Feyzullah Efendi 1195, copied in 696/1297 in Tabriz, which has two indications as to author and title by two different hands on its title page: *Sharḥ al-Talwīḥāt al-Suhrawardiyya taṣnīf al-fāḍil Muʿizz al-Dawla wa-huwa muʿāṣir al-ʿAllāma al-Shīrāzī wa-l-nasab ilā Ibn Kammūna al-yahūdī*, and: *Sharḥ al-Talwīḥ li-l-Shaykh al-fāḍil al-muḥaqqiq ʿIzz al-Dawla Saʿd b. Manṣūr al-maʿrūf bi-Ibn Kammūna al-Isrāʾīlī*; Lâleli 2522, undated: *Sharḥ al-Talwīḥāt li-l-ḥakīm al-mudaqqiq ʿIzz al-Dawla Saʿd b. Manṣūr al-maʿrūf bi-Ibn Kammūna al-Isrāʾīlī*; Lâleli 2516 written in 1086/1675: *Sharḥ al-Ishārāt mimmā allafahū mukammil ʿulūm al-awwalīn wa-l-ākharīn qudwat al-muḥaqqiqīn ʿIzz al-Dawla Saʿd b. Manṣūr b. Saʿd al-maʿrūf bi-Ibn Kammūna al-Isrāʾīlī*... See also infra, n. 193.

[96] See below, Chapter Two, *passim*.

Again, an exception among the testimonies is the aformentioned statement of an Imāmī copyist of Ibn Kammūna's *Maṭālib* that the author had converted to Twelver Shīʿism at the end of his life.[97] This statement may explain why Āghā Buzurg took his conversion to Shīʿism as historical fact; this is evidently why he included all writings of Ibn Kammūna he was aware of in his *Dharīʿa*—including the *Tanqīḥ*, although it is clear from the wording of the entry (*Dharīʿa*, vol. 4, p. 460 no. 2054) that he was unfamiliar with the text.[98] Initially, Āghā Buzurg also intended to include an entry on Ibn Kammūna in his *Ṭabaqāt aʿlām al-shīʿa*,[99] and it was only later that he or his son ʿAlī Naqī Munzawī decided not to do so. Munzawī was familiar with Perlmann's edition and translation of Ibn Kammūna's *Tanqīḥ* and also edited and studied the work.[100] Since he is the editor of most of the volumes (namely vols. 3–6) of the *Ṭabaqāt aʿlām al-shīʿa* of his father, it may be that it was he rather than Āghā Buzurg who decided not to include an entry on Ibn Kammūna.[101]

---

[97] See above n. 85.

[98] He also includes references to the various sophistries (*shubahāt*) that supposedly originated with Ibn Kammūna, ascribing them, however, to his ancestor whom he identified explicitly as Hibat Allāh Ibn Kammūna *al-Isrāʾīlī*. See *Dharīʿa*, vol. 13, pp. 24–25: *shubhat Ibn Kammūna wa-huwa Hibat Allāh b. Kammūna al-Isrāʾīlī min falāsifat al-yahūd kāna muʿāṣiran li-Ibn Sīnā*...; see also ibid., vol. 2, p. 286 no. 1160. For the *shubahāt*, see below, pp. 44 ff. Āghā Buzurg was familiar with *Kashf al-ẓunūn* and its various entries on Ibn Kammūna where he is regularly described as al-Isrāʾīlī, yet he apparently assumed that Ḥājjī Khalīfa confused ʿIzz al-Dawla Ibn Kammūna with his ancestor. See, e.g., *Dharīʿa*, vol. 13, pp. 152–53 no. 519.

[99] Āghā Buzurg al-Ṭihrānī: *Ṭabaqāt aʿlām al-shīʿa* 1–2, Beirut 1390–91/1971; 3–5, ed. ʿAlī Naqī Munzawī, Beirut 1392–95/1972–75; 6, ed. ʿAlī Naqī Munzawī, Tehran 1362/1983.

[100] ʿAlī Naqī Munzawī: *Ibn Kammūna wa-falsafatuhū fī l-nubuwwa*. PhD thesis, Université Saint Joseph, Beirut 1972. [*Al-Qism al-awwal: al-Madkhal wa-l-naṣṣ al-kāmil li-Tanqīḥ al-abḥāth; al-Qism al-thānī: al-Taʿlīqāt ʿalā Tanqīḥ al-abḥāth li-l-milal al-thalāth*]. See also his "Tanqīḥ al-abḥāth-i Ibn Kammūna." *Waḥīd* 6 v (1348/1969), pp. 369–71, 6 vi (1348/1969), pp. 503–14, 6 xii (1348/1969), pp. 1079–83.

[101] This is indicated in some of the footnotes to *Ṭabaqāt aʿlām al-shīʿa*, e.g., vol. 7, p. 69:

...مؤلف تنقيح الأبحاث الذي طبعه موسى برلمان في ١٩٦٧ وأثبت أنه يهودي ولذلك تركت ذكره مع أني ذكرت كتبه في الذريعة.

It cannot be decided conclusively whether Ibn Kammūna did or did not, at some stage, convert to Islam.[102] Yet the evidence against such a conclusion seems to be stronger than the evidence suggesting conversion. His position towards Judaism and Islam as far as his personal convictions are concerned is, on the basis of his *Tanqīḥ*, beyond doubt. It cannot be excluded that he became a Muslim for any of the reasons mentioned at the end of the chapter on Islam; yet it seems questionable whether he would have written a work such as *Tanqīḥ* following a conversion on the grounds of persecution or for opportunistic motives.[103] Another indicator against conversion is that he is invariably called ʿIzz al-Dawla by virtually all later scholars and scribes. Had he ever converted to Islam during his professional life, his title would have been modified into ʿIzz al-Dīn or ʿIzz al-Dawla wa-l-Dīn, as can be seen in the case of Rashīd al-Dawla (later: al-Dīn) Faḍl Allāh.[104] Had he converted to Islam

---

See also ibid., vol. 7, p. 176:

لقد ذكر آثار ابن كمونة ...ولكن لما تحقق أن ابن كمونة كان مشاركاً في تأليف كتاب تنقيح الأبحاث في الملل الثلاث الذي انحاز فيه نحو اليهودية، دين أبائه، حذفت ترجمته عن هذا الكتاب ...كل هذا أحيا فيه عرق اليهودية فانحاز إليهم في كتابه هذا الذي طبعه موسى برلمان في كاليفورنيا بأمريكا في 1967.

102 Contemporary scholars at times take Ibn Kammūna's conversion as historical fact without giving any reason for their assumption. See, e.g., Walbridge's and Ziai's "Translator's Introduction" to their edition of Suhrawardī's *Ḥikmat al-Ishrāq* (*The Philosophy of Illumination. A New Critical Edition of the Text of* Ḥikmat al-ishrāq *with English Translation, Notes, Commentary, and Introduction*. Provo, Utah 1999, p. xxii), in which they talk about "the Jewish convert to Islam, Ibn Kammūna."

103 This seems to be the view of Georges Vajda, who writes ("Le milieu juif à Baġdād." *Arabica* 9 (1962), p. 392): "Ce n'est pas à dire que la communauté juive de Baġdād n'ait pas produit encore, au XIIᵉ et au XIIIᵉ siècles, quelques personnalités intellectuelles marquantes comme les philosophes Hibat Allāh Abū l-Barakāt et Ibn Kammūna, qui ont d'ailleurs, au déclin de leur carrière, payé un tribut aux circonstances adverses plutôt qu'à une conviction intime en se convertissant à l'Islām."

104 See ʿAbbās Zaryāb-Khūʾi: "Sih nukta dar bara-yi Rashīd al-Dīn Faḍl Allāh." In *Proceedings of the Colloquium on Rashīd al-Dīn Faḍl Allāh. Tehran-Tabriz, 11–16 Aban 1348 (2–7 November 1969)*. Tehran 1971, pp. 123–35; Netzer: "Rashīd al-Dīn and His Jewish Background."—The only case we came across of the title ʿIzz al-Dīn being used for Ibn Kammūna by a Muslim is Moshe Perlmann's edition of an anonymous refutation of Ibn Kammūna's *Tanqīḥ* ("'Proving Muhammad's Prophethood', A Muslim critique of Ibn Kammuna." [Hebrew] In *Hagut Ivrit Ba'Amerika. Studies on Jewish Themes by Contemporary American Scholars*. Eds. Menahem Zohori, Arie Tartakover, and Haim Ormian. Vol. III. Tel Aviv 1974, p. 83:5). However, Perlmann clearly erred here, since the manuscript of the text has ʿIzz al-Dawla. For an improved edition of the text, see below, Appendix B: *Kitāb fī ithbāt al-nubuwwa li-Muḥammad ʿalayhi al-salām* (ms. Hamidiye 1447/3).

before or during the persecution he experienced, either the anonymous author of *al-Ḥawādith al-jāmiʿa*, or Ibn al-Fuwaṭī would have mentioned it. The possibility of his conversion to Twelver Shīʿism in Ḥilla during the last months of his life, as suggested by the copyist of the *Maṭālib*, cannot be ruled out altogether. Yet again, it would be hard to believe that the author of the *Tanqiḥ* would have converted to Islam, once he had escaped persecution.[105]

### 3. *Ibn Kammūna's Philosophical Thought*

A comprehensive assessment of Ibn Kammūna's philosophical thought can only be undertaken when critical editions of all his extant writings are available. In the following pages, only some basic outlines of his thought will be given.

As a philosopher, Ibn Kammūna was a follower of Ibn Sīnā. With some modifications, his cosmology is Avicennan. His distinction between the One as the Necessary Existent (*wājib al-wujūd*) and other existents as contingents (*mumkināt*) is also Avicennan, as is his notion of prophecy, explained in detail in the first chapter of his *Tanqīḥ*.[106]

One of the issues Ibn Kammūna was most concerned with is the human soul, and it is on this issue that he formulated his most distinct views. He wrote several independent treatises on the soul, three of which are extant.[107] His detailed discussions of the human soul in his other philosophical treatises and commentaries also testify to his interest in the issue. The main sources for his psychology are again Ibn Sīnā's writings.

[105] Although it is almost certain that Ibn Kammūna remained Jewish throughout his life, there are clear indications in his writings that attest to his openness to and respect for Islam and possibly Christianity. He freely adduces Qurʾānic verses in his writings and uses the relevant eulogistic formula when mentioning the Prophet Muḥammad. As was said earlier (see above, n. 91), Perlmann suggests that these eulogies were added by later scribes. This is unlikely since they are to be found in copies of his writings made by both Muslim and Jewish scribes. Moreover, in his *Kalimāt wajīza* (see below, Chapter Three, Text One), Ibn Kammūna uses the term *al-sharāʾiʿ al-ḥaqqa* which also indicates his respect for the three Abrahamic religions and their respective dispensations.

[106] Some instances where Ibn Kammūna follows Ibn Sīnā have been indicated in Perlmann's annotation to his edition of Ibn Kammūna's *Tanqīḥ*.

[107] See below, Chapter Two, Sections 1.3.8, 1.3.9, and 1.3.10.

However, he disagreed with Ibn Sīnā when he rejected the latter's view that a human rational soul comes into existence with the birth of the body in which it inheres. Instead, he maintained the pre-eternity of the human soul. His main argument is that the human soul is simple (*basīṭ*) while everything that is originated in time (*ḥadīth*) is a composite. Thus, the human rational soul must be pre-eternal (*qadīm*). His arguments for the simplicity of the human soul and for simples being necessarily pre-eternal seem to be original—at least this is what Ibn Kammūna claims. Following Suhrawardī, Ibn Kammūna criticizes the arguments against the transmigration of souls and seems to take a non-commital position on the issue. This is also the view of Suhrawardī in his *Ḥikmat al-ishrāq* and this is typical for later followers of Suhrawardī, such as Shams al-Dīn al-Shahrazūrī and Quṭb al-Dīn al-Shīrāzī.[108] However, since he is the only one of them maintaining the pre-eternity of the soul, his understanding of transmigration must have been different from the notion upheld by Suhrawardī, Shahrazūrī and presumably Quṭb al-Dīn al-Shīrāzī, who agreed on its temporal origin.[109] Apparently Ibn Kammūna was searching for Plato's explanation for the eternity of the soul, as is suggested by his use of Miskawayh's (d. 421/1030) *al-Fawz al-aṣghar* in one of his *maqālāt* devoted to the human soul. Ibn Kammūna was not entirely satisfied with the explanations he found there; he states that he was unable to grasp them fully as they lacked precision. Perhaps, he says, it was not originally presented as an apodictic proof (*burhān*) but rather rhetorically (*kalāman khiṭābiyyan*) and only later was it understood by the later philosophers, including Miskawayh, as a syllogism.[110]

Ibn Kammūna was not a typical Peripatetic philosopher, as he did not hesitate to criticize Aristotle openly and to maintain views that disagreed with what were perceived to have been Aristotelian positions. His detachment from the Aristotelian tradition was apparently due

---

[108] For the views of Suhrawardī and his early commentators on the issue of the transmigration of souls, see Sabine Schmidtke: "The Doctrine of the Transmigration of Soul according to Shihāb al-Dīn al-Suhrawardī (killed 587/1191) and his Followers." *Studia Iranica* 28 (1999), pp. 237–54.

[109] I (S.S.) take the opportunity to correct what I wrote earlier, viz. that Quṭb al-Dīn al-Shīrāzī was a follower of Ibn Kammūna on the issue of the pre-eternity of the soul ("The Doctrine of the Transmigration of Soul," pp. 249–50). See also below, p. 32.

[110] See his "Maqāla fī l-taṣdīq bi-anna nafs al-insān bāqiya abadan," pp. 115–16.

to the influence of the writings of Abu l-Barakāt al-Baghdādī and Suhrawardī on his thought. Ibn Kammūna repeatedly refers to *al-Kitāb al-Muʿtabar*, the *magnum opus* of the Jewish convert to Islam Abu l-Barakāt al-Baghdādī. The controversial views of Abu l-Barakāt in his *Muʿtabar*, which was a major critique of Aristotelian philosophy, were attacked by later scholars such as Suhrawardī, who disparagingly referred to him as *baʿḍ man tafalsafa min al-yahūd*, Fakhr al-Dīn al-Rāzī and Naṣir al-Dīn al-Ṭūsī. Ibn Kammūna attempted to justify Abu l-Barakāt's positions on a number of issues and refered to him respectfully even when he maintained different views.[111]

Shihāb al-Dīn al-Suhrawardī, the founder of the philosophy of illumination (*ḥikmat al-ishrāq*), was another critic of mainstream Aristotelian philosophy. When Ibn Kammūna wrote a commentary upon the latter's *Talwīḥāt*, Suhrawardī did not as yet figure among the higher ranks of philosophers, and it seems that Ibn Kammūna's *Sharḥ* was the first commentary ever written on any of the works of Suhrawardī and that the popularity of this commentary influenced subsequent reception of the *Shaykh al-ishrāq*. From his commentary on *Talwīḥāt*, as well as from some of his other writings, it is evident that Ibn Kammūna was familiar with and used Suhrawardī's other philosophical works in Arabic, namely his *Ḥikmat al-ishrāq, al-Mashāriʿ wa-l-Muṭāraḥāt, al-Muqāwamāt, al-Alwāḥ al-ʿimādiyya* and his *Kalimat al-taṣawwuf*. While he explicitly refers to the first three works mentioned, he never mentions *Kalimat al-taṣawwuf* by name although he used it extensively, particularly those parts containing moral advice and mystical instructions, many of which Ibn Kammūna adduced in his *Kalimāt wajīza* and some of his other writings.[112]

Ibn Kammūna followed Suhrawardī in a number of issues, such as the principle of the most noble contingency (*qāʿidat al-imkān al-ashraf*), his admission of movement in substance, his non-commital position on the issue of the transmigration of souls and his uncertainty about the

---

[111] See, e.g., his *Sharḥ al-Talwīḥāt*. Eds. Ziai and Alwishah, pp. 267–69. Ibn Kammūna also refers repeatedly to Abu l-Barakāt in his *maqāla*s on the human soul.

[112] See below, Chapter Two, Section 1.1.2 for references to other writings by Suhrawardī in his *Sharḥ al-Talwīḥāt*, and Chapter Three, Text One where adoptions from *Kalimat al-taṣawwuf* have been mentioned in the annotation.

number of spheres. Moreover, like Suhrawardī, Ibn Kammūna was not interested in mathematics, which he never addressed in his writings. Yet it would be inappropriate to consider Ibn Kammūna a follower of Suhrawardī, as he disagreed with him on certain issues such as the pre-eternity of the soul. There are aspects of Suhrawardī's thought and *œuvre* that Ibn Kammūna ignored, and there is no reference in his work to any of the allegorical writings of Suhrawardī, most of which are written in Persian and therefore may have been inaccessible to Ibn Kammūna. But he also never referred to those composed in Arabic, just as he ignored any writings by Ibn Sīnā in this genre. Moreover, Ibn Kammūna did not share Suhrawardī's veneration of the ancient philosophers of Persia, Egypt and Greece, and likewise he was not interested in Suhrawardī's specific terminology, which was derived from pre-Islamic Persian notions and figured prominently in his *Ḥikmat al-ishrāq*.

Ibn Kammūna was also familiar with, and made extensive use of, the writings of Abū Ḥāmid al-Ghazzālī (d. 505/1111) and Fakhr al-Dīn al-Rāzī. As has been pointed out by Perlmann, Ibn Kammūna used Ghazzālī's *Tahāfut al-falāsifa* and his *Munqidh min al-ḍalāl* in his *Tanqīḥ*,[113] and in his commentary on the *Talwīḥāt* he used Ghazzālī's *Mishkāt al-anwār*.[114] In his various *maqāla*s on *nafs* Ibn Kammūna rejected Ghazzālī's view on the soul, presumably having his *Tahāfut al-falāsifa* in mind.[115] Moreover, in his *Kalimāt wajīza* and *Ithbāt al-mabdaʾ* Ibn Kammūna adduced some pieces of moral advice which he seems to have paraphrazed from Ghazzālī's *Iḥyāʾ ʿulūm al-dīn*.[116] His familiarity with Rāzī's writings is evident from his superglosses on the latter's *Maʿālim*, in which he constantly refers to other writings by Rāzī, namely his *Nihāyat al-ʿuqūl*, his *Muḥaṣṣal* and his *Kitāb al-Arbaʿīn*, and the fact that

---

[113] See *Tanqīḥ al-abḥāth*. Ed. Perlmann, pp. 2–3.

[114] See *Sharḥ al-Talwīḥāt* [part III, *mawrid* 4, *mirṣād ʿarshī*]. Ed. Mūsawī, pp. 992–93.

[115] In his "Maqāla fī l-taṣdīq bi-anna nafs al-insān bāqiya abadan" (ed. Barkhʷāh, p. 124), Ibn Kammūna says that he was familiar with Ghazzālī's view on the soul, which he rejected.

[116] For example Ghazzālī's prohibition on cursing (*laʿn*) others and his argument that it is not a sin to refrain from cursing Satan (Iblīs), let alone others. See *Iḥyāʾ ʿulūm al-dīn* [*al-Juzʾ al-thālith. Kitāb āfāt al-lisān: al-āfāt al-thāmina*]. Beirut n.d., vol. 3, p. 118:23–24 which is to be found in Ibn Kammūna's *Kalimāt wajīza* and his *Ithbāt al-mabdaʾ*. See below, Chapter Three, Text One (p. 164) and Text Two (p. 190).

he quoted a great deal from his writings in the *Tanqīḥ*.[117] Ibn Kammūna had studied Rāzī's *Muḥaṣṣal*, together with the critical commentary on it by Ṭūsī, as is evident from the extant excerpts.[118] Ibn Kammūna was also familiar with Rāzī's view on the soul, which he criticized in his superglosses on his *Maʿālim* and in his *Maqāla fī l-taṣdīq bi-anna nafs al-insān bāqiya abadan*.[119] In addition, he was well-acquainted with Rāzī's views on logic, to which he refers in his commentaries on *Talwīḥāt* and *Ishārāt*.[120]

At some stage in his life Ibn Kammūna also became familiar with Najm al-Dīn Abū Bakr Muḥammad al-Nakhjawānī and his critical commentaries on Ibn Sīnā's *Ishārāt*. However, although he studied his writings closely, as is evident from the extant extracts that he copied,[121] there is no evidence that Nakhjawānī had any influence on his thought. Rather, Ibn Kammūna seems to have disapproved of his criticism.[122] Moreover, since Ibn Kammūna's epitomes of Nakhjawānī's critical commentaries on Ibn Sīnā's *Ishārāt* were written some eight years after he had completed his own commentary on the work, in which there is

---

[117] This is the case particularly in Chapters One and Four. See, e.g., *Tanqīḥ*. Ed. Perlmann, pp. 7, 8, 10, 18, 19, 69, 86, 88.

[118] See below, Chapter Two, Section 1.2.1.

[119] See ed. Barkhʷāh, p. 124 where he explicitly mentions Rāzī as an opponent of the philosophical notion of the soul.

[120] See, e.g., *Sharḥ al-Talwīḥāt* [part I, *marṣad* 1, *talwīḥ* 1: *fī gharaḍ al-manṭiq*]. Ed. Mūsawī, p. 9; *Sharḥ al-ishārāt*. Ms. Ragıp Paşa 845, f. 42b. In both instances, he specifically refers to Rāzī's notion of conception and consent.

[121] See below, Chapter Two, Sections 1.2.6 and 1.2.7.

[122] According to Āghā Buzurg (*Dharīʿa*, vol. 2, p. 97 no. 382), Ibn Kammūna writes at the end of his excerpts from Nakhjawānī's *al-Iʿtirāḍāt min Kitāb Zubdat al-naqḍ wa-lubāb al-kashf*:

إن أكثر هذه الاعتراضات يمكن الجواب عنها وينتصر لمصنف الإشارات عليه.

There is so far no study of Nakhjawānī, and none of his writings have as yet been edited. See the brief sketches provided by Shabībī: *Turāthunā*, pp. 100–4, Mūsā Mūsawī: *Min al-Suhrawardī ilā l-Shīrāzī*. Beirut 1979, pp. 79–87 (most of the material adduced by Mūsawī is based on Shabībī's account), and Mudarris Raḍawī: *Aḥwāl u āthār*, pp. 195, 565.—Nakhjawānī also has a work on medicine which is extant in manuscript: *Ajwiba ʿan iʿtirāḍāt Fakhr al-Dīn al-Rāzī li-l-qānūn* (ms. Ayasofya 3721, ff. 1a–66b). Another manuscript of the same text is extant in Paris; see Ferdinand Wüstenfeld: *Geschichte der arabischen Aerzte und Naturforscher. Nach den Quellen bearbeitet*. Göttingen 1840, p. 131 no. 233.—There are some indications that he may also be the author of a Mathnawī entitled *Ṭarīq al-taḥqīq*, the authorship of which is still disputed (ed. Bo Utas. Lund 1973).

no trace of Nakhjawānī, he seems to have become acquainted with the latter's writings at a relatively late stage of his life.

Among the philosophers of his own time, Ibn Kammūna had studied in depth the writings of Naṣīr al-Dīn al-Ṭūsī, Kātibī, and Athīr al-Dīn al-Abharī. In the introduction to his commentary on *Ishārāt*, Ibn Kammūna states that he has made extensive use of Ṭūsī's commentary. Kātibī's works and notions repeatedly served as a point of departure for his writings, and in his commentary on the *Talwīḥāt* Ibn Kammūna explicitly refers to Abharī's *Muntahā al-afkār*.[123] There are also indications that he used Abharī's *Kashf al-ḥaqā'iq*.[124]

### 4. *The Reception of Ibn Kammūna in later Muslim Thought*

Ibn Kammūna was one of the most popular and influential philosophers of the Eastern lands of Islam both during his lifetime and in the decades following his death. It was his commentary on Suhrawardī's *Talwīḥāt* that was copied and read most, followed by his independent philosophical work *al-Jadīd fī l-ḥikma*. Numerous manuscript copies of his *Sharḥ al-Talwīḥāt* were produced during this period, not only in Baghdad but also in places such as Isfahan, Tabriz, Mashhad, and possibly Najaf. Some of the copies made in Baghdad were transcribed in the Niẓāmiyya *madrasa*.[125] Given the quantity of manuscript copies of this commentary and the fact that it quickly became available in many of the cultural centres of the Eastern lands of Islam—as compared with Shahrazūrī's commentary on the *Talwīḥāt*—[126] Ibn Kammūna very soon became known as the "commentator on the *Talwīḥāt*" (*shāriḥ al-Talwīḥāt*).[127] Most

---

[123] See *Sharḥ al-Talwīḥāt* [part III, *mawrid* 1, *talwīḥ* 1: *fī dhātihi*]. Ed. Mūsawī, p. 701.

[124] Farāmarz Qarāmalikī states that the way Ibn Kammūna presented the liar paradox is apparently taken from Abharī's *Kashf al-ḥaqā'iq*. See his "Kitābshināsī taḥlīlī-yi paradox-i Durūghgū," p. 33.

[125] For details, see below, Chapter Two, Section 1.1.2.

[126] Ritter I, pp. 273–75 lists only four copies of Shahrazūrī's commentary.

[127] As reference is given to him by Jalāl al-Dīn al-Dawānī ("Shawākil al-ḥūr." In *Thalāth rasā'il*. Ed. Sayyid Aḥmad Tūysirkānī. Mashhad 1411/1991, p. 153:4–5), Quṭb al-Dīn al-Rāzī ("al-Risāla al-maʿmūla fī l-taṣawwur wa-l-taṣdīq." In *Risālatān fī l-taṣawwur wa-l-taṣdīq*. Ed. Mahdī Sharīʿatī. Qum 1416/[1995–96], p. 116:9), Najm al-Dīn Maḥmūd al-Nayrīzī (*Miṣbāḥ al-arwāḥ fī kashf ḥaqā'iq al-alwāḥ*. Ms. Ragıp Paşa 853,

of the extant copies of his *al-Jadīd fī l-ḥikma* were also produced during the fifty years following his death.[128]

The most immediate influence his philosophical writings and notions had was on his younger contemporary Quṭb al-Dīn Maḥmūd al-Shīrāzī (634–710/1236–1311).[129] It is not known whether or not the two ever met. Quṭb al-Dīn is known to have spent some time in Baghdad between 665/1266 and 667/1268, and he could have met Ibn Kammūna there.[130] Quṭb al-Dīn wrote two major philosophical works, a commentary on Suhrawardī's *Ḥikmat al-ishrāq* in Arabic, completed in 694/1295,[131] and an independent encyclopaedic work written in Persian, *Durrat al-tāj li-ghurrat al-dubbāj*, completed between 693/1294 and 705/1306, which contains an extensive section on philosophy.[132] Both works testify to the profound impact that the writings of Ibn Kammūna had on Quṭb al-Dīn al-Shīrāzī.

---

ff. 3b:15, 123a:19) and Ṣadr al-Dīn al-Shīrāzī ("Mullā Ṣadrā") ("Risālat al-taṣawwur wa-l-taṣdīq." In *Risālatān fī l-taṣawwur wa-l-taṣdīq*. Ed. Mahdī Sharīʿatī. Qum 1416/[1995–96], p. 67:7).

[128] See below, Chapter Two, Section 1.3.1.

[129] The influence of Ibn Kammūna's writings on those of his contemporary Shams al-Dīn al-Shahrazūrī still needs to be studied in depth. In his edition of vol. 2 [containing *al-Risāla al-rābiʿa fī l-ʿulūm al-ṭabīʿiyya*] of Shahrazūrī's *Rasāʾil al-Shajara al-ilāhiyya*, Najafqulī Ḥabībī has shown that Shahrazūrī has numerous adoptions therein from Ibn Kammūna's *Sharḥ al-Talwīḥāt* (Tehran 1384/2006).

[130] For the main stages of Quṭb al-Dīn al-Shīrāzī's biography, see Walbridge: *Science of Mystical Lights*, pp. 7–24.

[131] *Sharḥ Ḥikmat al-ishrāq* has been published at least twice: Lithograph edition by Asad Allāh Harātī. Tehran 1313–15/1895–97, containing also Mullā Ṣadrā's glosses on Quṭb al-Dīn's commentary [recently reprinted without place and date]. Critical edition by ʿAbd Allāh Nūrānī and Mahdī Muḥaqqiq. Tehran 1380/2001. If not stated otherwise, we will refer in the following to the critical edition only. Hans Daiber mentions another edition by Ibrāhīm Ṭabāṭabāʾī (Tehran 1934) which was not available to us (*Bibliography of Islamic Philosophy*, vol. 1, p. 751 no. 7316).

[132] No complete edition of *Durrat al-tāj* is yet available. Between 1317–20/1938–41 Sayyid Muḥammad Mishkāt published five portions of the work, namely *fātiḥa*, *jumla* 1: logic (*manṭiq*), *jumla* 2: first philosophy (*falsafa-yi ūlā*), *jumla* 3: physics (*ʿilm-i ṭabīʿī*), and *jumla* 5: metaphysics (*ʿilm-i ilāhī*). The third edition which was at our disposal (Tehran 1369/1990) has the original pagination for each separate volume, as well as consecutive pagination for the editor's introduction and all five volumes; in the following only the consecutive pagination will be given. Sayyid Ḥasan Mishkān Ṭabasī in 1324/1945 published portions of *jumla* 4 on mathematics (*ʿilm-i riyāḍī*), and in 1369/1991, Māhdukhtbānū Humāʾī published sections three and four of the *khātima*.

In the introduction to his *Sharḥ Ḥikmat al-Ishrāq*, Quṭb al-Dīn mentions without any further qualification that he had made use of other commentaries on Suhrawardī's writings when writing his own.[133] His main source is the one on the same text by Shahrazūrī, completed nine years before Quṭb al-Dīn wrote his.[134] The dependence of Quṭb al-Dīn's commentary on the *Sharḥ* of Shahrazūrī has been well established ever since the study on the two commentaries by Corbin.[135] Less well known hitherto has been that Quṭb al-Dīn also relied heavily on Ibn Kammūna's *Sharḥ al-Talwīḥāt* in his commentary which contains numerous quotations and borrowings from Ibn Kammūna's *Sharḥ*.[136]

Quṭb al-Dīn al-Shīrāzī offers some general elaborations on a variety of issues, taken from Ibn Kammūna's *Sharḥ al-Talwīḥāt* with only minor modifications. In none of these cases does he indicate his source. They include (1) elaborations on the philosophical principle that God does not act with specific objectives, for an agent who acts with an objective does so to reach self-perfection and must therefore be deficient. The entire passage in Quṭb al-Dīn's commentary is meant to clarify Suhrawardī's statement that the motion of the celestial spheres is not for the sake of what lies beneath them (*inna ḥarakat al-barāzikh al-ʿalawiyya laysat li-mā taḥtahā*);[137] (2) his critical remarks about the Ashʿarite notion of the

[133] Quṭb al-Dīn al-Shīrāzī: *Sharḥ Ḥikmat al-ishrāq*, p. 5:19.

[134] Ms. Esad Efendi 1932 contains a colophon at the end of Shahrazūrī's *Sharḥ Ḥikmat al-ishrāq* indicating that the author completed the work on 20 Rajab 685/11 September 1286; see f. 298a:10–14: *faragha min taʾlīfihī al-faylasūf al-kāmil... Shams al-Ḥaqq wa-l-Milla wa-l-Dīn... Muḥammad b. Muḥammad [sic] al-Shahrazūrī... fī ʿishrīn min Rajab min sanat 685/* [11 September 1286]. Cf. our "Some notes on a new edition of a medieval philosophical text in Turkey. Shams al-Dīn al-Shahrazūrī's *Rasāʾil al-Shajara al-ilāhiyya*," p. 77.

[135] See Shihāb al-Dīn al-Suhrawardī: *Opera Metaphysica et Mystica II*. Ed. Henry Corbin. Tehran 1952, Prolégomènes II, pp. 64–71 (6. Les deux grands Commentaires de Shamsaddîn Shahrazûrî et de Qotbaddîn Shîrâzî). The dependence of Quṭb al-Dīn al-Shīrāzī's commentary on Shahrazūrī's *Sharḥ* was first mentioned by Ḍiycʾ al-Dīn Durrī Iṣfahānī in the introduction to his Persian translation of Shahrazūrī's *Nuzhat al-arwāḥ* entitled *Kanz al-ḥikma* (Tehran 1316/1937, p. 12).

[136] Some of the borrowings have been mentioned by Sayyid Ḥusayn Sayyid Mūsawī: "Mullā Ṣadrā u Ibn Kammūna." In *Mullā Ṣadrā u muṭālaʿāt-i taṭbīqī. Majmūʿa-yi maqālāt-i hamāyish-i jahānī-yi ḥakīm Mullā Ṣadrā*. Vol. 5. Tehran 1381/2002, pp. 46–53. See also our "Quṭb al-Dīn al-Shīrāzī's (d. 710/1311) *Durrat al-Tāj* and Its Sources," pp. 309–28 for additional examples for the reception of Ibn Kammūna's writings in Quṭb al-Dīn al-Shīrāzī's *Sharḥ Ḥikmat al-ishrāq*.

[137] Quṭb al-Dīn al-Shīrāzī: *Sharḥ Ḥikmat al-ishrāq*, pp. 395:12–396:9 [commenting on

eternity of the divine attributes;[138] (3) his elaborations on the meaning of pleasure and pain and the difference between the two.[139]

There are other issues on which Quṭb al-Dīn al-Shīrāzī considered Ibn Kammūna's comments to be appropriate explanations since he included them in his commentary, such as (1) Ibn Kammūna's proof for the immortality of the soul following the death of the body;[140] (2) his explanation that the divine attributes can be thought of as either relative (*iḍāfiyya*), negative (*salbiyya*) or respective (*iʿtibāriyya*);[141] (3) his elaborations on conception (*taṣawwur*) and assent (*taṣdīq*),[142] and (4) his explanations as to which of the categories (*maqūlāt*) are receptive to change in intensity.[143] Quṭb al-Dīn, moreover, followed Ibn Kammūna in some instances where the latter criticized Suhrawardī, as in the case of Ibn Kammūna's argument for the absurdity of a body without inclination (*jism ʿadīm al-mayl*). Ibn Kammūna agreed with Suhrawardī's position but criticized his argument for it, and Quṭb al-Dīn incorporated the relevant passages in his *Sharḥ*.[144] Ibn Kammūna also criticized one of

part 2, discourse 3, chapter 4] which is taken from part III of Ibn Kammūna's *Sharḥ al-Talwīḥāt* [*mawrid* 2, *talwīḥ* 1: *fī l-ghanī*]. Ed. Mūsawī, pp. 805:12–806:14.

138 See Quṭb al-Dīn al-Shīrāzī: *Sharḥ Ḥikmat al-ishrāq*, p. 377:13–18 [commenting on part 2, discourse 3, chapter 1] which is taken from part III of Ibn Kammūna's *Sharḥ al-Talwīḥāt* [*mawrid* 1, *talwīḥ* 5: *fī kayfiyyat ibdāʿ al-wāḥid min jamīʿ al-wujūh*]. Ed. Mūsawī, p. 787:15–21.

139 Quṭb al-Dīn al-Shīrāzī: *Sharḥ Ḥikmat al-ishrāq*, p. 328:11–18 [commenting on part 2, discourse 2, chapter 5] which is taken from part III of Ibn Kammūna's *Sharḥ al-Talwīḥāt* [*mawrid* 3, *faṣl* 7: *al-ḥujja fī imtināʿ al-tanāsukh*]. Ed. Mūsawī, pp. 928:14–929:2.

140 Quṭb al-Dīn al-Shīrāzī: *Sharḥ Ḥikmat al-ishrāq*, pp. 24:21–243:5, 243:7, 243:11–17 [commenting on part 1, discourse 3, judgement on proofs of the immortality of the soul] which is taken from part III of Ibn Kammūna's *Sharḥ al-Talwīḥāt* [*mawrid* 3, *faṣl* 5: *iʿlam anna l-nafs bāqiya baʿda l-badan*...]. Ed. Mūsawī, pp. 908:21–909:5, 909:13–15, 910:7–12.

141 Quṭb al-Dīn al-Shīrāzī: *Sharḥ Ḥikmat al-ishrāq*, pp. 303:21–304:24 [commenting on part 2, discourse 1, chapter 9] which is taken from part III of Ibn Kammūna's *Sharḥ al-Talwīḥāt* [*mawrid* 1, *talwīḥ* 2: *kalām jumalī fī ṣifātihī*]. Ed. Mūsawī, pp. 732:5–733:8.

142 Quṭb al-Dīn al-Shīrāzī: *Sharḥ Ḥikmat al-ishrāq*, p. 41:7–9, 41:20–21 [commenting on part 1, discourse 1, rule two on the difference between conception and assent] which is taken from part I of Ibn Kammūna's *Sharḥ al-Talwīḥāt* [*marṣad* 1, *talwīḥ* 1: *fī gharaḍ al-manṭiq*]. Ed. Mūsawī, pp. 9:5–12, 9:19–10:2.

143 Quṭb al-Dīn al-Shīrāzī: *Sharḥ Ḥikmat al-ishrāq*, pp. 230:14–18, 230:22ff, 232:9–13, 232:17–20 [commenting on part 1, discourse 3, judgment concerning investigations related to prime matter and form] which is taken from part III of Ibn Kammūna's *Sharḥ al-Talwīḥāt*. Ed. Mūsawī, pp. 614:6–9, 614:22ff, 615:615:5–9, 615:18–20.

144 Quṭb al-Dīn al-Shīrāzī: *Sharḥ Ḥikmat al-ishrāq*, pp. 251:22–252:4 [commenting on

Suhrawardī's arguments against atomism as inadequate, and Quṭb al-Dīn quoted the relevant passages in his commentary.[145]

In addition to these borrowings, Quṭb al-Dīn mentioned some positions of Ibn Kammūna which he did not share, such as the view that the simple cannot emanate from the composite[146] and on the pre-eternity of the soul.[147] In his summaries of Ibn Kammūna's views on these issues, Quṭb al-Dīn evidently used other works by him other than *Sharḥ al-Talwīḥāt*. With regard to the first issue, Quṭb al-Dīn states that he formulated the argument himself with minimal modifications (*maʿa adnā taghyīr*), using various writings by Ibn Kammūna. Various passages with similar wording to Quṭb al-Dīn's, but none of which seems to have been Quṭb al-Dīn's exclusive source, can be found in his *Maqāla fī l-taṣdīq bi-anna nafs al-insān bāqiya abadan*,[148] *Maqāla fī anna l-nafs laysat bi-mizāj al-badan wa-lā kāʾina ʿan mizāj al-badan*,[149] and *al-Jadīd fī l-ḥikma*,[150] of which the first text mentioned was apparently his main source.[151] In regard to the second issue, viz. Ibn Kammūna's view on the pre-eternity of the soul, Quṭb al-Dīn partly quotes and partly summarizes his *Maqāla fī l-taṣdīq bi-anna nafs al-insān bāqiya abadan*.[152] In both cases Quṭb al-Dīn ascribed the views mentioned to *baʿḍ al-muʿāṣirīn min al-afāḍil* or *baʿḍ al-afāḍil min al-muʿāṣirīn* in order to make it clear that he did not share them. Such references were apparently understood by the readers of the time to refer to Ibn Kammūna.[153]

---

part 1, discourse 3, principle stating that the simple may emanate from the composite] which is taken from part II of Ibn Kammūna's *Sharḥ al-Talwīḥāt* [*Faṣl fī l-jism al-mutaḥarrik wa-l-mayl*]. Eds. Ziai and Alwishah, pp. 119:5–7, 120:6–7, 120:9–11, 120:13–14.

145 Quṭb al-Dīn al-Shīrāzī: *Sharḥ Ḥikmat al-ishrāq*, pp. 234:3–11 [commenting on part 1, discourse 3, principle on the denial of the individual atom] which is taken from part II of Ibn Kammūna's *Sharḥ al-Talwīḥāt*. Eds. Ziai and Alwishah, pp. 29:10–30:1, 30:3–4, 30:6–7.

146 *Sharḥ Ḥikmat al-ishrāq*, pp. 249:10–250:5.

147 *Sharḥ Ḥikmat al-ishrāq*, pp. 431:12–432:14.

148 Ibn Kammūna: "Maqāla fī l-taṣdīq bi-anna nafs al-insān bāqiya abadan," p. 97.

149 Ibn Kammūna: "Maqāla fī anna l-nafs laysat bi-mizāj al-badan wa-lā kāʾina ʿan mizāj al-badan," pp. 138–39.

150 Ibn Kammūna: *Al-Jadīd fī l-ḥikma*, p. 249.

151 *Sharḥ Ḥikmat al-ishrāq*, pp. 431:12–432:7 which is presumably taken from Ibn Kammūna: "Maqāla fī l-taṣdīq bi-anna nafs al-insān bāqiya abadan," pp. 91–92.

152 Ibn Kammūna: "Maqāla fī l-taṣdīq bi-anna nafs al-insān bāqiya abadan," pp. 91–92.

153 There is, for example, a marginal note in ms. Ragıp Paşa 854 (f. 112b) on a

The impact of Ibn Kammūna on Quṭb al-Dīn al-Shīrāzī is even more evident in the latter's *Durrat al-Tāj*. The work consists of a *fātiḥa*, five portions (sg. *jumla*), and an epilogue (*khātima*). The five portions constituting the main body of the work deal with 1. logic, 2. first philosophy (*falsafa-yi ūlā*), 3. physics, 4. mathematics and 5. metaphysics. The sections dealing with first philosophy, physics and metaphysics (*jumla* 2, 3, and 5) are translations of chapters 2, 3, 4, 5, and 6 of Ibn Kammūna's *al-Jadīd fī l-ḥikma*. Quṭb al-Dīn even kept editorial statements that disagree with the context of *Durrat al-tāj*. In the section on first philosophy, for example, he translated Ibn Kammūna's statement that the discussion on *al-kayfiyyāt bi-l-kam al-munfaṣil* belonged to arithmetics, which is beyond the purpose of the book, i.e., *al-Jadīd fī l-ḥikma* (pp. 281–82), although Quṭb al-Dīn treats arithmetics at length in *jumla* 4.

The text of *jumla* 1 of *Durrat al-tāj*, dealing with logic, is far less dependent on the *Jadīd fī l-ḥikma*. The text of the first three chapters (*maqālat*) of this section is independent from Ibn Kammūna's *Jadīd*, although Quṭb al-Dīn was clearly inspired by the work, as the number of chapters and the respective chapter headings correspond exactly with those in the *Jadīd*. Chapters 4, 5, and 7 of the portion on logic are again dependent on Ibn Kammūna's work; here Quṭb al-Dīn translated, with minor exceptions, the corresponding chapters of *al-Jadīd fī l-ḥikma*, supplementing them with copious commentary, exemplifying or other additions. In some cases Quṭb al-Dīn adds critical remarks directed against *baʿḍī mutaʾakhkhirān* or *baʿḍī az mutaʾakhkhirān*, whose identity still needs to be clarified.[154] Throughout chapters 4, 5, and 7, moreover, he makes a number of editorial changes. He further subdivides the material presented in each chapter of the logic section into several *taʿlīm*, even when taking material from Ibn Kammūna. Chapter 6 is a verbatim translation of the corresponding Chapter (*faṣl*) Six of *al-Jadīd fī l-ḥikma*.[155]

---

passage where Quṭb al-Dīn refers to Ibn Kammūna as *baʿḍ al-afāḍil min al-muʿāṣirīn* (= *Sharḥ Ḥikmat al-ishrāq*, p. 431:12) stating: *qīla al-murād bihī faylasūf al-yahūd Ibn Kammūna mawlawī*. For further examples, see below, pp. 136–37, and n. 188.

[154] Cf. *Durrat al-tāj*, vol. 2, pp. 431:17–432:5; vol. 3, p. 424:5–18.

[155] For further details, see our "Quṭb al-Dīn al-Shīrāzī's (d. 710/1311) *Durrat al-Tāj* and Its Sources."

Quṭb al-Dīn al-Shīrāzī had a number of Ibn Kammūna's writings in his possession, among them a copy of Ibn Kammūna's *Sharḥ al-talwīḥāt*, which in 692/1292–93 he gave as a gift to a prince.[156] He also wrote a codex, copied around 685/1286 in Konya, containing Ibn Kammūna's *Kalimāt wajīza*, *Tanqīḥ* and possibly *Jadīd fī l-ḥikma*, which is partially preserved.[157] In addition, two incomplete copies of the codex are extant. One of them (ms. Majlis 593) was written in 1060/1650.[158] This copy does not contain *al-Jadīd fī l-ḥikma*, which is included, however, in a

[156] Ms. Ahmed III 3244 (= Topkapı 6707). In the dedication written by Quṭb al-Dīn's hand (f. 1a), Ibn Kammūna is referred to as *al-Imām al-fāḍil wa-l-ḥakīm al-kāmil*, see below, Chapter Two, Section 1.1.2.

[157] Ms. Marʿashī 12868. The codex contains the following items: (1) Fragment from an unidentified philosophical work (f. 1 [heavily destroyed]); (2) various poems in Arabic (f. 2); (3) various poems in Arabic and Persian (f. 3a); (4) fragment of report about a conversation between Aristotle and Alexander (f. 3b); (5) fragment of *Majlis-i maktūb-i Khʷārazm* by Shahrastānī (ff. 4a–22a); (6) an anonymous Persian history of the Mongols (ff. 22b–39b); (7) Ibn Kammūna: *Kalimāt wajīza* (ff. 40a–59b); (8) Ibn Kammūna: *Tanqīḥ al-abḥāth* (ff. 59b–126b); (9) *min kalām Arisṭū* (ff. 126b–128b); (10) various poems in Persian (f. 129); (11) Samawʾal al-Maghribī: *Ifḥām al-yahūd* (ff. 130a–144b); (12) *Nuskhat masʾala waradat ʿalā Samawʾal min baʿd al-zanādiqa al-mutafalsifa* (ff. 144b–145b); (13) *Nuskhat al-jawāb* [li-Samawʾal al-Maghribī] (ff. 145b–147a); (14) fragment from an unidentified work apparently on medicine (f. 147). See also Aḥmad al-Ḥusaynī, Sayyid Maḥmūd Marʿashī Najafī [et al.]: *Fihrist-i Kitābkhāna-yi ʿUmūmī-yi Haḍrat-i Āyat Allāh al-ʿuẓmā Najafī Marʿashī* 1–. Qum 1975–, vol. 32, pp. 637–43, 1151–57. We are grateful to Sayyid Maḥmūd Marʿashī Najafī for having provided us with a copy of this codex.

[158] Ms. Majlis 593 consists of 203 pp. [from p. 100 onwards, the codex is also foliated, p. 100 being f. 17a], 19 lines to a page, 24,5 × 15,5 cm; it contains the following items: (1) Fragment from an unidentified work, apparently on philosophy (p. 1); (2) Fragment from an unidentified philosophical work (p. 2); (3) Ibn Kammūna: *Kalimāt wajīza* (pp. 3–40); (4) *Majlis-i maktūb-i Khʷārazm* by Shahrastānī (pp. 40:5–69); (5) Ibn Kammūna: *Tanqīḥ al-abḥāth* (pp. 69–178); (6) Samawʾal al-Maghribī: *Ifḥām al-yahūd* (pp. 178–199:16); (7) *Nuskhat masʾala waradat ʿalā Samawʾal min baʿd al-zanādiqa al-mutafalsifa* (pp. 199:17–200:16); (8) *Nuskhat al-jawāb* [li-Samawʾal al-Maghribī] (pp. 200:17–202); (9) Excerpt from an unidentified philosophical work (pp. 202:15–203). A microfilm of ms. Majlis 593 is preserved in the Central Library of Tehran University (no. 2320); see Muḥammad Taqī Dānishpazhūh: *Fihrist-i mīkrūfīlmhā-yi khaṭṭī-yi kitābkhāna-yi markazī-yi dānishgāh-i Tihrān* 1–3. Tehran 1348–63/1969–84, vol. 1, p. 673.—The codex contains the following colophons: p. 40 : *wa-qad tamma taswīduhū fī shahr Ṣafar li-sanat 1060 min al-hijra al-muqaddasa* (at the end of Ibn Kammūna's *Kalimāt wajīza*); p. 68: *nuqila min khaṭṭ al-ʿAllāma al-Shīrāzī* (at the end of the *Majālis* al-Shahrastānī); p. 178 (at the beginning of *Ifḥām al-yahūd*): *nuqilat min khaṭṭ al-ʿAllāma al-Shīrāzī*; p. 199 (at the end of *Ifḥām al-yahūd*): *ṣūrat khaṭṭ al-ʿAllāma al-Shīrāzī*.

second copy taken from the Shīrāzī codex (ms. Fâtih 3141).[159] The Fâtih codex was written by Tāj al-Dīn al-Kirmānī, a student of Quṭb al-Dīn al-Shīrāzī.[160] Although the scribe does not mention his source, there are indications which make it patently clear that he had the original Shīrāzī codex at his disposal. His codex contains most of the writings also included in the original codex, and the similarity of the versions of Ibn Kammūna's *Tanqīḥ* and *Kalimāt wajīza* preserved in both codices likewise points to the Shīrāzī codex as its *Vorlage*. It is likely that Quṭb al-Dīn also possessed a copy of Ibn Kammūna's *Maqāla fī l-taṣdīq bi-anna nafs al-insān bāqiya abadan* which he used in his commentary on *Ḥikmat al-ishrāq*.

Presumably Quṭb al-Dīn al-Shīrāzī encouraged his students to read the writings of Ibn Kammūna. Numerous copies of the works of Ibn Kammūna were made between 690/1291 and 710/1309 when Quṭb al-Dīn was active as a teacher.[161] It is possible that apart from Tāj al-Dīn al-Kirmānī there were other students of Quṭb al-Dīn, whose connection to him as a teacher has not been as yet established, who copied the works of Ibn Kammūna.[162]

---

[159] Ms. Fâtih 3141 (293ff, 20/25 lines to a page, 8 x 16,5 cm) contains the following items: (1) Ibn Kammūna: *Al-Jadīd fī l-ḥikma* ff. 1–158 (= ed. Ḥamīd Marʿīd al-Kabīsī (Baghdad 1403/1982), pp. 145:1–498:7), ff. 160–172 (= pp. 498:7–531:7), f. 159 (= pp. 531:8–533:3), f. 210 (= ed. pp. 533:3–536:22), ff. 173–201 (= pp. 536:22–598:4); (2) Ibn Kammūna: *Tanqīḥ al-abḥāth*, ff. 202–209 (= ed. Perlmann, pp. 2:1–16:4), ff. 211–263 (= pp. 17:2–108:8, 111:9–11); (3) Correspondence of Samawʾal al-Maghribī with an anonymous accuser, ff. 264–265; (5) Ibn Kammūna: *Kalimāt wajīza* ff. 266–286; (6) Excerpts from various unidentified historical, philosophical and poetic works, partly Arabic, partly Persian, ff. 286–293.

[160] The manuscript of the *ijāza* issued by Quṭb al-Dīn al-Shīrāzī to Tāj al-Dīn al-Kirmānī (dated 696/1296–97; ms. Yūsuf Agha (Konya) 6624, fol. 281) which may contain more information on what Kirmānī studied with Quṭb al-Dīn was unavailable to us.

[161] See below, Chapter Two, Sections 1.1.2 and 1.3.1 and *passim*.

[162] A preliminary list of his students is given by Walbridge (*Science of Mystic Lights*, pp. 172–74: Appendix B: Quṭb al-Dīn's Students). An in-depth study of Quṭb al-Dīn's teaching activities would require, among other things, a thorough investigation of all extant manuscripts of his writings. Such an investigation would also bring to light some of the erroneous attributions, such as his presumably alleged commentaries on Ibn Sīnā's *Ishārāt* and his *Najāt*. For his commentary on Ibn Sīnā's *Najāt*, reference is given to ms. Damat İbrahim Paşa 817, and for his commentary on the *Ishārāt* to ms. Ragıp Paşa 861; see John Walbridge: *The Philosophy of Quṭb al-Dīn Shīrāzī. A Study in the Integration of Islamic Philosophy*. PhD thesis: Harvard University Cambridge, Mass., 1983,

For over a century and a half after the death of Quṭb al-Dīn al-Shīrāzī, knowledge of Ibn Kammūna among Muslims remained limited to his commentary on Suhrawardī's *Talwīḥāt*. In the first half of the 8th/14th century, glosses were written on the commentary by Naṣīr al-Dīn al-Kāshī (or al-Kāshānī) al-Ḥillī.[163] Another philosopher of the 8th/14th century who made use of Ibn Kammūna's *Sharḥ al-Talwīḥāt* was Quṭb al-Dīn Muḥammad al-Rāzī al-Buwayhī (d. 766/1365), who quotes a long passage from the commentary in his *Risāla fī l-taṣawwur wa-l-taṣdīq*, including Ibn Kammūna's definitions of conception and assent, indicating his source.[164] He was presumably also aware that Quṭb al-Dīn al-Shīrāzī followed Ibn Kammūna on this issue, for immediately after quoting Ibn Kammūna's words he refers to Quṭb al-Dīn as "one of them" (*baʿḍahum*) and quotes a few lines from his *Sharḥ Ḥikmat al-Ishrāq*,

pp. 244, 247. An examination of these two manuscripts shows, however, that the former is Fakhr al-Dīn al-Isfarāyinī's commentary on the *Najāt*, while the latter is Quṭb al-Dīn al-Rāzī's *al-Muḥākamāt bayn Sharḥay al-Ishārāt*. For a preliminary overview of manuscripts of writings of Quṭb al-Dīn, see ibid., pp. 235–76: Appendix C: Quṭb al-Dīn's Works. The list was compiled on the basis of reference works and manuscript catalogues, and Walbridge apparently did not have the opportunity to inspect the manuscripts himself.

[163] Extant in the margins of the commentary in ms. Ragıp Paşa 852 (completed on 17 Rabīʿ I 752/14 May 1351). See below, Chapter Two, Section 1.1.2.—Kāshī also wrote glosses on some other philosophical and theological writings, viz. Ṭūsī's supercommentary on Ibn Sīnā's *Ishārāt* and on Shams al-Dīn al-Iṣfahānī's commentary on Ṭūsī's *Tajrīd al-ʿaqāʾid*. On Kāshī's life and work, see *Dharīʿa*, vol. 2, p. 223 no. 879; vol. 6, p. 112 no. 601; vol. 13, p. 365 no. 1361; Āghā Buzurg: *Ṭabaqāt aʿlām al-shīʿa*, vol. 3, pp. 149–50; Abū ʿAbd Allāh b. ʿĪsā al-Afandī: *Riyāḍ al-ʿulamāʾ* 1–6. Ed. Aḥmad al-Ḥusaynī. Qum 1401/1981, vol. 4, pp. 180–82; vol. 5, p. 247; Muḥsin al-Amīn: *Aʿyān al-shīʿa* 1–11. Beirut 1403/1983, vol. 8, p. 309; Muḥammad b. Ḥasan al-Ḥurr al-Āmilī: *Amal al-āmil* 1–2. Ed. Aḥmad al-Ḥusaynī. Baghdad 1385/1965, vol. 2, p. 202 no. 612; ʿAbbās al-Qummī: *Al-Kunā wa-l-alqāb* 1–3. Tehran 1368/1989, vol. 3, p. 218; Nūr Allāh Shushtarī: *Majālis al-muʾminīn* 1–2. Ed. Aḥmad ʿAbd al-Munāfī. Tehran 1375–76/1955–56, vol. 2, pp. 216–17; ʿAbd Allāh Niʿma: *Falāsifat al-shīʿa. Ḥayātuhum wa-ārāʾuhum*. Beirut 1962, pp. 314–16.—Among Kāshī's students were Sayyid Ḥaydar Āmulī (d. after 787/1385), ʿAbd Allāh Sharafshāh to whom Kāshī issued an *ijāza* in 759/1357–58 (cf. Āghā Buzurg: *Ṭabaqāt aʿlām al-shīʿa*, vol. 5, pp. 122, 149, 153, 185), and al-Miqdād al-Suyūrī (d. 826/1423), who repeatedly refers to his teacher's views in his writings; cf. Sabine Schmidtke: *Theologie, Philosophie und Mystik im zwölferschiitischen Islam des 9./15. Jahrhunderts. Die Gedankenwelten des Ibn Abī Ǧumhūr al-Aḥsāʾī (um 838/1434–35—nach 906/1501)*. Leiden 2000, p. 46.

[164] Quṭb al-Dīn al-Rāzī: "al-Risāla al-maʿmūla fī l-taṣawwur wa-l-taṣdīq," pp. 116:9ff, taken from Ibn Kammūna: *Sharḥ al-Talwīḥāt*. Ed. Mūsawī, pp. 9:5–10:15. Quṭb al-Dīn al-Rāzī introduces the quote with the words: *qāla shāriḥ al-Talwīḥāt*.

which are Quṭb al-Dīn's concluding words to the same quotation taken from Ibn Kammūna's commentary on the *Talwīḥāt* with no mention of its origin.[165] Shams al-Dīn Muḥammad Ibn Mubārakshāh al-Bukhārī (d. after 760/1359) also used Ibn Kammūna's *Sharḥ al-Talwiḥāt* in his commentary on Kātibī's *Ḥikmat al-ʿayn* and explicitly referred to it.[166] Like Quṭb al-Dīn al-Rāzī, he was aware of Ibn Kammūna's impact on Quṭb al-Dīn al-Shīrāzī's *Sharḥ Ḥikmat al-Ishrāq*, and he is the first to mention it explicitly.[167]

In the 9th/15th century, Jalāl al-Dīn Muḥammad al-Dawānī (d. 908/1502) departed from the pattern of his predecessors by using a wider range of writings by Ibn Kammūna.[168] In his commentary on *Hayākil al-nūr*, completed on 11 Shawwāl 872/4 May 1468, Dawānī refers to Ibn Kammūna's discussion of the proof for the uniqueness of God (*istiḥālat wujūd wājibayn / waḥdat wājib al-wujūd*).[169] Ibn Kammūna had dealt with this issue first in his *Maṭālib*, completed in Shaʿbān 657/ July-August 1259. There he sets out to prove that the species (*nawʿ*) of the Necessary Existent is specific to Him, i.e., that it is not shared by any other being (*wājib al-wujūd yajibu an yakūna nawʿuhū fī shakhṣihī ay lā yakūnu min dhālika l-nawʿ siwā shakhṣ wāḥid*),[170] and the subsequent proof

---

165 Quṭb al-Dīn al-Rāzī: "al-Risāla fī l-taṣawwur wa-l-taṣdīq," p. 121:5: *qāla baʿḍuhum fī l-farq bayn al-ḥukm wa-l-taṣdīq...* The quote (p. 121:5–9) is taken from Quṭb al-Dīn al-Shīrāzī: *Sharḥ Ḥikmat al-ishrāq*, pp. 41:24–42:3.

166 Shams al-Dīn Muḥammad Ibn Mubārakshāh al-Bukhārī: *Sharḥ Hikmat al-ʿayn*. Kazan 1319/1901, pp. 275:8–11 (*qāla ʿIzz al-Dawla fī sharḥihī li-l-Talwīḥāt...*), 286:30 (*ʿIzz al-Dawla fi sharḥihī li-l-Talwīḥāt*).

167 Ibn Mubārakshāh al-Bukhārī: *Sharḥ Hikmat al-ʿayn*, pp. 286:30–287:1: *hakadhā dhakarahū ʿan ʿIzz al-Dawla fī sharḥihī li-l-Talwīḥāt wa-ṭabaʿahū jamʿ min al-ʿulamāʾ ḥattā ṣāḥib al-ḥawāshī ṭāba tharāhu fī sharḥihī li-l-Ishrāq*. The quotation preceding this final note (p. 286:26–30) is taken from *Sharḥ al-Talwīḥāt*. Eds. Ziai and Alwishah, pp. 118:17–119:4, and it is quoted verbatim by Quṭb al-Dīn al-Shīrāzī in his *Sharḥ Ḥikmat al-ishrāq*, p. 252:21–24.—The only extant version of Quṭb al-Dīn al-Shīrāzī's glosses ("*al-ḥawāshī al-quṭbiyya*") on *Ḥikmat al-ʿayn* are the quotations from them included by Ibn Mubārakshāh in his commentary.—On Ibn Mubārakshāh, see Ḥasan Ṭārimī: "Bukhārī, Muḥammad b. Mubārakshāh." In *Dānishnāma-yi jahān-i Islām*, vol. 2, pp. 384–85.

168 For Jalāl al-Dīn al-Dawānī, see Stephan Pohl: *Zur Theosophie im nachmongolischen Iran. Leben und Werk des Ǧalāladdīn ad-Dawwānī (gest. 902/1502)*. Bochum 1997 [unpublished manuscript]. For his inventory of his works, see Reza Pourjavady: "Kitābshināsī-yi āthār-i Jalāl al-Dīn Dawānī." *Maʿārif* 15 i/ii (1377/1998), pp. 81–138.

169 Jalāl al-Dīn al-Dawānī: "Shawākil al-ḥūr fī sharḥ Hayākil al-nūr," p. 166:11–13.

170 Ibn Kammūna: "al-Maṭālib al-muhimma," p. 77:21. In his *al-Jadīd fī l-ḥikma*

he adduces is meant to show the impossibility of His species being shared by two necessary beings (*law ḥaṣala min nawʿihī shakhṣān…*).[171] Ibn Kammūna concludes the proof by pointing out that all it shows is that there cannot be two necessary existents sharing the same essence (*fa-qad bāna min hādhā istiḥālat wujūd wājibayn mutashārikayn fī l-māhiyya*). It is not an adequate proof that there cannot be two necessary existents, each being the only representative within its own specific species (*wa-lā yantiju hādhā l-burhān istiḥālat wujūd wājibayn kayfa kānā li-annahū min al-jāʾiz fī l-ʿaql an yakūna fī l-wujūd mawjūdān nawʿ kull wāḥid minhumā fī shakhṣihī wa-yakūnān mutashārikayn fī wujūb al-wujūd*). To rule this out, he continues, would require a different proof, and he admits that so far he has been unable to work out an appropriate proof (*fa-hādhā l-iḥtimāl in imtanaʿa kāna imtināʿuhū bi-burhān ghayr hādhā l-burhān wa-lam aẓfar bihī ilā l-ān*).[172] Dawānī inserts in his *Shawākil al-ḥūr fī sharḥ Hayākil al-nūr* the whole passage that contains Ibn Kammūna's criticism, together with the latter's confession that he had been unable to formulate an appropriate proof to settle the issue. Dawānī does not give the title of the *Maṭālib*, but refers to his source as "one of his writings" (*baʿḍ taṣānīfihī*).[173] Ibn

---

(p. 535:5–6), he formulates this principle as follows: *kull mā huwa wājib al-wujūd li-dhatihī fa-inna nawʿahū lā budda wa-an yanḥaṣira fī shakhṣihī…* Quṭb al-Dīn al-Shīrāzī translates this passage as follows (*Durrat al-tāj*. Ed. Mishkāt, p. 815:21–22): *har chih ān wājib wujūd ast (li-dhātihi) nawʿ-i ān lā budd bāshad kī munḥaṣir bāshad dar shakhṣ-i ū.*

[171] Cf. Ibn Kammūna: "al-Maṭālib al-muhimma," p. 77:21–31. The complete proof reads as follows:

وبرهان ذلك أنه لو حصل من نوعه شخصان لكانا متشاركين في الماهية مختلفين في الهوية، إذ لو لم يقرن بأحد منهما هوية لزم حصول الاثنينية بلا امتياز وهو بديهي البطلان. فتلك الهوية إما أن يكون نفس ماهية واجب الوجود أو أمر زائد عليها، فإن كانت نفس ماهيته وجب انحصار نوعه في شخصه وهو المطلوب. وإن لم يكن كذلك لزم المحال من وجهين: أحدهما أنها لو كانت زائدة لكانت محتاجة إلى الماهية فتكون ممكنة، فافتقرت إلى علة. فإن كانت تلك العلة نفس ماهية الواجب لزم من تحقق تلك الماهية تحققِ تلك الهوية، فلا يوجد من نوعه شخصان وهو المطلوب. وإن كانت علة الهوية غير ماهية الواجب، كان الواجب مفتقراً في هويته إلى غيره وكل ما هو كذلك فليس بواجب، فالواجب ليس بواجب هذا خلف. والوجه الآخرِ أن اختصاص كل واحِد من الهويتين بأحد الشخصين إن كان لأجل الماهية أو لوازمها وجب في المفارق أن يكون لازماً وهو محال، ولزم أيضا أن يكونِ نوع الواجب في شخصه وهو خلاف الفرض. وإن كان لأجل أمر خارج عن الذات وما يلزمها فيكون الواجب محتاجاً في هويته إليه وذلك باطل.

[172] Ibn Kammūna: "al-Maṭālib al-muhimma," p. 78:2–3.

[173] Dawānī: "Shawākil al-ḥūr," p. 166:11–16. The complete passage reads as follows:

ولذلك قال ابن كمونة في بعض تصانيفه أن هذا البرهان ينتج استحالة وجود واجبين متشاركين في الماهية ومن الجائز في العقل أن يكون في الوجود واجبان ينحصر نوع كل واحد منهما في شخصه ويكونان مشتركين في وجوب الوجود. فهذا الاحتمال إن امتنع كان امتناعه ببرهان غير هذا البرهان ولم أظفر به إلى الآن. هذا ما ذكره في هذا الكتاب.

Kammūna seems to have pondered the problem intensively after having completed his *Maṭālib*, for he supplies a proof in his *Taqrīb al-maḥajja* completed in Dhu l-Ḥijja 657/November 1259, i.e., some three months after the composition of the *Maṭālib*. Here he argues that all parts of the world are inseparably linked to each other so that the entirety of creation constitutes one composite. If there were two necessary existents they would both necessarily be perfect causes (*ʿilla tāmma*). This, however, is incompatible with creation being one composite.[174] Dawānī does not seem to have known Ibn Kammūna's *Taqrīb al-maḥajja*, but he was familiar with his *al-Jadīd fī l-ḥikma* (completed in 676/1278), containing a detailed discussion of this issue.[175] Here Ibn Kammūna's main argument is that if there were two necessary existents, regardless of whether they are of the same or different species, they would share the necessity of their existence (*wujūb al-wujūd*), and since the two necessary existents are not the same, the necessity of existence could not be part of their essence (*māhiyya*) or reality (*ḥaqīqa*) but would have to be something additional, accidental (*ʿaraḍī*), which is absurd.[176] Dawānī adduces this

---

[174] See below, Chapter Three, Text Three, p. 206.—The complete argument reads as follows:

> ومدبر العالم جلّت عظمته واحد لأن أجزاء العالم مرتبط بعضها بالبعض، بمعنى انتفاعه به كما يظهر مما مضى. فهو شخص واحد مركب منها، فلو وجد إلهان فإن كان أحدهما مستبدّاً بتدبيره اجتمع علتان تامتان على معلول واحد بالشخص أو ببعضه والآخر بالآخر وجب استبداده بتدبيرهما لخلقه أحدهما لأجل الآخر. فلو كان الثاني مؤثراً في شيء منهما عاد المحال، وإن لم يكن واحد منهما مستبدا به ولا شيء منه لم يكونا إلهين. وهذا برهان غريب في التوحيد اخترعت تقريره على هذا الوجه وختمت به هذا المختصر الذي من اطلع على فوائده وأسراره وتأمل ما ذكروه في المطولات تثبتت له منزلته.

[175] Ibn Kammūna: *Al-Jadīd fī l-ḥikma*, pp. 535–42. The argument that he had formulated in his *Taqrīb al-maḥajja* is to be found here on p. 540:14ff.

[176] Ibn Kammūna: *Al-Jadīd fī l-ḥikma*, pp. 537:3–538:4. The complete argument runs as follows:

> وإذ قد ثبت أن نوع الواجب لا يدخل تحته شخصان فصاعدا فنقول الآن أنه يمتنع وجود شخصين هما واجبا الوجود سواء كانا من نوع واحد أو من أكثر. أما إن كان نوعهما واحداً فلما مر. وأما إن كان نوع كل واحد منهما مغايرا لنوع الآخر فوجوب الوجود يجب إذ ذاك ألا يكون نفس حقيقتها وإلا لكان نوعهما واحداً. فإن مفهوم وجوب الوجود لا يختلف وألا يكون داخلاً في حقيقتهما. وإلا لكان الواجب مركباً، إما من أمرين وجوديين إن كان وجوب الوجود وجوديا أو من وجودي وعدمي إن كان هو أو الجزء الآخر عدميا، أو من عدميين. والكل يقتضي أن لا يكون الواجب واجباً. فإن ما يفتقر إلى جزئه الذي هو غيره فليس بواجب، والذي يتقوم بأمر عدمي فليس بموجود فضلاً عن أن يكون واجب الوجود لا سيما إذا لم يكن في أجزائه ما هو وجودي البتة. فلو صح وجود واجبين من نوعين لكان وجوب الوجود عرضياً لازماً لكل واحد منهما. فيكون كل واحد منهما يشارك الآخر في وجوب الوجود ويمتاز عنه بتمام ماهيته. وحينئذ لا يكون معروض وجوب الوجود في ذاته واجباً لا بمعنى انفكاكه عن الوجود الواجبي، بل بمعنى أن العقل يمكن أن يلاحظه وحده من غير ملاحظة ذلك الوجود، فلا تكون ماهية المعروض هي المؤثرة فيه، إذ اللاشيء لا يؤثر إلا إذا كان في الأعيان فيلزم أن يتقدم على وجوده تقدماً بالذات. وليس ذلك العارض المشترك بينهما واجباً في نفسه، لأنه لا

proof in his *Shawākil al-ḥūr*, quoting from it the most significant passages which, he explicitly states, are Ibn Kammūna's words in his *Kāshif* (i.e. *al-Jadīd fī l-ḥikma*).[177] Dawānī adduces some critical remarks against this proof and argues that regardless of whether the existence of God and of contingent beings is the same only by analogy (*bi-l-tashkīk*) (as is upheld by the Peripatetics) or whether existence is understood as light (*nūr*) as maintained by the people of illumination (*ʿalā dhawq ahl al-ishrāq*), with regard to God there is nothing but pure existence. Thus the entire discussion about whether there can be two necessary existents and the subsequent considerations of the possible ontological foundations of the necessity of their respective existence is superfluous.[178] In his *Risālat Ithbāt al-wājib al-jadīda* Dawānī discusses the issue again in more detail.[179] Here again he quotes Ibn Kammūna's remarks from the *Maṭālib* and makes the same arguments he presented in his *Shawākil al-ḥūr*. He adds that this issue is the most subtle and most significant problem in the discipline of metaphysics and that he is the first to have dealt with it adequately.[180] The problem of the proof for the uniqueness of the Necessary Existent Dawānī derived from Ibn Kammūna's writings subsequently developed into one of the most popular philosophical problems, and it was later

---

يوجد في الخارج من غير تخصيص يزيل اشتراكه وإذا لم يكن واجباً فهو ممكن فيفتقر إلى علة هي غير معروضة فيفتقر الواجب الوجود في وجوده إلى علة خارجة عنه فلا يكون الواجب واجباً هذا خلف.

[177] Dawānī: "Shawākil al-ḥūr," p. 166:15–19. The complete passage which is taken from Ibn Kammūna's *al-Jadīd fī l-ḥikma*, p. 537:6–9, 15–16, reads as follows:

وذكر في كتاب الكاشف أن تعدد الواجب محال مطلقاً أما إن كان نوعهما واحداً فلما مرّ وأما إن كان نوع كل منهما مغايراً لنوع الآخر فلأن وجوب الوجود يجب إذ ذاك أن لا يكون نفس حقيقتهما وإلا لكان نوعهما واحداً فإن مفهوم وجوب وجود لا يختلف وأن لا يكون داخلاً في حقيقتهما وإلا لكان الواجب مركباً فلو صح وجود واجبين من نوعين لكان وجوب الوجود عرضياً لازماً لكل واحد منهما.

[178] Dawānī: "Shawākil al-ḥūr," pp. 167:4–171:4.

[179] Jalāl al-Dīn al-Dawānī: "Risālat ithbāt al-wājib al-jadīda." In *Sabʿ Rasāʾil li-l-ʿAllāma Jalāl al-Dīn Muḥammad al-Dawānī (al-mutawaffā 908 h.) wa-l-Mullā Ismāʿīl al-Khʷājūʾī al-Iṣfahānī (al-mutawaffā 1173 h.)*. Ed. Sayyid Aḥmad Tūysirkānī. Tehran 1381/2002, pp. 125–37.

[180] Dawānī: "Risālat ithbāt al-wājib al-jadīda," pp. 127:15–128:5. The relevant passage reads as follows:

وهذا مجمل ما ذكرناه مفصلاً وهو برهان متين مختصر لا يرد عليه ما ذكره ابن كمونة في بعض تصانيفه من أن البراهين التي ذكروها إنما تدل على امتناع تعدد الواجب مع اتحاد الماهية وأما إذا اختلف فلا بد له من برهان آخر ولم أظفر به إلى الآن. أقول: لأن هذا المطلوب أدق المطالب الإلهية وأحقها بأن يصرف فيها الطالب وكد وكده ولم أر في كلام السابقين ما يصفو عن شوب ريب ولا في كلمات اللاحقين ما يخلو عن وصمة عيب فلا عليّ إلا أن أشبع فيه الكلام حسبما يبلغ إليه فهمي القاصر وإن كنت موقنا بأنه سيصير عرضة لملام اللئام ...

named after Ibn Kammūna as "Ibn Kammūna's Sophistry regarding the Uniqueness of God" (*shubhat Ibn Kammūna fī l-tawḥīd*).

Dawānī also referred briefly to Ibn Kammūna's *Sharḥ al-Talwīḥāt* in his commentary on Suhrawardī's *Hayākil al-nūr* when dealing with a passage on the pre-eternity of the soul.[181] Moreover, in his second supercommentary on Ṭūsī's *Tajrīd al-ʿaqaʾid*, and later on in his independent treatise on the liar paradox, *Nihāyat al-kalām fī ḥall shubhat jadhr al-aṣamm*, he quotes from a correspondence between Ibn Kammūna and Kātibī on this problem.[182]

There is only little indication that philosophers who came after Dawānī used any work by Ibn Kammūna other than his commentary on Suhrawardī's *Talwīḥāt*. Moreover, it also seems that after Dawānī most scholars no longer learned of Ibn Kammūna through his own writings, but mostly through secondary sources. The most important sources for the later reception of Ibn Kammūna were the writings of Dawānī and to some extent also those of Quṭb al-Dīn al-Shīrāzī, although later philosophers were often no longer aware of the influence that Ibn Kammūna had on Quṭb al-Dīn's writings. Dawānī's contemporary, Ṣadr al-Dīn al-Dashtakī (d. 903/1498), for example, quoted Ibn Kammūna's solution to the liar paradox from his correspondence with Kātibī in his second supercommentary on ʿAlī Qūshchī's (d. 879/1474–75) commentary on Ṭūsī's *Tajrīd al-ʿaqāʾid* and in his independent treatise on the problem, *Risāla fī shubhat jadhr al-aṣamm*.[183] Since Dashtakī did not add anything to what had been quoted from the correspondence by Dawānī, it is likely that the quotations in the latter's writings were

---

[181] Dawānī: "Shawākil al-ḥūr," p. 153:4–5, where he refers to *shāriḥ al-Talwīḥāt*.

[182] See Jalāl al-Dīn al-Dawānī: *Ḥāshiya jadīda ʿalā l-Tajrīd*. Ms. Āstān-i Quds 19827, ff. 195b:21–196a:8; idem: "Nihāyat al-kalām fī ḥall shubhat 'kull kalāmī kādhib'." Ed. Aḥad Farāmarz Qarāmalikī. *Nāme-ye Mufīd* 2 i (1375/1996), pp. 113:21–114:6.—See also below, p. 44.

[183] Ṣadr al-Dīn al-Dashtakī: *Ḥāshiya jadīda ʿalā Sharḥ al-Tajrīd*. Ms. Majlis 1998, f. 116b:15–20; idem: "Risāla fī shubhat jadhr al-aṣamm." Ed. Aḥad Farāmarz Qarāmalikī. *Kheradnāme-ye Ṣadrā* 2 v/vi (1375/1997), pp. 79–80.—For another edition of Dashtakī's treatise on the liar paradox and a study, see Afsānah Muḥammadī: *Risāla fī shubhat jadhr al-aṣamm 'kull kalāmī kādhib'*. MA Thesis, University of Tehran, 1381/2002, esp. edition (with separate pagination), pp. 9–11.

his sole source.[184] Ghiyāth al-Dīn al-Dashtakī (d. 949/1542–43) put Dawānī's quotations from Ibn Kammūna's *Maṭālib* and *al-Jadīd fī l-ḥikma* on the arguments for the uniqueness of God in his supercommentary of Dawānī's commentary on Suhrawardī's *Hayākil al-nūr*.[185] Approving of the solution of Ibn Kammūna as quoted by Dawānī from his *Jadīd fī l-ḥikma*, Ghiyāth al-Dīn then reproached Dawānī for having criticized it.[186] In addition, Ghiyāth al-Dīn also had direct access to Ibn Kammūna's *Sharḥ al-Talwīḥāt*, as can be seen from his reference to a comment made by Ibn Kammūna in it on the *burhān al-musāmata* for the limitation of dimensions (*tanāhī abʿād*), which he quotes in part, and of which he approves.[187] On other issues, Ghiyāth al-Dīn disagreed with Ibn Kammūna. He rejected Ibn Kammūna's notion of the pre-eternity of the soul as explained by Quṭb al-Dīn al-Shīrāzī and singled out Ibn Kammūna's main argument in favour of this position, viz. that the cause of a simple can only be caused by a simple whereas the cause of the temporal is always composite; and since the soul is simple it cannot be temporal and therefore must be eternal,[188] pointing out briefly that temporal origination of simples is possible, as is the case

[184] See above n. 182.—The quotation from the correspondence supplied by Dawānī is encountered repeatedly in later works, such as Shihāb al-Dīn Maḥmūd al-Ālūsī's (d. 1270/1854) *Gharāʾib al-ightirāb wa-nuzhat al-albāb fī l-dhahāb wa-l-iqāma wa-l-iyāb*. Ed. Khayr al-Dīn al-Ālūsī. Baghdad 1317/1899–1900, p. 210.

[185] Ghiyāth al-Dīn Manṣūr al-Dashtakī al-Shīrāzī: *Ishrāq hayākil al-nūr li-kashf ẓulumāt shawākil al-ghurūr*. Ed. ʿAlī Awjabī. Tehran 1382/2003, pp. 180:6–15: *wa-ʿlam anna l-shāriḥ naqala ʿan Ibn Kammūna annahū qāla fī baʿḍ taṣānīfihī*...

[186] Dashtakī: *Ishrāq hayākil al-nūr*, pp. 180:16–181:2ff (Dawānī's criticism quoted from his "Shawākil al-ḥūr," p. 166:20–24 ), p. 181:3ff (Dashtakī's commentary).

[187] Dashtakī: *Ishrāq hayākil al-nūr*, p. 291:12–15: *wa-yaqrubu minhu mā raʾaytuhū baʿda dhālika fī Sharḥ al-Talwīḥāt li-Ibn Kammūna fa-innahū qāla baʿda taqrīr burhān al-musāmata*.... The entire passage is based on Ibn Kammūna: *Sharḥ al-Talwīḥāt*. Eds. Ziai and Alwishah, p. 71:4–8; ed. Mūsawī, p. 330:6–9.

[188] Dashtakī: *Ishrāq hayākil al-nūr*, p. 167:11–12: *wa-li-Ibn Kammūna dalīl istadalla bihī ʿalā qidam al-nafs wa-ḥarrarahū al-ʿAllāma al-Shīrāzī*... Dashtakī paraphrases in the following (pp. 167:13–169:3) Qutb al-Dīn al-Shīrāzī's *Sharḥ Ḥikmat al-ishrāq*, pp. 431:12–432:14. At the beginning of this passage, Quṭb al-Dīn refers to Ibn Kammūna as *baʿḍ al-afāḍil min al-muʿāṣirīn*. Dashtakī introduces the argument with *qāla* (p. 167:13) which refers back to Ibn Kammūna whom he mentioned two lines before by name (p. 167:11). This indicates that it was evident to Dashtakī that the phrase *baʿḍ al-afāḍil min al-muʿāṣirīn* referred to Ibn Kammūna.

with simple accidents (*annahū manqūḍ bi-ḥudūth aʿrāḍ basīṭa*).[189] Najm al-Dīn Maḥmūd al-Nayrīzī (Ḥājjī Maḥmūd, d. after 943/1536) repeatedly adduced explanations provided in Ibn Kammūna's *Sharḥ al-Talwīḥāt* in his commentary on Suhrawardī's *al-Alwāḥ al-ʿimādiyya*.[190] His reliance on Ibn Kammūna is much more extensive than he admits to, and concerns issues such as the proof for the existence of the soul[191] and divine knowledge.[192]

In his supercommentary on the section on metaphysics in Qūshchī's commentary on Ṭūsī's *Tajrīd*, Shams al-Dīn al-Khafrī (d. 942/1535–36) discusses in detail the problem of the uniqueness of God.[193] Khafrī refers to it as "Ibn Kammūna's sophistry" (*shubhat Ibn Kammūna*), and he seems to have been the first to use this label and presumably coined it. Moreover, he explicitly blames Ibn Kammūna for having formulated the *shubha*, calling him "The Pride of the Devils" (*iftikhār al-shayāṭīn*).[194] Although Khafrī does not adduce any of the relevant passages from Dawānī's writings, it is likely that his knowledge of the paradox came from the latter's writings. Had he had direct access to Ibn Kammūna's works, it is unlikely that he would have focussed so much on this issue or that he would have judged him on it exclusively. On this issue, moreover, Khafrī also criticized the commentator, ʿAlī Qūshchī, for having followed, or so at least is Khafrī's understanding, Ibn Kammūna.[195] However, there are no indications that Qūshchī was using Ibn Kammūna's writings or even that he adopted any of his positions when discussing the issue of divine uniqueness in his commentary.[196] This also suggests that Khafrī

[189] Dashtakī: *Ishrāq hayākil al-nūr*, p. 169:4–5.

[190] Najm al-Dīn Maḥmūd al-Nayrīzī: *Miṣbāḥ al-arwāḥ fī kashf ḥaqāʾiq al-alwāḥ*. Ms. Ragıp Paşa 853, ff. 3b:15ff, 10bb:16ff, 118b:12ff, 123a:19–123b:21.—A comprehensive study of Maḥmūd al-Nayrīzī is currently being prepared by Reza Pourjavady.

[191] Nayrīzī: *Miṣbāḥ al-arwāḥ*, f. 118b.

[192] Nayrīzī: *Miṣbāḥ al-arwāḥ*, ff. 123–b. This passage is taken from Ibn Kammūna: *Sharḥ al-Talwīḥāt*. Ed. Mūsawī, pp. 879:14–880:14.

[193] Shams al-Dīn Muḥammad b. Aḥmad al-Khafrī: *Taʿlīqa bar Ilāhiyyāt Sharḥ al-tajrīd-i Mullā ʿAlī Qūchshī*. Ed. Fīrūzah Sāʿatchiyān. Tehran 1382/2004, pp. 208:13–209:4.

[194] Khafrī: *Taʿlīqa*, p. 209:2–3: *wa-bi-dhālika yandafiʿu shubhat Ibn Kammūna alladhī bi-ibdāʾihā ṣāra iftikhār al-shayāṭīn*. According to the editor, two out of four manuscripts give his name as *Ibn Kammūna al-yahūdī*; see ibid., p. 209 n. 3.

[195] Khafrī: *Taʿlīqa*, p. 209:15.

[196] See ʿAlī Qūshchī: *Sharḥ Tajrīd al-kalām*. Lithograph edition. Tehran 1301/[1883–84], pp. 359–60.

was unfamiliar with the writings of Ibn Kammūna, for otherwise he would not have made such a claim.

As a result of the popularity of the writings of Dawānī and of Khafrī's supercommentary on Ṭūsī's *Tajrīd* among the philosophers of subsequent generations, Ibn Kammūna now became primarily known as a formulator of philosophical and logical sophistries (*shubahāt*). In addition to *shubhat al-tawḥīd*, two other *shubha*s were now ascribed to him. One of them was the liar paradox, known among Muslim philosophers as *shubhat jadhr al-aṣamm* or *shubhat 'kull kalāmī kādhib'*. ʿIṣmat Allāh b. Kamāl al-Dīn al-Bukhārī (10th/16th c.) was one of the earliest scholars to claim that this paradox originated with Ibn Kammūna. He even states that it is called *aṣamm* ("deaf") because, so he maintained, Ibn Kammūna was deaf.[197] Ibn Kammūna had corresponded about the liar paradox with Kātibī[198] and had written about it in his *al-Jadīd fī l-ḥikma*.[199] However, the attribution of this *shubha* to Ibn Kammūna as its alleged originator is without foundation, as this paradox was well known and had already been dealt with by Muslim scholars before him.[200]

---

197 In his *Sharḥ al-Mughālaṭāt al-Bukhāriyya* which is a commentary on the *Risāla fī l-Mughālaṭāt al-gharība* written by his father Kamāl al-Dīn Maḥmūd b. Niʿmat Allāh al-Bukhārī (d. 947/1540–41). See Aḥad Farāmarz Qarāmalikī: "Kitābshināsī-yi taḥlīlī-yi pārādox-i durūghgū nazd-i dānishmandān-i musalmān," pp. 32–33, 40. For an analysis of Kamāl al-Dīn's treatise, see idem: "Kamāl al-Dīn Maḥmūd Bukhārī u muʿammā-yi jadhr-i aṣamm." *Kheradnāme-ye Ṣadrā* 4 xiv (1377/1999), pp. 53–57.

198 See below, Chapter Two, Section 1.5.4 (Correspondence with Najm al-Dīn al-Kātibī II).

199 Ibn Kammūna: *Al-Jadīd fī l-ḥikma*, pp. 205–6.

200 See Larry B. Miller: "A Brief History of the Liar Paradox." In *Of Scholars, Savants, and their Texts. Studies in Philosophy and Religious Thought. Essays in Honor of Arthur Hyman*. Ed. Ruth Link-Salinger. Frankfurt/Main 1989, pp. 173–82; Farāmarz Qarāmalikī: "Kitābshināsī-yi taḥlīlī-yi pārādox-i durūghgū nazd-i dānishmandān-i musalmān."—The various editions and studies by Farāmarz Qarāmalikī (see below, ABBREVIATIONS AND BIBLIOGRAPHY) constitute a valuable point of departure for a "Catalogue of the *Insolubilia* Literature in Medieval Muslim Philosophy" (the term being borrowed from Paul Vincent Spade: *The Medieval Liar. A Catalogue of the* Insolubilia-*Literature*. Toronto 1975). It is striking that the liar paradox became relevant to Muslim philosophers (apparently only in the Eastern lands of Islam) roughly at the same time as the problem started to be dealt with intensively in Medieval Europe, i.e., at the turn of the 12th/13th century. The methodological questions, that have been formulated for Medieval Europe

The second sophistry ascribed to Ibn Kammūna was the so-called "problem of necessary requirement" (*shubhat al-istilzām*). This *shubha*, which challenged the philosophical notion of the eternity of the world (*qidam al-ʿālam*), apparently originated with Kātibī. In one of his exchanges with Ṭūsī, Ibn Kammūna asked for Ṭūsī's view on this problem which, as he states here explicitly, had been formulated by Kātibī.[201] The first scholar who attributed this *shubha* to Ibn Kammūna is Shams al-Dīn al-Ḥusaynī al-Astarābādī ("Mīr Dāmād", d. 1041/1631). He fails to provide any reference for his claim, simply stating that this problem "has been transmitted from Ibn Kammūna" (*shubhat al-istilzām al-manqūla ʿan Ibn Kammūna*) or labelling it "the famous sophistry of Ibn Kammūna" (*al-mughālaṭa al-mashhūra li-Ibn Kammūna*). The significance the problem had in his eyes is evident, as he devoted several of his writings to it.[202] Mīr Dāmād also discussed the problem of the uniqueness of God, stating that the ascription to Ibn Kammūna was not accurate, as he was not the first to discuss it. Mīr Dāmād's derogatory description of Ibn

---

and that are still being discussed, have so far not been touched upon in Islamic studies: "What were the origins of the discussion of the insolubilia in the Middle Ages? . . . first, *when* did the mediaevals begin to discuss such paradoxes, and second, *where* did they learn about them? Did they inherit the Liar, or did they discover it for themselves?" (Paul Vincent Spade: "The Origins of the Mediaeval *Insolubilia* Literature." *Franciscan Studies* 33 (1973), p. 293). See also Christopher J. Martin: "Obligations and Liars." In *Sophisms in Medieval Logic and Grammar. Acts of the Ninth European Symposium for Medieval Logic and Semantics, held at St Andrews, June 1990*. Ed. Stephen Read. Dordrecht/Boston/London 1993, pp. 357–81.

201 See below, Chapter Two, Section 1.5.1 (Correspondence with Naṣīr al-Dīn Ṭūsī I).—The problem as formulated by Ibn Kammūna in the correspondence runs as follows: "the [temporal] originatedness of the world does not necessitate the negation of reality, and whatever does not necessitate the negation of the reality is itself a reality that resulted from the originatedness of the real world" (*ḥudūth al-ʿālam lā yastalzimu irtifāʿ al-wāqiʿ wa-kull mā lā yastalzimu irtifāʿ al-wāqiʿ fa-huwa wāqiʿ yuntiju anna ḥudūth al-ʿālam wāqiʿ*).

202 Mīr Dāmād: "Taʿlīqāt manṭiqiyya." Ed. ʿAbd Allāh Nūrānī. In *Collected Texts and Papers on Logic and Language*. Eds. Mehdi Mohaghegh and T. Izutsu. Tehran 1974, pp. 303–304 (6. *Shubhat al-istilzām al-manqūla ʿan Ibn Kammūna*), pp. 305–7 (7. *Ḥall mughālaṭat Ibn Kammūna*). Editions of both texts are also included in *Muṣannafāt-i Mīr Dāmād mushtamil bar dah ʿunwān az kitāb-hā u risāla-ha u ijāza-hā u nāma-hā*. Vol. 1. Ed. ʿAbd Allāh Nūrānī. Tehran 1381/2003, pp. 528–29 (12. *Dafʿ shubhat al-istilzām al-manqūla ʿan Ibn Kammūna*), pp. 529–31 (13. *Ḥall mughālaṭat Ibn Kammūna*). Included in the volume is a third treatise on the same issue, entitled *Jawāb al-mughālaṭa al-mashhūra li-Ibn Kammūna* (p. 534 no. 15).

Kammūna in this context as *rajul min al-mutafalsifīn al-muḥdathīn yuʿrafu bi-Ibn Kammūna* suggests that he knew him only on the basis of secondary writings, most likely those of Dawānī and Khafrī.[203]

Ṣadr al-Dīn al-Shīrāzī ("Mullā Ṣadrā", d. 1050/1640) discussed the *shubhat al-tawḥīd* on numerous occasions. The most elaborate version is to be found in his *magnum opus al-Ḥikma al-mutaʿāliya fī l-asfār al-arbaʿa*. Here, he presents his own solution to the *shubha* which, he states, is "ascribed to Ibn Kammūna" (*mimmā yunsabu ilā Ibn Kammūna*).[204] As was the case with Dawānī and particularly with Khafrī, his solution is based on the analysis of the notion of pure existence (*wujūd baḥt*). To conclude his discussion, Mullā Ṣadrā clearly has Khafrī's words in mind when he states that in this way a solution is found for the problem that has been ascribed to Ibn Kammūna, and that "one of them" (*baʿḍahum*)—he refers back to the aforementioned *ahl al-ʿilm wa-l-ʿirfān*—called him the "Pride of the Devils" (*iftikhār al-shayāṭīn*) for having invented the problem.[205] The entire section of Mullā Ṣadrā's *al-Ḥikma al-mutaʿāliya* is entitled *rajm al-shayṭān* ("Stoning Satan") indicating his approval of Ibn Kammūna's designation as *iftikhār al-shayāṭīn*.[206] However, Mullā Ṣadrā's reservation about Khafrī's judgment is evident. For, as was the case with Mīr Dāmād, Mullā Ṣadrā is aware that Ibn Kammūna was not the first to express doubts about the conclusiveness of the proofs for the

---

[203] Mīr Dāmād: "Al-Taqdīsāt." In *Muṣannafāt-i Mīr Dāmād*, p. 115.

[204] Ṣadr al-Dīn al-Shīrāzī: *Al-Ḥikma al-mutaʿāliya fī l-asfār al-ʿaqliyya al-arbaʿa* 1–9. Beirut [5]1419/1999, vol. 1, pp. 131:17–132:3.

[205] Ṣadr al-Dīn al-Shīrāzī: *Al-Ḥikma al-mutaʿāliya*, vol. 1, p. 132:2–4: *wa-bihā yandafiʿu mā tashawwashat bihī ṭabāʾiʿ al-aktharīn wa-taballadat adhhānuhum mimmā yunsabu ilā Ibn Kammūna wa-qad sammāhu baʿḍuhum bi-iftikhār al-shayāṭīn li-ishtihārihī bi-ibdāʿ hādhihī l-shubha al-ʿawīṣa.*

[206] Ṣadr al-Dīn al-Shīrāzī: *Al-Ḥikma al-mutaʿāliya*, vol. 1, pp. 131–32. Cf. also his *Sharḥ al-Hidāya al-athīriyya*, completed in 1029/1620 (Lithograph edition. Tehran 1313/[1895–96], pp. 291–92). He refers to the *shubha* also in his glosses on the section on metaphysics of Ibn Sīnā's *Kitāb al-Shifāʾ* (*Taʿlīqāt ʿalā l-Shifāʾ*. Lithograph edition. Qum 1399/[1978–79], p. 304), in his "Masāʾil al-Qudsiyya" (in *Sih risāla falsafī*. Ed. Sayyid Jalāl Āshtiyānī. Qum 1378/1999, p. 214:16: *al-shubha al-mashhūra al-mansūba ilā Ibn Kammūna*) and in his "Risālat al-Shawāhid al-Rubūbiyya" (in *Majmūʿa-yi rasāʾil-i falsafī-yi Ṣadr al-mutaʾallihīn*. Ed. Ḥāmid Nājī Iṣfahānī. Tehran 1375/1996, p. 285:2: *al-shubha al-mashhūra al-mansūba li-Ibn Kammūna*). See also Fazlur Rahman: *The Philosophy of Mullā Ṣadra (Ṣadr al-Dīn al-Shīrāzī)*. Albany 1975, pp. 134–36.

uniqueness of the Necessary Existent.[207] Elsewhere, he sets this out more specifically, explaining that the *shubha* had explicitly been formulated by Suhrawardī in his *Muṭāraḥāt* and was implied in his *Talwīḥāt*. It was again taken up by Ibn Kammūna, a commentator on Suhrawardī, "in one of his writings" (*fī baʿḍ muṣannafātihī*), Mullā Ṣadrā continues to explain. It has become associated with his name because he was unable to solve it. This was because he followed Suhrawardī in his understanding of existence as a mental notion with no counterpart in reality (*li-iṣrārihī ʿalā iʿtibāriyyat al-wujūd wa-annahū lā ʿayn lahū fī l-khārij*). Mullā Ṣadrā then quoted the relevant passage from Ibn Kammūna's *Maṭālib*, which he presumably took from Dawānī's *Shawākil al-ḥūr* or his *Risālat Ithbāt al-wājib al-jadīda*.[208]

Mulla Ṣadrā's severe attitude to Ibn Kammūna is comprehensible only if one assumes that he did not have immediate access to his writings. As with Khafrī and Mīr Dāmād, he presumably misjudged him, since everything he knew about Ibn Kammūna came from secondary sources. Had he read his works, particularly those written after the *Maṭālib* where Ibn Kammūna offers his own solutions to the problem, he would have taken these into account.[209]

Moreover, all clear quotations from Ibn Kammūna's works in Mullā Ṣadrā's writings seem to have been found in the writings of later scholars,

---

[207] Saḍr al-Dīn al-Shīrāzī: *Al-Ḥikma al-mutaʿāliya*, vol. 1, pp. 132:5–133:2:

فإني قد وجدت هذه الشبهة في كلام غيره ممن تقدمه زماناً وهي أنه لم لا يجوز أن يكون هناك هويتان بسيطتان مجهولتان الكنه مختلفتان بتمام الحقيقة يكون كل واحد منهما واجب الوجود بذاته ويكون مفهوم واجب الوجود منتزعاً منهما مقولاً عليهما قولا عرضياً فيكون الاشتراك بينهما في هذا المعنى العرضي المتنزع عن نفس ذات كل منهما والافتراق بصرف حقيقة كل منهما.

[208] Saḍr al-Dīn al-Shīrāzī: *Al-Ḥikma al-mutaʿāliya*, vol. 6, p. 63:7–12:

والشبهة مما أوردها هو أولاً في المطارحات تصريحاً وفي التلويحات تلميحاً ثم ذكرها ابن كمونة وهو من شراح كلامه في بعض مصنفاته واشتهرت باسمه ولإصراره على اعتبارية الوجود وأنه لا عين له في الخارج تبعاً لهذا الشيخ الإشراقي قال في بعض كتبه: إن البراهين التي ذكروها إنما تدل على امتناع تعدد الواجب مع اتحاد المهية وأما إذا اختلفت فلا بد من برهان آخر ولم أظفر به إلى الآن.

The passages in Suhrawardī's *Muṭāraḥāt* and *Talwīḥāt* that Mullā Ṣadrā presumably had in mind are Suhrawardī: *Opera Metaphysica et Mystica I*, p. 393:9–15 (*Muṭāraḥāt*) and pp. 386, 394–95 (*Talwīḥāt*); see also Sayyid Ḥusayn Sayyīd Mūsawī: "Mullā Ṣadrā u Ibn Kammūna," pp. 55–57.

[209] Ibn Kammūna moreover insists, contrary to Mullā Ṣadrā's claim, that God's existence alone is not a mental notion. See *Sharḥ al-Talwīḥāt* [part III, *mawrid* 1, *talwīḥ* 1]. Ed. Mūsawī, p. 706:21–22: …*fa-lā wujūd ghayr iʿtibārī illā wujūdihī.*

which confirms that he did not have direct access to Ibn Kammūna's ideas. In his treatise on *al-Taṣawwur wa-l-taṣdīq*, Mullā Ṣadrā obviously used Quṭb al-Dīn al-Rāzī's treatise on the same subject when quoting the passage from Ibn Kammūna's *Sharḥ al-Talwīḥāt*.[210] When commenting on a passage concerned with conception and assent in his glosses on Quṭb al-Dīn al-Shīrāzī's *Sharḥ Ḥikmat al-Ishrāq*, Mullā Ṣadrā remarked that Quṭb al-Dīn followed Ibn Kammūna on that issue. To support this, Mullā Ṣadrā quoted the entire passage from *Sharḥ al-Talwīḥāt* which is also to be found in Quṭb al-Dīn al-Rāzī's *Risālat al-taṣawwur wa-l-taṣdīq*.[211]

There are further indications that Mullā Ṣadrā did not have direct access to the writings of Ibn Kammūna. He followed Ibn Kammūna in his view that the simple cannot emanate from the composite, which was part of the latter's argument for the pre-eternity of the soul. Mullā Ṣadrā became familiar with this argument through Quṭb al-Dīn al-Shīrāzī's commentary on *Ḥikmat al-ishrāq*, where Quṭb al-Dīn quoted Ibn Kammūna's argument and criticized him for it.[212] Since Quṭb al-Dīn did not mention Ibn Kammūna by name referring to him only as "one of the contemporary scholars" (*baʿḍ al-muʿāṣirīn min al-afāḍil*),[213] the identity of the holder of this view was not clear to Mullā Ṣadrā, and so he referred to him in his glosses to Quṭb al-Dīn's *Sharḥ Ḥikmat al-ishrāq* as the "one who maintains this position", *hādhā l-qāʾil*, or *al-qāʾil al-madhkūr*.[214] Mullā Ṣadrā quotes Ibn Kammūna's argument in its entirety, together with Quṭb al-Dīn's criticism and his own counter critique, again in his *al-Ḥikma al-mutaʿāliya*, where he is likewise ignorant of the

[210] Saḍr al-Dīn al-Shīrāzī: "Risālat al-taṣawwur wa-l-taṣdīq," pp. 67:7–68:12. See the identical passage in Quṭb al-Dīn al-Rāzī: "al-Risāla al-maʿmūla fī l-taṣawwur wa-l-taṣdīq," pp. 116:9–117:11.

[211] Saḍr al-Dīn al-Shīrāzī: [Glosses on Quṭb al-Dīn al-Shīrāzī's *Sharḥ Ḥikmat al-ishrāq*], p. 41: *kamā qarrarahū al-shāriḥ al-ʿAllāma wa-qad akhadha hādhā l-taḥqīq min kalām Ibn Kammūna fī Sharḥ al-Talwīḥāt ḥaythu qāla fī taqsīm al-ʿilm ilā l-taṣawwur wa-l-taṣdīq: inna ḥuṣūl ṣūrat al-shayʾ*...

[212] Quṭb al-Dīn al-Shīrāzī: *Sharḥ Ḥikmat al-ishrāq*. Lithograph edition, pp. 255–56 [= ed. Muḥaqqiq and Nūrānī, pp. 249:10ff]. See also above, p. 32.

[213] Quṭb al-Dīn al-Shīrāzī: *Sharḥ Ḥikmat al-ishrāq*. Lithograph edition, p. 255:17.

[214] Ṣadr al-Dīn al-Shīrāzī: [Glosses on Quṭb al-Dīn al-Shīrāzī's *Sharḥ Ḥikmat al-ishrāq*], p. 256.

identity of the holder of this position which he ascribes here to *baʿḍ al-muḥaqqiqīn*.[215] It was also through Quṭb al-Dīn that Mullā Ṣadrā learned about the doctrinal result of Ibn Kammūna's argument, the pre-eternity of the soul. Again, the identity of Ibn Kammūna is unknown to him.[216] However, although he approved of the main argument defending this position, Mullā Ṣadrā says that such a conclusion, i.e., the pre-eternity of the soul, cannot be derived from this premise because the soul is simple.[217]

With few exceptions, Ibn Kammūna's philosophical writings remained unknown to other scholars of the Safavid period and beyond.[218] However, mainly because of the *shubhat al-tawḥīd* and the two other *shubha*s mentioned above, the name of Ibn Kammūna continued to appear in philosophical writings in Iran.

In the late Safavid period, *shubhat al-tawḥīd* became one of the most challenging philosophical issues and numerous independent tracts were written on it. One of the earliest examples is a treatise written by ʿAlī b. Faḍl Allāh al-Jīlānī (Gīlānī) al-Fūmanī, presumably from the end of the 10th/16th century. The author quotes the relevant passages from Ibn Kammūna's writings, most probably on the basis of Dawānī's quotations. Moreover, Jīlānī was aware of Khafrī's response, as he used

[215] Ṣadr al-Dīn al-Shīrāzī: *Al-Ḥikma al-mutaʿāliya*, vol. 2, pp. 194:17–197:1, which is a complete quotation from Quṭb al-Dīn's *Sharḥ Ḥikmat al-ishrāq*. Ed. Muḥaqqiq and Nūrānī, pp. 249:10–251:4. Thereupon, Mullā Ṣadrā repeats his approval of the position of *hādhā l-qāʾil* (i.e. Ibn Kammūna) (p. 197:2ff): *wa-innī dhakartu fī l-ḥawāshī allatī ʿallaqtuhā ʿalā dhālika l-kitāb wa-sharḥihī bi-anna kalām hādhā l-qāʾil qawī jiddan fī jamīʿ mā dhakarahū illā fī qidam al-nafs*...; see also Sayyid Ḥusayn Sayyīd Mūsawī: "Mullā Ṣadrā u Ibn Kammūna."

[216] See Ṣadr al-Dīn al-Shīrāzī: *Al-Ḥikma al-mutaʿāliya*, vol. 2, p. 198:5 where he writes again about *hādhā l-qāʾil al-madhkūr* when discussing his view of the pre-eternity of the soul.

[217] Ṣadr al-Dīn al-Shīrāzī: *Al-Ḥikma al-mutaʿāliya*, vol. 2, pp. 197:2–198:17; vol. 8, pp. 379:1–380:2; idem: [Glosses on Quṭb al-Dīn al-Shīrāzī's *Sharḥ Ḥikmat al-ishrāq*], p. 458.

[218] One of the few scholars who was familiar with the writings of Ibn Kammūna was Sayyid Aḥmad ʿAlawī (d. between 1054/1644 and 1060/1650) who quotes from Ibn Kammūna's *Sharḥ al-Talwīḥāt* in his commentary on Mīr Dāmād's *Qabasāt* when discussing the issue of time (*zamān*). See Sayyid Aḥmad ʿAlawī: *Sharḥ al-Qabasāt*. Ed. Ḥāmid Nājī Iṣfahānī. Tehran 1376/1997, p. 456 which is taken from part II of *Sharḥ al-Talwīḥāt*. Ed. Mūsawī, p. 403:13–16; eds. Ziai and Alwishah, pp. 189:15–190:2.

the title Khafrī gave to Ibn Kammūna.[219] Shams al-Dīn Muḥammad b. Niʿmat Allāh al-Jīlānī ("Mullā Shamsā Gīlānī", d. 1064/1654), one of the renowned philosophers at that time, was deeply involved in this discussion. He devoted five works to the uniqueness of God, two of which are specifically concerned with the *shubhat al-tawḥīd* of Ibn Kammūna, namely *Ibṭāl wujūd al-wājibayn* and *Dafʿ shubhat Ibn Kammūna*.[220] Mullā Ṣadrā's student and son-in-law ʿAbd al-Razzāq al-Lāhījī (d. 1072/1661) discusses the *shubha*, formulating his own solution in a philosophical work written in Persian, *Gawhar-i murād*.[221] Āqā Ḥusayn b. Jamāl al-Dīn al-Khʷānsārī (d. 1098/1687) also wrote on the issue in glosses on Ibn Sīnā's *Shifāʾ*.[222] Muḥammad b. ʿAbd al-Fattāḥ Sarāb Tunkābunī (d. 1124/1712) wrote a treatise on the *shubhat al-tawḥīd* (completed in 1103/1691–92) in which he commented on the solutions formulated by Mullā Shamsā Gīlānī and ʿAbd al-Razzāq al-Lāhījī.[223] Moreover,

---

[219] A manuscript of the text is preserved in the Majlis Shūrā-yi Islāmī, ms. Majlis Sinā 602, ff. 129b–130a (cat. 1/389).

[220] Cf. Ibrāhīm Dībājī: "Aḥwāl u āthār-i Mullā Shamsāyī-yi Gīlānī." In *Collected Papers on Islamic Philosophy and Mysticism*. Eds. Mehdi Mohaghegh and Hermann Landolt. Tehran 1971, pp. 53–96. See also Daiber: *Bibliography*, vol. 1, p. 831 no. 8080; *Muʿjam al turāth al kalāmī*, vol. 3, p. 148 no. 5428.

[221] See ʿAbd al-Razzāq Lāhījī: *Gawhar-i murād*. Ed. Muʾassassa-yi taḥqīqāti-yi Imām Ṣādiq. Qum 1383/2004, pp. 204–6; the section is introduced as follows: *ammā shubha-yi mashhūra bih shubhat Ibn Kammūna*.

[222] Āqā Ḥusayn b. Jamāl al-Dīn Khʷānsārī: "Ḥāshiya ʿalā l-Shifāʾ (Ilāhiyyāt)." Ed. Ḥāmid Nāji Iṣfahānī. In *Dabīrkhāna-yi kongiri-yi Āqā Ḥusayn Khʷānsārī*. Qum 1378/1999, pp. 289ff [*taʿlīqa* 46/18]. For the glosses, see also *Dharīʿa*, vol. 13, pp. 24–25 no. 68. Khʷānsārī does not mention Ibn Kammūna explicitly but states that the *shubha* perplexed the thought of most scholars and their minds were obsessed with finding a solution (*al-shubha al-mashhūra allatī tashawwashat bihā ṭibāʿ al-aktharīn wa-taballadat adhhānuhum ʿan dafʿihā*). He then says that the originator of this *shubha* was called "Pride of the Devils" for having produced so intricate and complicated a problem that intelligent people were unable to solve it (*li-ishtihārihī bi-ibdāʿ shubha ʿawīṣa wa-ʿuqda shadīda ʿajazat al-adhkiyāʾ ʿan ḥallihā*). Khʷānsārī presents his own version of the *shubha* and cites various solutions to it including Mullā Ṣadrā's. He rejects them all, stating that because of this *shubha* he is not aware of any rational argument for the unity of God; while there are numerous rational proofs for the existence of the Necessary Existent and for prophecy, the unity of God can be proved only on the basis of the Qurʾān and the *aḥādīth*.—Khʷānsārī also wrote on the liar paradox in his superglosses on Dawānī's glosses (*ḥawāshī qadīm*) on Ṭūsī's *Tajrīd al-ʿaqāʾid*. See Aḥad Farāmarz Qarāmalikī: "Āghā Ḥusayn Khʷānsārī u muʿammā-yi jadhr-i aṣamm." *Kheradnāme-ye Ṣadrā* 3 x (1376/1998), pp. 77–89. Khʷānsārī did not refer to Ibn Kammūna's view on the issue nor did he attribute the paradox to him.

[223] *Risāla fī taḥrīr shubhat al-tawḥīd*. The autograph is preserved in ms. Mahdawī 592/2, ff. 3a–6a.

Tunkābunī wrote an epistle on the liar paradox (completed in 1097/1686) in which he quotes Ibn Kammūna's view, evidently on the basis of the quotation cited by Dawānī in his writings, and then comments on it.[224]

In contrast to his philosophical writings, Ibn Kammūna's comparison of the three religions, *Tanqīḥ al-abḥāth*, never achieved much attention in the Islamic world and indeed remained unknown to many scholars, despite the scandal the work had provoked in Baghdad four years after its completion. Compared with the number of manuscripts of Ibn Kammūna's philosophical works, there are few extant copies of the *Tanqīḥ* transcribed by Muslims. The work is included in the above-mentioned codex written by Quṭb al-Dīn al-Shīrāzī as well as in the two incomplete preserved copies of the codex.[225] These three contain the same version of Ibn Kammūna's *Tanqīḥ* which differs significantly from all other extant manuscript copies of the text containing numerous additions and variants. There is a fourth manuscript of the work of Muslim provenance, which is held in the Chester Beatty Library (Ar. 4965). This copy varies from the three other manuscripts of Muslim provenance and is undated. However, it contains a reading colophon written by the owner of the copy, Shihāb al-Dīn Aḥmad b. Abi l-Qādir b. al-Hādī (?) stating that he read it to his Shaykh al-ʿAllāma Badr al-Dīn b. al-Ṭabbākh al-Ashʿarī at the beginning of 965/1557.[226]

A further indication of how little attention the work received in the Muslim world is the fact that the two tracts which it is known were written to refute it are lost. The first reaction is entitled *al-Durr al-manḍūd fī l-radd ʿalā faylasūf al-yahūd* and was written by Muẓaffar al-Dīn Aḥmad b. ʿAlī b. al-Sāʿātī al-Baghdādī (651–694/1254–1294–5). The second was composed by Zayn al-Dīn Sarījā b. Muḥammad al-Malaṭī (d. 788/1386–7) and is entitled *Nuhūḍ ḥathīth al-nuhūd ilā duḥūḍ khabīth al-yahūd*.[227]

---

[224] *Risālat shubhat jadhr al-aṣamm*. The autograph is perserved in ms. Mahdawī 592/1, ff. 8b–11a.

[225] See above, nn. 157, 158, 159.

[226] We were unable to inspect a fifth manuscript of Muslim provenance, ms. Shīrāz 538.

[227] See below, Chapter Two, Section 1.4.1.—For Ibn al-Sāʿātī, a Ḥanafī scholar who was teaching at the Mustanṣiriyya *madrasa*, see Ibn al-Fuwaṭī: *Majmaʿ al-ādāb*, vol. 5, pp. 260–1 no. 5040. Presumably the earliest reference to his refutation is in Zayn al-Dīn

The only extant Muslim reaction to the *Tanqīḥ* is a comparatively brief treatise by an anonymous author which was apparently composed in the 8th/14th century. In his introduction the author states that it was Ibn Kammūna's *Tanqīḥ* that caused him to write the tract; yet the subsequent main text of the treatise, consisting of two chapters, is rather a polemical treatise against Judaism in its own right, than a refutation of the *Tanqīḥ*. Instead the author uses the *Tanqīḥ*, as well as Samawʾal's *Ifḥām al-yahūd*, as a source for his own treatise.[228] The only other extant refutation of Ibn Kammūna's *Tanqīḥ* was written by a Christian (presumably Jacobite), Ibn al-Maḥrūma, who around 733/1333 composed lengthy glosses (*ḥawāshī*) on Chapters Two and Three of the *Tanqīḥ* dealing with Judaism and Christianity.[229]

Towards the end of the 18th century, Ibn Kammūna's *Tanqīḥ* came to the attention of Āqā Muḥammad ʿAlī Bihbahānī (1144–1216/1732–1801), apparently through the aforementioned Shīrāzī codex[230] or the codex copied in 1060/1650 (ms. Majlis 593).[231] Bihbahānī, a son of the famous Waḥīd Bihbahānī (1118–1207/1706–1792), is mostly known for persecuting and killing some of the leaders of the Niʿmatullāhī order and for his profound hostility towards the Niʿmatullāhīs which has found literary expression in his *Khayratiyya dar ibṭāl-i ṭarīqa-yi ṣūfiyya*.[232] Shortly

---

Qāsim b. Quṭlūbughā: *Tāj al-tarājim fī ṭabaqāt al-Ḥanafiyya*. Ed. Gustav Fluegel. Leipzig 1862, p. 4 no. 10 (= *Tāj al-tarājim fī man ṣannafa min al-Ḥanafiyya*. Ed. Ibrāhīm Ṣāliḥ. Beirut 1412/1992, p. 16 no. 14). Sarījā al-Malaṭī's refutation of the *Tanqīḥ* is often not listed in bio-bibliographical works; see, e.g., Khayr al-Dīn al-Ziriklī: *Al-Aʿlām* 1–8. Beirut 1980, vol. 3, p. 82; ʿUmar Riḍā Kaḥḥāla: *Muʿjam al-muʾallifīn* 1–14. Beirut n.d., vol. 4, p. 209.

228 The text was introduced and edited on the basis of the only extant manuscript (ms. Hamidiye 1447/3) by Perlmann: "'Proving Muhammad's Prophethood', A Muslim critique of Ibn Kammuna," pp. 75–97. For an improved edition, see below, Appendix B: *Kitāb fī ithbāt al-nubuwwa li-Muḥammad ʿalayhi al-salām*.

229 For details, see below, Chapter Two, Section 1.4.1.

230 See above, n. 157.

231 See above, n. 158.

232 Ed. Sayyid Mahdī Rajāʾī. Qum 1412/1991. For manuscripts, see *Dharīʿa*, vol. 7, p. 286 no. 1407; Munzawī: *Fihrist-i nuskha-hā-yi khaṭṭī-yi fārsī*, vol. 2 i, p. 940; *Muʿjam al-turāth al-kalāmī*, vol. 3, p. 233 no. 5820.—On the work and Bihbahānī's relation to the Sufis, see Nasrollah Pourjavady and Peter Lamborn Wilson: *Kings of Love. The Poetry and History of the Niʿmatullāhī Sufi Order*. Tehran 1978, pp. 128–31; Nasrollah Pourjavady:

before his death (in 1216/1801), he completed *Rādd-i shubahāt al-kuffār*, a polemical work against Judaism and Christianity.[233] In his introduction, the author says that he will go into detail about the three religions (*dar laṭā'if-i abḥāth-i milal-i thalāth*) and that he will present his material in an introduction (*muqaddamāt*), three chapters (*sih baḥth*) and an epilogue (*khātima*). He further gives an enlarged title of the work: *Rādd-i shubahat-i kuffār dar dhikr abḥāth-i milal-i thalāth*,[234] which already alludes to Ibn Kammūna's *Tanqīḥ al-abḥāth li-l-milal al-thalāth*. A comparison of *Rādd-i shubahāt al-kuffār* and *Tanqīḥ* reveals that not only the overall structure of Bihbahānī's work is modelled according to Ibn Kammūna's *Tanqīḥ*; both the introduction and the subsequent three chapters of the *Rādd* consist, to a large extent, of translations of extensive portions of the *Tanqīḥ*, although Bihbahānī refrains from explicitly mentioning either the title of this work or its author throughout the entire text of his *Rādd*. Bihbahānī's introduction is, with few omissions, a translation of Ibn Kammūna's first chapter dealing with prophecy in general[235] with only some additional remarks by the author. Bihbahānī's first chapter, dedicated to Judaism, is basically a translation of most of Ibn Kammūna's second chapter on Judaism, with a few additions from other sources.[236] Bihbahānī's second

---

"Opposition to Sufism in Twelvershiism." In *Islamic Mysticism Contested. Thirteen Centuries of Controversies and Polemics*. Eds. Frederick de Jong and Bernd Radtke. Leiden 1999, pp. 614–23.

233 Ed. Sayyid Mahdī Rajā'ī. Qum 1413/1992.

234 *Rādd*, p. 24:18–19. On p. 5:7 he mentioned the title under which the work has been published: *Rādd-i shubahāt al-kuffār*.

235 *Tanqīḥ*, pp. 2–21; transl. *Examination*, pp. 14–39.—Bihbahānī translated the following portions of this chapter: *Tanqīḥ*, pp. 2:3–14 (*wa-l-mumtana'āt*) [= *Rādd*, p. 6:12–7:5], 2:16–4:17 [= 7:6–10:4], 5:3–5 [= 10:4–6], 6:1 [= 10:6–7], 7:13–20 [= 10:10–23], 8:3–22 [= 11:1–12:13], 9:22–23 (*makhṣūṣ*) [= 12:14–16], 10:1 (*wa-hādhā*)-2 [= 12:16–17], 8:23–9:4 [= 12:18–13:2], 10:3–5 [= 13:3–8], 9:5 [= 14:13–15], 10:6–7 [= 14:16–20], 9:6–21 [= 14:21–15:22], 10:8–18 [= 15:23–16:17], 15:18–16:19 [= 16:19–18:8], 17:1–20:16 [= 18:9–24:7], 20:19–22 [= 24:7–13], 21:1–2 (*wa-Aghātādīmūn*), 21:3 (*wa-āmanat*)-5 (*'alayhi al-salām*), 21:6 (*wa-āmana*)-7 (*bi-nubuwwatihī*), 21:10–13 [= 24:13–16].

236 *Tanqīḥ*, pp. 22–50; transl. *Examination*, pp. 40–77.—Bihbahānī translated the following portions of this chapter: *Tanqīḥ*, pp. 22:5–23:20 (*shamsiyya*) [= *Rādd*, pp. 27:10–28:19], 24:10–22 [= 28:20–29:12], 25:11–27:14 [= 29:14–31:17], 27:17–24 (*mulūkihim*) [= 31:17–32:4], 28:1 (*bal*)-12 (*'adaduhā*) [= 32:5–22], 28:13–20 (*li-tawātur*) [= 32:23–33:11], 28:22–32:3 [= 33:11–38:4], 32:4 (*wa-kānat*)-32:17 (*bihā*) [= 38:4–21], 32:18–19 (*nusiyat*), 32:22–33:13 (*al-tawrāh*), 33:16–20 [= 38:22–40:7], 35:7–13 (*sulṭānihī*), 35:13

chapter devoted to Christianity is, with a few omissions and with the exception of the first two sections and part of section eight, a translation of Ibn Kammūna's third chapter on Christianity.[237] The third chapter, dedicated to Islam, contains a translation of Ibn Kammūna's fourth chapter, though with considerable omissions and numerous additions from other sources.[238]

## 5. *The Reception of Ibn Kammūna in Judaism*

Compared with the significance of Ibn Kammūna's philosophical writings and thought for the later development of philosophy among Muslims, the reception of his writings within Jewish circles was limited. The fact that there are manuscripts of his writings in Hebrew characters preserved in European libraries, in the various collections of material from the Ben Ezra Genizah in Fustat/Old Cairo and in the Abraham Firkovitch collection in St. Petersburg, which partly originates from the Genizah of the Karaite synagogue of Cairo,[239] show that some of his philosophical writings, as well as his two comparative works, circulated among Jews and particularly Karaites. Out of his philosophical writings, the following were copied and read by later generations of Jews: *Maqāla*

(*wa-yu'akkidu*)-15 (*bihī*), 35:16–36:2 (*al-ma'mūra*), 36:4–8 (*al-ān*), 36:11–12 (*sana*), 37:1–9 (*al-amr*), 37:9 (*wa-qad*)-10 (*dhalika*), 37:11 (*wa-man*)-20, 40:7–42:4, 43:9–18, 45:5–47:5 (*al-dawām*), 47:10–14 (*ʿanhu*), 47:17–18 (*naskh*) [= 40:15–49:18], 49:15–50:3 [= 49:19–50:1].

237 *Tanqīḥ*, pp. 51–66; transl. *Examination*, pp. 78–99. Bihbahānī translated the following portions of this chapter: *Tanqīḥ*, pp. 51:4–52:16 (*al-ʿaqīda*) [= *Rādd*, pp. 57:5–59:13], 52:16 (*wa-ittafaqū*)-54:7 (*bi-maṣnūʿ*), 54:10–15 (*lā ghayr*) [= 60:8–62:22], 54:16–55:20, 55:23–57:7 [= 63–67:4], 57:8–14 [= 67:6–16], 57:14–21 (*al-nabī*), 57:23–58:1 [= 68:3–21], 58:5–14 (*ajlika*), 58:15 (*wa-kayfa*)-20 (*baʿdahum*), 59:1–7 [= 69:1–70:13], 59:8–61:17, 61:22–62:20, 62:22 (*wa-naḥnu*)-23 (*ḥikmatihī*), 63:3–5 (*wuqūʿihī*), 63:6–64:9, 64:13–65:1 (*al-anājīl*), 65:4–5 (*muḥtamil*) [= 70:14–78:12].

238 *Tanqīḥ*, pp. 67–108; transl. *Examination*, pp. 100–57.—For Bihbahānī and his *Rādd-i shubahāt al-kuffār*, see our "Muslim Polemics against Judaism and Christianity in 18th Century Iran. The Literary Sources of Āqā Muḥammad ʿAlī Bihbahānī's (1144/1732–1216/1801) *Rādd-i shubahāt al-kuffār*." *Studia Iranica* 35 (2006), pp. 69–94.

239 On the various collections, see David Sklare: "A Guide to Collections of Karaite Manuscripts." In *Karaite Judaism. A Guide to its History and Literary Sources*. Ed. Meira Polliack. Leiden 2003, pp. 893–924, esp. pp. 895, 905–9 with further references.

*fī l-taṣdīq bi-anna nafs al-insān bāqiya abadan* and *al-Maṭālib al-muhimma*, of which incomplete copies are preserved in the Firkovitch collection,[240] and his commentary on Suhrawardī's *Talwīḥāt*, three fragments of which are preserved in the same collection. Two of the fragments are evidently from the same copy of the work,[241] and another substantial fragment of this copy belonged to the *Nachlass* of Moses Wilhelm Shapira (1830–1884) and came into the possession of the State Library, Berlin in 1885.[242] The three fragments taken together constitute (with the exception of at least four folios) a complete copy of parts I (logic) and II (physics) of Ibn Kammūna's *Sharḥ al-Talwīḥāt*.[243]

Ibn Kammuna's comparative works were also read and copied in Jewish circles. The Firkovitch collection contains three fragments of his treatise on the differences between the Rabbanites and Karaites, two of which originally formed parts of the same copy of the work.[244] The only complete manuscript of the text, which is in the State Library in Berlin, was copied in 1341 C.E. in Baghdad.[245] Moreover, three complete copies of his *Tanqīḥ* written in Hebrew script are preserved in the State Library, Berlin,[246] the Bodleian Library, Oxford,[247] and the Magnes Museum, Berkeley.[248] Two fragments which are evidently from the same copy of the *Tanqīḥ* are preserved in the Firkovitch collection,[249] and a

---

240 RNL II Firk. Yevr.-Arab. I 1416 (*Maqāla fī l-taṣdīq*) and Yevr.-Arab. II 73 (*Maṭālib*).

241 RNL II Firk. Yevr.-Arab. I 3137 and 3138. RNL II Firk. Yevr.-Arab. I 857 is written by a different hand.

242 Berlin Ms. Or. Fol. 1321.

243 See below, Appendix A: The Firkovitch/Shapira Codex (*Sharḥ al-Talwīḥāt*).—This is not the only case of a work of which fragments were collected by both Firkovitch and Shapira. See, e.g., Gregor Schwarb: "Découverte d'un nouveau fragment du *Kitāb al-muġnī fī 'abwāb al-tawḥīd wa-l-'adl* du Qāḍī 'Abd al-Ǧabbār al-Hamaḏānī dans une collection karaïte de la British Library." *Mélanges de l'Institut Dominicain d'Etudes Orientales du Caire* 27 (2006) (in press).

244 RNL II Firk. Yevr.-Arab. I 913 and 914. Yevr.-Arab. I 1703 is a fragment from a different copy of the tract.—Yevr.-Arab. II 807 contains a one-page fragment (f. 33b) with some lines of a paraphrastic summary of the introduction to the tract.

245 Ms. Or. Oct. 256/2, f. 180.

246 Ms. Or. Oct. 256/1.

247 Ms. Huntington 390.

248 Ms. Magnes Museum KC 24.

249 RNL II Firk. Yevr.-Arab. I 1702 and Yevr.-Arab. II 490.

third fragment is part of the Adler collection at the Jewish Theological Seminary, New York.[250]

There is, so far, only one case of a later Jewish thinker who is known to have studied Ibn Kammūna's writings and to have used them in his own writings, namely David ben Joshua Maimonides, a late descendant of Maimonides active circa 1335–1410 C.E..[251] In his Judaeo-Arabic commentary on Maimonides' commentary on the *Mishneh Tōrāh* he quotes from Ibn Kammūna's *al-Jadīd fī l-ḥikma*;[252] he was apparently also familiar with his *Sharḥ al-Talwīḥāt*.[253] It is not unlikely that the copy of Ibn Kammūna's commentary on Suhrawardī's *Talwīḥāt*, portions of which were collected by Abraham Firkovitch and by Moses Wilhelm Shapira, is the manuscript that was available to David ben Joshua.[254]

---

250 Ms. Jewish Theological Seminary 5519 (ENA 95).

251 On David ben Joshua, see Paul B. Fenton: "The Literary Legacy of David ben Joshua, last of the Maimonidean Negīdim." *The Jewish Quarterly Review* 75 i (1984), pp. 1–56; idem: "A Judaeo-Arabic Commentary on Maimonides' *Mišnē Tōrā* by Rabbi David Ben Joshua Maimonides (circa 1335–1414)." [Hebrew] In *Heritage and Innovation in Medieval Judaeo-Arabic Culture. Proceedings of the Sixth Conference of the Society for Judaeo-Arabic Studies*. Eds. Joshua Blau and David Doron. Ramat Gan 2000, pp. 145–60. Fenton has also edited and translated one of David ben Joshua's writings: *Al-Murshid ilā l-tafarrud wa-l-murfid ilā l-tajarrud*. Edited for the first time with a Hebrew translation and notes by Paul B. Fenton. Jerusalem 1985.

252 Cf. Fenton: "Literary Legacy," p. 33 no. 38; idem: "A Judaeo-Arabic Commentary on Maimonides' *Mišnē Tōrā*," p. 154. The quote (ms. RNL II Yevr.-Arab. I 1497, ff. 47b:15–48a:21) is taken from Ibn Kammūna's *al-Jadīd fī l-ḥikma*, pp. 403:14–404:6 and includes the passage where Ibn Kammuna quotes Mu'ayyad al-Dīn al-ʿUrḍī. Other than was suggested by Fenton ("Literary Legacy," p. 37 no. 48) who did not identify *al-Jadīd* as the source of David, this second part of the quote is therefore not taken directly from a writing by ʿUrḍī.—For Ibn Kammūna's use of ʿUrḍī, see below, Chapter Two, Section 1.2.3.

253 Suggested by Fenton ("Literary Legacy," p. 37) for ms. West. Coll. Ar. II 75 which is a portion of David ben Joshua's *al-Murshid ila l-tafarrud*.

254 There is also an extensive paraphrastic quotation from Ibn Kammūna's *Sharḥ al-Talwīḥāt* contained in a copy in Hebrew script of a partial commentary on Maimonides' *Sefer ha-maddaʿ* (i.e., the first book of his commentary on the *Mishneh Tōrāh*) by the Muslim author ʿAlā' al-Dīn Abu l-Ḥasan ʿAlī b. Ṭaybugha al-Ḥalabi al-Ḥanafī al-Muwaqqit (d. Aleppo 793/1391). The text of ms. BL Add. 27,294, ff. 60b–61b is introduced as follows פצל מן אלתלויחאת ללשיח שהאב אלדין אלסהרורדי קדס אללה רוחה ושרחה ללשיח אלאמאם אלעלאמה עז אלדולה בן כמונה סקא אללה צובה אלגמאם See also G. Margoliouth: *Catalogue of the Hebrew and Samaritan Manuscripts in the British Museum* 1–4. London 1965–77, vol. 2, pp. 108–11 for a description of this manuscript. Another fragment of the same work is preserved in ms. RNL Firk. II

Yevr.-Arab. I 1426 which does not include the chapter based on Suhrawardī and Ibn Kammūna.—On the work and its author, see also the studies of Moritz Steinschneider: "Schriften der Araber in hebräischen Handschriften, ein Beitrag zur arabischen Bibliographie." *Zeitschrift der Deutschen Morgenländischen Gesellschaft* 47 (1893), pp. 338–41; Y. Tzvi Langermann: "Arabic writings in Hebrew manuscripts. A preliminary relisting." *Arabic Sciences and Philosophy* 6 (1996), pp. 137–60; G. Margoliouth: "A Muhammadan Commentary on Maimonides' Mishneh Torah. Chs. I–IV." *The Jewish Quarterly Review* 13 (1901), pp. 488–507; Fenton: "A Judaeo-Arabic Commentary on Maimonides' *Mišnē Tōrā*," pp. 150ff; David Blumenthal: "An Arabic Translation to 'Hilkhot Yesode ha-Torah'." [Hebrew] *Da'at* 14 (1985), pp. 113–47; Gregor Schwarb: "The commentary of 'Alā' al-Dīn Abū l-Ḥasan 'Alī b. Ṭaybugha (d. 793/1391) on Hilkhot Yesodey ha-Torah, I–IV: a scientific encyclopaedia." (Forthcoming).—For the Hebrew original of *Sefer ha-madda'*, see Maimonides: *Mishneh Torah. The Book of Knowledge* [*Sefer ha-Madda'*]. Edited according to the Bodleian (Oxford) Codex with Introduction, Biblical and Talmudical References, Notes and English Translation by Moses Hyamson. Jerusalem 1962.—Ibn Ṭaybugha was a contemporary and close friend of David ben Joshua, and the manuscript contains "various additions by at least one Jewish redactor of the work." Cf. Margoliouth: *Catalogue*, vol. 2, p. 108; see also idem: "A Muhammadan Commentary," pp. 490–92. Certainly, one of these "Jewish redactor[s]" was David ben Joshua.

CHAPTER TWO

# INVENTORY OF THE WRITINGS OF IBN KAMMŪNA

The purpose of this inventory is to serve as a reference tool for further research on Ibn Kammūna. Included are all works of which he is known to be the author or that have been attributed to him. Each entry gives the title of the work, together with a brief description of its contents, and contains references to relevant bibliographical sources. As far as can be established, the date of composition is also given. Editions, translations, commentaries and studies are indicated, and the location of manuscripts of extant works, together with brief descriptions is also given. We have tried to trace and describe all extant manuscripts of Ibn Kammūna's writings and to inspect as many manuscript copies as possible. Where this has not been feasible, the information provided is based on the descriptions in the respective manuscript catalogues. References to manuscripts of Ibn Kammūna in secondary literature that proved to be wrong, as well as misattributions in catalogues, have been omitted. In many cases, Ibn Kammūna did not give his writings specific titles. The first title mentioned for each entry is either the one that was given to the work by Ibn Kammūna, or that can be deduced from his wording in the respective works and from cross references in other works. Alternative titles, as found in quotations of later scholars, in manuscript catalogues, or in modern scholarship are indicated in parentheses. Given the sometimes confusing state of the titles, the respective *incipit* and *explicit* are quoted for each work in order to facilitate future identifications of as yet unknown manuscripts. For those works that have either not been edited as yet or that are difficult to trace, the quoted *incipit* includes the text of the entire introduction. The respective sources for *incipit* and *explicit* have been indicated. Misreadings in the editions and manuscripts have been emended without specific notice. The abbreviations used in this chapter for manuscript collections and catalogues and for bibliographical references are explained below (see Abbreviations and References).

## 1. *Authentic Writings*

### 1.1. *Commentaries and Glosses*

#### 1.1.1. *Sharḥ al-uṣūl wa-l-jumal min muhimmāt al-ʿilm wa-l-ʿamal (Sharḥ al-Ishārāt)*

A commentary on *al-Ishārāt wa-l-tanbīhāt* of Ibn Sīnā, dedicated to Sharaf al-Dīn Hārūn, one of the sons of *ṣāḥib al-dīwān* Shams al-Dīn al-Juwaynī. As Ibn Kammūna says in the introduction, the commentary is heavily indebted to the commentary of Naṣīr al-Dīn al-Ṭūsī. It was composed in less than two months and completed on 28 Shaʿbān 671/19 March 1273. A copy of the work was sent by Ibn Kammūna to the library of Sharaf al-Dīn Hārūn and his father.

(*Dharīʿa*, vol. 13, p. 94 no. 301; GAL, vol. 1, p. 592; GALS, vol. 1, p. 769; d'Herbelot: *Bibliothèque orientale*, p. 476; Ḥujjatī & Dānishpazhūh: *Fihrist*, vol. 1, p. 473; *Kashf al-ẓunūn*, vol. 1, p. 303; Mahdawī: *Fihrist*, p. 35; Rescher: *Development*, p. 205; Steinschneider: *Arabische Literatur*, p. 239 no. 4)

Introduction (Ragıp Paşa 845, ff. 1b:1–3a:22):—

أحمد الله على حسن توفيقه لما يتنزل من السفر إليه منزلة التقدم بالاعداد للراحلة والزاد وأسأله هداية طريقه حتى لا أضل في شعبها المتوعرة عن سنن الرشاد كما ضل كثير من العباد وإلهام الحق بتحقيقه إذ بإلهامه ينتهي ذلك السفر إلى اليقين في الاعتقاد لا بفكرة العبد الذي لا ينسب إليه من ذلك إلا مجرد القبول والاستعداد وأن نصلي على المصطفين من عباده لرسالته خصوصاً على محمد وآله وعترته. أيها الحريصُ على تحقق الحق إني مهدٍ إليك في هذا الكتاب الذي هو كالشرح لكتاب الإشارات والتنبيهات مع زيادة ما استخرجته من براهين قواعدهم وإفادة ما استنتجته من قوانين قواعد جمة أصولاً وجُملاً من الحكمة إن اجدت الفطانة بيدك سهل عليك تفريع تلك الأصول إلى جزئياتها وتفصيل تلك الجمل إلى أجزائها ومبتدىء من علم المنطق ومنتقل عنه إلى العلم الطبيعة وما قبله ومقدم على جميع ذلك ذكر ماهية الحكمة وأقسامها الأصلية وما يتضمنه هذا الكتاب منها. وأما الحكمة فصناعة نظرية يستفيد بها الإنسان تحصيل ما عليه الوجود في نفسه وما عليه الواجب مما ينبغي أن يكتسبه بعلمه ليستكمل نفسه صائرةً عالماً معقولاً مضاهياً للعالم الموجود ويستعد للسعادة القصوى الأخرية بحسب الطبيعة البشرية. وأما أقسامها الأصلية فهي أنها إن كانت علماً هو آلة لما عداه من العلوم فهو المنطق

ولهذا وجب الابتداء به وإن لم يكن كذلك فإن كانت علماً بما لا يتعلق وجوده بفعل الإنسان فهو العلم النظري وإلا فهو العلم العملي، والمبحوث عنه في النظري إن كان مفتقراً إلي المادة الجسمية في الوجود فهو علم الطبيعة، والعلم الطبيعي إن كان حده متعلقاً بها أيضاً، وإلا فهو العلم الرياضي، وإن لم يكن مفتقراً إليها فهو علم ما قبل الطبيعة باعتبار اشتماله علي بيان أكثر مبادئ علمها. ولنقدم ما يبحث فيه من المجردات علتها بالعلية والشرف وما بعد الطبيعة باعتبار التعلم كون إدراك المحسوسات قبل إدراك المعقولات. وأما العملي فإن كان علماً بالتدبير الذي يختص بالشخص الواحد فهو علم الأخلاق، وإلا فهو علم تدبير المنزل إن كان علماً بما لا يتم إلا بالاجتماع المنزلي، وعلم السياسة إن كان علماً بما لا يتم إلا بالاجتماع المدني، فهذه أمهات العلوم وكل علم جزئي فلا بد وأن ينسب إلى واحد منها. وقد اشتمل هذا الكتاب من كل علم من هذه على ما هو من أهم مطالبه وأقرب إلى تزكية النفس به. أما المنطق فلأنه آلة النظر، وأما علم الطبيعة وما قبله فلأنهما يبحثان عن أحوال نفس الإنسان وسائر المجردات الممكنة وعن واجب الوجود وصفاته وكيفية صدور سائر الموجودات عنه على ترتيبها. ولا شك أن العلم بذلك من أعظم المهمات التي يجب الاعتناء بتحصيلها. وهذه العلوم الثلاثة هي أجلّ ما فيه لكون مسائلها أهم من مسائل غيرها، فأما الرياضي فلما لم يكن كذا لم نتعرض منه إلا لقدر نزر وقع التلويح إليه في بعض فصول علم ما بعد الطبيعة. وأما العلم العملي فأشرف ما فيه كيفية رياضات العارفين. قد ثبت في نمط مقاماتهم من العلم المذكور. فهذا ما يتعلق بشرح الخطبة والمقدمة وعلى هذا الأسلوب آتي على شرح باقي الكتاب مع الزيادات التي منها ما التقطتُه من كتب الحكماء الأكابر، ومنها ما استنبطتُه بفكري الفاتر مختصراً أكثره من شرح مولانا علامة العالم جامع علوم المتقدمين والمتأخرين ملك العلماء المحققين سلطان الحكماء المدققين نصير الحق والدولة والدين فإنه أعلى الله محله وأسبغ ظله جمع فيه محاسن الشروح الموجودة وبلغ الغاية في تحقيق المعاني المقصودة، فهو بما استوعب فيما حقق من المقاصد المقال قد ضيق على غيره في الزيادة المجال والاكتفاء به. وإن كان الأولى بأن لا أتعرض لشرح الكتاب لولا أمر ورد إليّ ممن حكمه جازم وطاعة أمره، إذ هو من أولى الأمر، أمر لازم لا سيما ورياسته طبعية لا وضعية وحقيقية لا اتفاقية وهو المولى المالك المعظم شرف الدنيا والدين هرون سيد الملوك والأكابر رب المناقب والمفاخر الجامع بين الفضيلتين العلمية والعملية الحاوي للرياستين الدينية والدنيوية ذو الفطنة الوقادة والدرية والعبادة الممنوح بنقاء السريرة واستقامة السيرة المستحق بما جمع من هذه الخلال الرضية بمقتضى ما اشترطه الرئيس في عهد الوصية أن يُتحَف بنفائس الحقائق العلمية ويُبذل له أسرار الدقائق الحكمية ولد الصاحب الاعظم ملك الوزراء العالم العادل المنصف الكامل صاحب ديوان الممالك دستور الشرق والغرب شمس الدنيا والدين محمد بن محمد صاحب الديوان الجويني

المحيي بكرمه معالم العلوم الدارسة وآثار الفضائل الطامسة المتخصص بفعل الخيرات المنفرد باقتناء الكمالات أنار الله سعدهما وأعلى جدهما وقمع ضدهما. وقد عملت هذا الكتاب برسم خزانتهما العامرة وأتيت فيه بجميع ألفاظ الرئيس أبي علي في الإشارات من غير إخلال بشيء منها محافظةً على فصاحتها وعذوبتها إلا بما هو لضرورة اندراج الكلام مثل واو عطفٍ وفاء تعقيب ولإظهار المعنى كالتصريح بضمير أو ما اختلفت فيه النسخ والروايات، وإن كنتُ أجمع بينهما في الأغلب، فمن أحب أن يميز ألفاظ الإشارات عن باقي الكلام فليميز ذلك بعلامة الحمرة أو غيرها وسميتُه شرح الأصول والجمل من مهمات العلم والعمل وقسمتُه مجارياً لقسمة الإشارات إلى قسمين أولهما في المنطق وهو عشرة أنهاج والثاني في علمي الطبيعة وما قبله مما يندرج فيهما من غيرهما وهو عشرة أنماطٍ والقسمان يشتملان على ثلاث مائة وأربعة وخمسين فصلاً معنونة بسبعة وثلاثين عنواناً معنونٍ بكل واحد منها، إما فصل واحد أو أكثر وهذه أسماؤها على ترتيب ذكرها في الكتاب ... فلأشرع الآن في ذكر ذلك مستعيناً بمن لا يخيب قاصده ولا يضل مسترشده مستمداً منه الهداية وحسن الخاتمة وسعادة الدارين إنه لطيف كريم رؤوف رحيم.
القسم الأول في علم المنطق . . .

End (Ragıp Paşa 845, f. 198a:3–16):—

فهذا ما تيسّر إيراده على حكم العجلة في أوقات مختلسة من الشواغل الدنيوية من غير معاودة تنقيح ولا مراجعة تهذيب وأرجو مع اعترافي بالتقصير أن يكون ما أوردتُه موافقاً للغرض ملحوظاً بعين الرضا على أن الذي فيه من قبلي هو شيء يسير وقدر قليل وما عداه فمأخوذ من أقاويل أكابر العلماء وأساطين الحكماء ومن الله تعالى أسال وإليه أضرع وأرعب أن يجعلني من الداخلين في رحمته المشمولين بنعمته اللائذين ببابه الآمنين من عذابه المبتهجين بلقائه الواصلين إلى سعادة الأبد في زمرة أوليائه إنه أرحم الراحمين وأكرم الأكرمين. والحمد لله رب العالمين وصلواته على ملائكته المقربين وعلى أنبيائه وأوليائه أجمعين وخصوصاً على محمد وآله الطاهرين وأختم بخير انتجز تأليفه في الاثنين الثامن والعشرين من شعبان من سنة إحدى وسبعين وستمئة هجرية في قريب من شهرين.

*Editions*

Eds. Ibrahim Marazka, Reza Pourjavady & Sabine Schmidtke. Tehran: Iranian Institute of Philososophy (forthcoming).

*Manuscripts*

Awqāf ʿāmma 6474, 181 ff., 12 × 22 cm, "*nuskha jayyida qadīmat al-khaṭṭ*" (cat. vol. 4, p. 60 no. 6261).

British Library (India Office) 4545 (= Arabic Loth 484), 174 ff., 11.4 × 20.3 cm, 29 lines to a page. Copied by Ḥasan (or ʿAlī ?) b. ʿAbd al-Raḥmān b. ʿAbd Allāh... al-Mālikī and dated 12 Rabīʿ II (?) 734/20 December 1333 (cf. colophon f. 174a). The leaves at the beginning and the end of the manuscript are in disorder; the correct order is ff. 1, 4, 3, 6, 5, 2, 7... and 168, 170, 169, 172, 171, 173, 174 (cat. 1877, p. 135).

Lâleli 2516, 292 ff., 15 × 21 cm, 18/19 lines to a page, copy completed on 3 Rabīʿ II 1086/27 June 1675 by Darwīsh ʿAbd al-Raḥmān Ibn Kamāl al-ʿĪthāwī al-Shāmī (cf. colophon f. 198a). The title page has: *Sharḥ al-Ishārāt mimmā allafahū mukammil ʿulūm al-awwalīn wa-l-ākhirīn qudwat al-muḥaqqiqīn ʿIzz al-Dawla Saʿd b. Manṣūr b. Saʿd al-maʿrūf bi-Ibn Kammūna al-Isrāʾīlī*... (cat. p. 193).

Lâleli 2487/6, ff. 222b–223a, 16.5 × 12 cm, 16 lines to a page. Excerpt from Ibn Kammūna's commentary in *namaṭ* 9, *ishāra* 15 that is introduced as follows:

من كلام الشيخ الإمام الفاضل العالم المختار عز الدولة بن كمونة رحمه الله تعالى في شرح لكتاب الإشارات عند مقامات العارفين شرح بعض مصطلحات الصوفية في أحوالهم وملكاتهم الشريفة التي هي مقاماتهم قال الخاطر عندهم هو ما يرد على النفس...

The corresponding passage in the commentary is as follows:

وأنا شارح في هذا الموضع بعض مصطلحات الصوفية في أحوالهم وملكاتهم الشريفة التي هي مقاماتهم وإن كان جميع ذلك داخلاً فيما سبق ذكر بالخاطر عندهم هو ما يرد على النفس...

The text of the excerpt continues up to the beginning of the following *tanbīh* and corresponds to ms. Ragıp Paşa 845, ff. 184a:13–184b:18 (cf. Ritter II, p. 76).

Ragıp Paşa 845, 198 ff., 13 × 25 cm, 23 lines to a page, copied in Ramaḍān 674/February–March 1276 by Muḥammad b. Abī Lājik al-Nīlī al-Turkī and collated with a copy transcribed from the autograph and read to the author (cf. colophon f. 198a: *balagha muqābalatan wa-taṣḥīḥan bi-l-aṣl al-manqūl minhu ḥasab al-jahd wa-l-ilṭāq wa-kāna al-aṣl qad nuqila min khaṭṭ al-muṣannif wa-quriʾa ʿalā muṣannifihī*). The title page, which has numerous ownership statements, indicates the title of the work as follows with no hint that Ibn Kammūna is the author: *Sharḥ Ishārāt al-Shaykh Abī ʿAlī Ibn Sīnā*. There is a note by a different hand at the upper margin of the title page stating: *min taṣānīf al-Mawlā al-sharīf*...*Abu l-Faraj al-madʿūw bi-Ibn Kammūna* (cat. p. 64).

### 1.1.2. *Sharḥ al-Talwīḥāt (al-Tanqīḥāt fī Sharḥ al-Talwīḥāt)*

A commentary on the *Talwīḥāt* of Shihāb al-Dīn al-Suhrawardī produced in response to a request by some unidentified scholars, whom the author refers to as *jamāʿa min al-sāda al-nubalāʾ wa-l-aṣḥāb al-fuḍalāʾ*, and completed at the beginning of 667/1268.[1] There are no indications in the text that Ibn Kammūna was aware of a commentary on the same work by Shams al-Dīn al-Shahrazūrī (if the latter's commentary in fact predates Ibn Kammūna's). In the introduction he says that he used other works by Suhrawardī. Throughout the text, he explicitly refers to *Ḥikmat al-ishrāq* (eds. Ziai & Alwishah, pp. 221, 224, 382, 388, 424–25; ed. Mūsawī, p. 997), *al-Mashāriʿ wa-l-Muṭāraḥāt* (eds. Ziai & Alwishah, pp. 180, 296; ed. Mūsawī, p. 997), *al-Alwāḥ al-ʿimādiyya* (ed. Mūsawī, p. 994) and *Muqāwamāt* (eds. Ziai & Alwishah, p. 452); although he refrains from explicitly mentioning it, he also used Suhrawardī's *Maqāmāt al-ṣūfiyya* (*Kalimat al-taṣawwuf*), particularly with respect to Sufi terminology; especially his definitions of *tawakkul*, *shukr*, and *riḍā* (ed. Mūsawī, pp. 1020–21) are based on Suhrawardī's *Kalimat al-taṣawwuf* (ed. Habibi, pp. 133–34). Following the structure of Suhrawardī's *Talwīḥāt*,

[1] Ziai and Alwishah state that Malik 934 contains a colophon on f. 288b according to which "the date of the completion of the work by Ibn Kammūna *at the Niẓāmiyya Madrasa in Baghdād* [was] in the beginning of the year 667 A.H." (emphasis ours). See the introduction to their edition, p. 47. This information is, however, incorrect; the reference in the colophon to the Madrasa Niẓāmiyya refers exclusively to the copyist. The complete text of the colophon reads as follows:

وقع الفراغ من نسخه يوم الاثنين أول الصبح العاشر ربيع الأول سنة أربع وسبعمائة هجرية في المدرسة النظامية بدار السلام مدينة بغداد حماها الله تع عن الآفات وقع الفراغ من تصنيفه في أوائل سنة سبع وستين وستمائة.

the commentary consists of three parts: logic (part I), physics (part II), metaphysics (part III).

(*Dharīʿa*, vol. 13, pp. 152–53 no. 519; GAL, vol. 1, p. 437; GALS, vol. 1, pp. 769, 782; *Kashf al-ẓunūn*, vol. 2, p. 420; Rescher: *Development*, p. 205; Ritter I, pp. 273–75; Suhrawardi: *Opera Metaphysica et Mystica I*, Prolégomènes, pp. lxiii ff; Steinschneider: *Arabische Literatur*, p. 240 no. 5)

Introduction (ed. Mūsawī, pp. 2:1–3:8):—

بعد حمد الله تعالى على آلائه المتواترة والرغبة أن يوفّق للهداية في الدنيا والسعادة في الآخرة والصلاة على ملائكته أولي الذوات الزكية الطاهرة وعلى من اصطفاه من أنبيائه وأوليائه المؤيّدين بالمعجزات الباهرة والكرامات الظاهرة خصوصاً على محمد وآله. إنه لما كان المختصر الموسوم بالتلويحات للإمام العلامة شهاب الدين السهروردي قدّس الله روحه ونوّر ضريحه مشتملاً على الحقائق العلمية من أسناها وأرفعها ومن الدقائق الحكمية على أولاها وأنفعها متضمّناً للمهمات الموصلة إلى الكمال المسعدة في المآل وكان مع ما هو عليه من شدة الإيجاز المنتهي في كثير من المواضع إلى ما يكاد أن يكون من قبيل الإلغاز غير موجود له فيما بلغنا بشرح يرجع في فهم مقاصده إليه ويعول في إبانة غوامضه عليه، التمس مني جماعة من السادة النبلاء والأصحاب الفضلاء أحسن الله تعالى توفيقهم وسهّل إلى كل خير طريقهم أن أشرح المختصر المذكور شرحاً لا يخرج إلى الإطناب ولا يخلّ بشيء من أغراض الكتاب فدافعتهم عن ذلك معتذراً بكثرة الأشغال والهموم وقلّة بضاعتي من هذه العلوم وبأن الشرح الملتمس يستغني عنه أرباب الفطانة وأولو الألباب إذ لا يتعذر عليهم بتأملهم للكتب المبسوطة الاطلاع على مقاصد هذا الكتاب فألحوا إلحاحاً موجباً للموافقة مقتضياً ترك المناقشة والمحاققة. فعملت هذا الشرح في أثناء اشغالي على حكم العجلة من غير مراجعة لتنقيح المعاني ولا معاودة لتهذيب الألفاظ جامعاً فيه مباحث التلويحات على ترتيبها مجتهداً في تبيينها وتهذيبها مع تفسير مقاصدها وتقرير قواعدها وتكثير فوائدها وتحرير معاقدها وبسط موجزها وحلّ ملغزها وتقييد مرسلها وتفصيل مجملها منقحاً لما أتى به صاحب الكتاب على سبيل المساهلة وموضحاً لما سلك فيه طريق المراعاة والمجاملة وإن كنت مستفيداً معظم ذلك من بواقي كتبه ومقتدياً في الأكثر بطريقه في سلوك الحق ومذهبه مع أني في المواضع المستغلقة أذكر بحث الكتاب بعبارة يتبين من أثنائها المراد، وأورده على وجه يتحقق شرح المتن من نفس ذلك الإيراد ملتزماً في جميع ذلك شريطة الاختصار غير متعرض للتعريفات المؤدية إلى الإسهاب والإكثار متضرعاً إلى الله تعالى أن يلهمني الصواب ويجعلني في زمرة عباده الأبرار في دار الثواب، إنه جواد كريم رئوف رحيم. قال الإمام صاحب الكتاب رضي الله عنه . . .

End (ed. Mūsawī, pp. 1022:20–1023:2):—

...ومثل هذا في فن الخطابيات غير مستنكر ولا خال من زيادة فائدة. على أنها مع كونها جارية على قانون الخطابة هي مشتملة على لطائف كثيرة علمية ولولا علمي بأنها لا تخفى على من أتقن المباحث السالفة لكنت أنبّه على كل واحد منها وليس في هذه الوصايا ما هو خارج عن آداب الشرائع الحقّة واجبها ومندوبها، وبالجملة يجب على طالب السعادة الأبدية المواظبة عليها والتمسّك بها فأمن من التزم امتثال أوامرها وترك مناهيها لسيادة السرمد حائزاً وبسعادة الأبد فائزاً.

Most manuscripts contain an additional remark by Ibn Kammūna following the end of the text (Yeni Câmi 766, ff. 214a:23–214b:20; Majlis 1850, f. 291b; Feyzullah Efendi 1194, ff. 245b:18–246a:12; Feyzullah Efendi 1195, f. 205b:12–27; Ayasofya 2392, ff. 261a:19–261b:12; Chester Beatty 3598, ff. 216a:14–216b:4; Leiden Or. 137, ff. 281b:12–282a:2):—

وهذا آخر شرح التلويحات على الوجه المناسب لغرض السادة المقترحين وإن كان قد لا يناسب غرض غيرهم من بعض الطالبين والمستشرحين وأنا اقترح على من ينظر فيه أن يقيم عذري فيما طفيء فيه العلم أو زلّت به القدم فإن استكشاف أسرار الدقائق اللطيفة واستشفاف أنوار الحقائق الشريفة مما يتعذر مع العوائق الدنيوية ويتعسر مع العلائق البدنية لا سيما إذا انتهت كثرة الشواغل إلى حد المنع من معاودة التنقيح والتهذيب واستغرقت الوقت فلم يتأت التأني في اختيار الألفاظ وجودة الترتيب هذا على تقدير حدة الخاطر وغزارة العلم فكيف إذا كانت الفكرة كليلة والبضاعة من العلم قليلة على أن من يحكم بالتخطئة لا لأجل الحسد والعناد ولا عن هوى يعدل به عن سنن الرشاد لعلّه يجد مخرجاً صالحاً لو دقق النظر منهجاً واضحاً لو لاحظ المقصد المعتبر مع أنه كثيراً ما يقع الغلط من قبل الناسخين فيظن من المصنفين فإني قد وجدت من أغلاطهم حتى في تغيير كلمة أو حرف ما استحال به المعنى وفسد، وشاهدت وقوع ذلك في هذا الشرح قبل إتمامه، فكيف إذا طال به الأمد. ولهذا ما يجب أن تهمل المقابلة التامة للكتب العلمية بالنسخ المعول عليها الموثوق في الصحة إليها ولا تكلان إلا على الواحد القوي الهادي إلى الصراط السوي هو ربّ العزة والجلال وله المنّ ومنه الإفضال. فعليك أن تعتصم به إذ هو موجدك وأن تتعرض لروح نفحاته إذ هو هاديك ومرشدك ومن لم يستنصر بعناية موجده ذلّ ومن لم يستنصر بهداية مرشده ضلّ. وأسألك يا من لا يخيب قاصده ولا يضلّ مسترشده أن تعضدني بإدراك عنايتك وترشدني بأنوار هدايتك وأن تجعلني لالتزام التأدب بقويم آدابك من المنتهجين وبدوام التقرب إلى كريم جنابك من المبتهجين وأن تحشرني مع الذين زينت قلوبهم

بالتقوى فتمسكوا بالسبب الأقوى فارتفعوا عن حضيض الشقاوة والبلوى فاتصلوا بمحل السعادة القصوى برحمتك يا أكرم الأكرمين وأرحم الراحمين.

Ibn Kammūna's colophon (Ayasofya 2423, f. 333a; Ayasofya 2392, f. 261b:13–14):—

وقع الفراغ من تصنيف هذا الكتاب في أوائل سنة 667 هجرية

*Glosses*

Extant in the margin of Ragıp Paşa 852 (completed on 17 Rabīʿ I 752/ 14 May 1351). The reference to "Naṣīr al-Dīn al-Kāshī" in the margin of f. 194a suggests that the glosses signed throughout the text as ن or ن ك were written by Naṣīr al-Dīn al-Kāshī (or al-Kāshānī) al-Ḥillī.

*Editions*

Quotations from part I of the commentary are included in the annotation of ʿAlī Akbar Fayyāḍ's edition of the logic section of Suhrawardī's *Talwīḥāt* (*Manṭiq al-talwīḥāt*. Tehran 1334/1955). The editor refrains from mentioning Ibn Kammūna by name and identifies the quotes simply as *sharḥ*.[2]

*Al-Tanqīḥāt fī sharḥ al-Talwīḥāt ta'līf-i Ibn Kammūna*. Ed. Sayyid Ḥusayn Sayyid Mūsawī. PhD thesis: Dānishgāh-i Tihrān, 1375–76/1996–97.
[The edition is based on the following manuscripts: Ahmed III 3244; Malik 934; Majlis 1849/5; Āstān-i Quds 700; Āstān-i Quds 703].

Ibn Kammūna: *Al-Tanqīḥāt fī Sharḥ al-Talwīḥāt. Refinement and Commentary on Suhrawardī's* Intimations. *A Thirteenth Century Text on Natural Philosophy and Psychology*. Critical Edition, with Introduction and Analysis by H. Ziai & A. Alwishah. Costa Mesa, California 2003.

[2] Daiber seems to suggest (*Bibliography of Islamic Philosophy*, vol. 1, p. 445 no. 4463) that Fayyāḍ's edition includes an edition of the entire commentary. This is not the case.

[The edition, containing part II only, is based on ms. Malik 934. The editors collated the text with the following manuscripts: Yahuda 201/2; Leiden Or. 137; Mishkāt 273; Majlis 1849/5]. On this edition, see the review by John Walbridge (*Iranian Studies* 38 iii (2005), pp. 520–1).

*Manuscripts*

Ahmed III 3208 (= Topkapı 6705), 386 ff., 25.5 × 11.5 cm, 29 lines to a page, copied by Muḥammad b. al-Ḥasan al-Qāshī (or al-ʿĀshī), completed on 14 Ṣafar 696/11 December 1296 in the Madrasa Niẓāmiyya in Baghdad (Ritter I, p. 274; cat. vol. 3, p. 639).
[Microfilm copy: Maʿhad 231 (cat. vol. 1, pp. 222–23)].

Ahmed III 3219 (= Topkapı 6706), 265 ff., 24 × 13 cm, 35 lines to a page, copied by "Aḥmad", completed at the end of Jumādā I 691/ May 1292 (Ritter I, p. 274; cat. vol. 3, p. 639).
[Microfilm copy: Maʿhad 232 (cat. vol. 1, p. 223)].

Ahmed III 3243 (= Topkapı 6708), 169 ff., 25.6 × 18 cm, 25 lines to a page, containing part III. Completed on 2 Rajab 729/2 May 1329 (Ritter I, p. 274; cat. vol. 3, p. 640).
[Microfilm copy: Maʿhad 235 (cat. vol. 1, pp. 222–23)].

Ahmed III 3244 (= Topkapı 6707), 308 ff. [ms. has two paginations; only the modern pagination is referred to here], 25 × 16.3 cm, 27 lines to a page, containing parts I (ff. 1b–85b), II (ff. 86b–167a), and III (ff. 168b–307a). The codex also contains two pieces of Naṣīr al-Dīn al-Ṭūsī written by the same hand (ff. 307b–308a). Copy completed on 16 Jumādā II 687/17 July 1288 (see colophon f. 307a:3–4). The copy belonged to the library of Quṭb al-Dīn al-Shīrāzī. On *laylat al-qadr* of Ramaḍān 692/August 1293, he presented the copy as a gift to the son of a certain *amīr* Kamāl al-Dīn, as is indicated in the dedication written by his hand on f. 1a:

انتقل مني بحق الهبة الشرعية هذا الكتاب وهو شرح التلويحات للإمام الفاضل والحكيم
الكامل عز الدولة ابن كمونة أسقي الله ثراه وجعل الجنة مثواه إلى الولد الأعز خلف الفضلاء
شرف الأمراء كمال الملة والدين وعز ومجد الإسلام والمسلمين متعنى الله بطول بقائه ودوام
لقائه تذكرة له ولمن طلبه منه وهو الأخ الصالح الملك الصالح رزقني الله شرف مناقشته

ويمن مشافهته وأدامه على طريقته المشهورة وسيرته المستورة (؟) حرره محبه وهو أحوج خلق الله إليه محمود بن مسعود الشيرازي ختم الله له بالحسنى...ليلة القدر قبيل الصبح من رمضان سنة اثنتين وتسعين وستمائة.

The manuscript contains Ibn Kammūna's "critical edition" of Suhrawardī's text; the scribe used several manuscripts and frequently notes the variants (cat. vol. 3, pp. 639–40; Ritter I, pp. 273–74; Suhrawardī: *Opera Metaphysica et Mystica I*, Prolégomènes, pp. lxiv–lxv).

[Microfilm copies: Maʿhad 230 (cat. vol. 1, p. 222); Mīnuwī 65 (cat. p. 80); Ritter/Uppsala MF 15:2701–2710 (cat. pp. 179–80)].

Ahmed III 3284 (= Topkapı 6687), 233 ff., 24 × 12.5 cm, 25 lines to a page, containing parts I (ff. 1–114a, completed on 8 Dhu l-Ḥijja 684/4 February 1286) and II (ff. 114b–233b, completed between 11–20 Rajab 685/2–11 September 1286), copied by Ismāʿīl b. Ibrāhīm b. al-Zāhid for his own use (*li-nafsihī*). Part III of this copy is extant in Yeni Câmi 766 (Ritter I, p. 273; cat. vol. 3, p. 633).

[Microfilm copy: Maʿhad 233 (cat. vol. 1, p. 223)].

Ākhūnd 591, 69 ff., 21 lines to a page, 24/26 × 15 cm, containing part I only, incomplete in the beginning and the end (cat. p. 1333).

[Microfilm copy: Āstān-i Quds microfilm 141].

Āstān-i Quds 700, 324 ff., 30 × 20 cm, 21 lines to a page, containing parts II and III; copy completed on 8 Dhu l-Qaʿda 1031/14 September 1622 by Wajīh al-Dīn b. Ismāʿīl b. ʿAbbās al-Muʿallim al-Makkī (cat. 1926–85, vol. 4, pp. 170–72; cat. 1990, p. 330).

[Microfilm copy: Āstān-i Quds microfilm 9683].

Āstān-i Quds 701, 144 ff., 19 × 13 cm, 25 lines to a page; containing parts II and III, incomplete in the beginning and the end (cat. 1926–85, vol. 4, pp. 170–72; cat. 1990, p. 330).

Āstān-i Quds 702, 208 ff., 22 × 17 cm, 12 lines to a page, containing part II; copied in 1297/1880–1 (cat. 1926–85, vol. 4, pp. 170–72; cat. 1990, p. 330).

[Microfilm copy: Āstān-i Quds microfilm 9771].

Āstān-i Quds 703, 336 ff., 19 × 26 cm, 25 lines to a page, containing parts I, II and III, copied by ʿAbd al-Raḥmān Maḥmūd al-Qirimī (?); at the end of part II there is a colophon of the copyist dated beginning of Muḥarram 705/July 1305 (cat. 1926–85, vol. 4, pp. 170–72; cat. 1990, p. 330).

[Microfilm copies: Dānishgāh microfilms 1522 and 2909 (cat. 1969–84, vol. 1, pp. 340–41)].

Āstān-i Quds 11523, 287 ff., 16 × 23.5 cm, 22 lines to a page, containing parts I and II; copied by Maḥmūd al-Ḥusaynī in Muḥarram 1298/December 1880–January 1881 from a corrupt manuscript (cat. vol. 11, p. 186).

Atıf Efendi 1587, 265 ff., 25.2 × 16.5 cm, 27 lines to a page, containing parts I (ff. 1–111) and II (ff. 111–266). Copied at the beginning of the 8th/14th century by Ḥabīb b. Rasūl b. Ibrāhīm b. Zakariyyā; collation completed on 20 Dhu l-Qaʿda 711/29 March 1312 (cat. 1892, p. 93; cat. 1987, p. 256; cf. also *Nawādir*, vol. 1, pp. 164–65; *Mukhtārāt*, p. 143; Ritter I, p. 274).

Awqāf ʿāmma 5342, 415 ff., 19 × 25 cm, containing part III (cat. vol. 2, pp. 291–92 no. 3401; in the catalogue the manuscript is misidentified as a *Sharḥ Hayākil al-ḥikma* by an anonymous author).

Ayasofya 2392, ff. 44–261, 25 × 17.5 cm, 17–22 lines to a page, containing parts II (ff. 1b–44a [*Talwīḥāt*], 45b–125b [commentary]) and III (ff. 126b–261b). Copied by al-Fatḥ b. ʿAbd Allāh b. al-Fatḥ b. ʿAlī b. Muḥammad al-Bundārī *al-mutaṭabbib*; part II completed on 20 Jumādā I 732/18 February 1332 (see colophon f. 44a), part III completed on 25 Ramaḍān 732/20 June 1332 in the Madrasa Niẓāmiyya in Isfahan (see colophon f. 261b:13–19) (cat. p. 143; Ritter I, p. 274).

Ayasofya 2423, 336 ff., 13 × 19 cm, 23 lines to a page, part I (ff. 1–94a) completed on 17 Rabīʿ I 763/15 January 1362 by Ḥasan b. Muḥammad b. al-Ḥasan al-Muṭahhar, part II (ff. 95b–183b) on 18 Jumādā I 764/5 March 1363 by Ḥasan b. Muḥammad b. al-Muṭahhar for his own use (*li-nafsihī*), part III (ff. 184b–333a) completed on 5 Rabīʿ II 762/12 February 1361 by the same scribe *li-nafsihī*. The title page has: *Kitāb Sharḥ al-Talwīḥāt li-l-Faylasūf al-muḥaqqiq Hibat Allāh b. Kammūna* (cat. p. 145; Ritter I, p. 274).

Ayasofya 2424, 129 ff., 19.4 × 13.3 cm, 21 lines to a page. Contains part II (ff. 2b–114a) and the beginning of part III (ff. 115b–130a) of the commentary. Copied by Ṣalāḥ al-Dīn b. Dāwūd al-Ṭanḥūmī (?) (cf. undated colophon on f. 114a). Presumably 8th/14th century (cat. p. 145; Ritter I, p. 274).

Bankipore 2356, 323 ff., 10 × 6.5 cm, 22 lines to a page, containing parts II (ff. 1–116a) and III (ff. 116b–323). The copy is not dated (apparently 11th/17th century) and transcribed from a copy dated 706/1306–7 (cat. 1908–, vol. 21, pp. 100–1).

Bankipore 2357, 163 ff., 9.7 × 6.5 cm, 23 lines to a page, containing part I. Missing in the beginning (beginning = ed. Mūsawī, p. 2:22) and the end (end = ed. Mūsawī, p. 114:6). The copy is not dated and was apparently written in 8th/14th century. Ff. 83–103 are written in a later *Nastaʿlīq* hand and were evidently later added to the copy. The manuscript belongs to the same copy as Bankipore 2358 (cat. 1908–, vol. 21, p. 102).

Bankipore 2358, 170 ff., 9.7 × 6.5 cm, 23 lines to a page The copy is not dated and was apparently written in the 8th/14th century. Ff. 100–103 and 122 are written in a later *Nastaʿlīq* hand and were evidently later added to the copy. The manuscript belongs to the same copy as Bankipore 2357 (cat. 1908–, vol. 21, p. 102).

Berlin Ms. Or. Fol. 1321, 18.2 × 25.8 cm, 99 ff., 25 lines to a page, fragment containing the following portions of the text: ff. 1a:1–30b (= ed. Mūsawī, pp. 228:19–297:4), ff. 31–49 (= 298:26–340:25/ eds. Ziai & Alwishah, pp. 13:2–88:4), ff. 50–99 (= 364:1–496:3/ 125:3–339:6). The manuscript which belonged to the *Nachlass* of Moses Wilhelm Shapira was originally part of the same copy as St. Petersburg RNL II Firk. Yevr.-Arab. I 3137 and RNL II Firk. Yevr.-Arab. I 3138 (cat. 1878–97, vol. 2, pp. 65–66 no. 217; Steinschneider: "Schriften der Araber," pp. 364–65). See also below Appendix A: The Firkovitch/ Shapira-Codex of Ibn Kammūna's *Sharḥ al-Talwīḥāt*.
[Microfilm copy: IMHM no. 1809].

British Library OR 6348, 228 ff., 17 × 25 cm. Contains parts II (ff. 1b–77b) and III (ff. 78a–225b). Ff. 5b–78a written in black ink (instead of brown) and possibly by another hand; ff. 224a–225b written by a different hand; ff. 226–228 are from a different manuscript. Copy completed in Muḥarram 900/October 1494 (cf. colophon f. 225b) (cat. 1912, p. 12).

British Library OR 7728, 384 ff., 18.5 × 25.3 cm. Contains parts I (ff. 2b–119a), II (ff. 119b–221b) and III (ff. 222b–384a). Copy completed in 1037/1627–28 by Muḥammad b. ʿAbd al-Raḥīm (colophon f. 384a) (cat. 1912, p. 12).

British Library Delhi Arabic 2105, 135 ff., 13 × 24 cm, ff. 1–41: 33

lines to a page, ff. 42–135: 35 lines to a page. The manuscript is in disorder and contains quires from at least two different works written by different hands: ff. 1–41 which is a fragment from another unidentified philosophical work (possibly Ghazzālī's *Maqāṣid*), and ff. 42–135, containing fragments of parts II and III of Ibn Kammūna's *Sharḥ al-Talwīḥāt*.

Carullah 998, 101 ff., 17.7 × 12.5 cm, 15 lines to a page. Contains parts II (ff. 1–43a) and III (ff. 43b–98a) of Suhrawardī's *Talwīḥāt*. Copied by Dāwūd b. Ḥasan at the end of Ramaḍān 720/October–November 1320. The copy contains two variants of the complete name of the scribe; according to the dated colophon his name is as follows:

بقلم العبد المفتقر إلى رحمة ربه داؤود بن حسن بن علي العرماني (؟)

The scribe collated his copy with Ibn Kammūna's copy of the text and noted throughout the copy selected portions of Ibn Kammūna's commentary in the margins. In the collation mark on the lower margin of the last folio, he gives his name as follows:

... مقابلة بحسب الاستطاعة من نسخة نقلت من أصل المصنف الشيخ ابن كمونة مصنف شرح هذا الكتاب وأنا أقل خلق الله تعالى مسطر هذا الكتاب بقلمه ... داؤود بن حسن بن علي بن أبي الفتح المعروف بابن المسلح (؟)

(Ritter I, p. 272; Suhrawardī: *Opera Metaphysica et Mystica I*, Prolégomènes, pp. lxvi–lxvii).

Chester Beatty 3598, ff. 20–216, 23.5 × 16 cm, 29 lines to a page; containing parts II (ff. 20–74b) and III (ff. 75a–216); incomplete in the beginning (f. 20a:1 = ed. Mūsawī, p. 371:20); part II completed at the end of Jumādā II 729 (?)/April 1329 (cf. colophon f. 74b). On f. 20a title and author are written by a different hand as follows: *Sharḥ al-Talwīḥāt fī ʿilm al-manṭiq wa-l-ḥikma wa-ghayrihimā li-l-Ḥakīm ʿIzz al-Dawla Saʿd b. Manṣūr al-mashhūr bi-Ibn Kammūna* (cat. vol. 6, p. 34).

Chester Beatty 4612, 227 ff., 24.5 × 17.5 cm, 25/31 lines to a page, containing parts II (ff. 1a–83b—incomplete, the beginning of the manuscript corresponds to ed. Mūsawī 295:21 [*an yakūna maʾkhūdhan*]) and III (ff. 84a–227b); written by at least two different hands; undated, presumably 8th/14th century (cat. vol. 6, p. 34).

Damat İbrahim Paşa 820, 234 ff., 21 × 15.5 cm, 23 lines to a page, contains parts I (ff. 1b–119a) and II (ff. 119b–234b) of the commentary. The title is given as *al-qism al-awwal min Sharḥ al-Talwīḥāt wa-l-thānī minhu wa-huwa kāmil al-kitāb li-Mawlānā Mullā ʿIzz al-Dawla wa-huwa kāmil sharḥ al-fannayn al-manṭiqiyyāt wa-l-ṭabīʿiyyāt*... (f. 1a). Ownership statement dated 886/1481–2 (f. 1a) (Ritter I, p. 275).

Dānishkadah 194b, 194 pp., 21.3 × 34 cm, containing part III only; copy completed on 15 Muḥarram 1349/13 June 1930 by Muḥammad Kāẓim; it was later collated with another manuscript by al-Ḥakīm and Sayyid Islām al-Dīn. It contains an ownership statement by Ḍiyāʾ al-Dīn Durrī al-Iṣfahānī (cat. vol. 1, p. 572).

Feyzullah Efendi 1194, iv + 246 ff., 12.5 × 17.6 cm, 27 lines to a page, containing parts I (ff. 1b–58b), II (ff. 59a–120b) and III (ff. 121a–296a). There is a scribe's colophon at the end of part I (f. 58b:28–29), dated beginning of Shaʿbān 700/April 1301:

فرغت من تعليقها يوم الخميس أوائل شعبان سنة خصص

The manuscript contains conflicting and partly mistaken information on the author of the commentary. On f. IIIa is written:

شرح التلويحات لأفضل المحققين وأكمل المدققين أبي البركات البغدادي قدس الله نفسه

On f. 1a it is written:

شرح التلويح للشيخ الفاضل المحقق عز الدولة سعد بن منصور المعروف بابن كمونة وهو
معاصر لصاحب التلويح شهاب الدين السهروردي ووفاته سنة اثنين وثلثين وستمائة...

Feyzullah Efendi 1195, 205 ff., 16.7 × 24 cm, 29 lines to a page, containing parts I (ff. 2b–61b), II (ff. 62a–113b) and III (ff. 114a–205b); copied by ʿAbd Allāh b. Mubārak b. Hadiyya, completed in a *madrasa* in Tabriz on 22 Shawwāl 696/12 August 1297 (see colophon f. 205b:27–29). The title page (f. 2a) contains two partly conflicting indications as to title and author by two different hands:

شرح التلويحات السهروردية تصنيف الفاضل معز الدولة وهو معاصر العلامة الشيرازي
والنسب إلى ابن كمونة اليهودي

شرح التلويح للشيخ الفاضل المحقق عز الدولة سعد بن منصور المعروف بابن كمونة الإسرائيلي
وهو شرح ممزوج والشارح معاصر الماتن توفي الماتن اثنين وثلاثين وستمائة

(cf. *Nawādir*, vol. 1, p. 165; *Mukhtārāt*, p. 143).

Flügel 1331 (= Codex Vindobenensis Palatinus Mxt. 453), 510 ff., 8.5 × 5.7 cm, 17 lines to a page, containing parts I (ff. 1b–279b) and II (ff. 280a–510a) (cat. vol. 2, pp. 606–7).

Gharawiyya (cat. 1959, p. 25 no. 24: *Sharḥ al-Talwīḥāt al-Suhrawardiyya Saʿd b. Manṣūr b. Saʿd al-maʿrūf bi-Ibn Kammūna*). According to Āghā Buzurg, who inspected this copy (cf. *Dharīʿa*, vol. 13, pp. 152–53 no. 519), it is incomplete and may contain only part II.

Hekimoğlu Ali Paşa 854/2, ff. 46b–275b, 31.5 × 19 cm, 32 lines to a page, copied ca. 1100/1688–89. Containing parts I (ff. 46b–100a), II (ff. 100a–165a) and III (ff. 165a–275b). The codex contains no indication that the commentary was composed by Ibn Kammūna (Ritter I, p. 275; *Nawādir*, vol. 1, p. 165; *Mukhtārāt*, p. 143).

Istanbul University al-qism al-ʿarabī 2775, 312 ff., copied in 1101/1689–90 (*Nawādir*, vol. 1, p. 165).

John Rylands 340, 104 ff., 11 × 20 cm, 30 lines to a page. The manuscript was copied around 1680 C.E. in India and is deficient both at the beginning and the end (cat. columns 674–76 no. 400).

Lâleli 2521, 310 ff., 22 × 16.5 cm, 25 lines to a page, contains parts I (ff. 1b–157a) and II (ff. 157a–310a). The title page has: *Sharḥ Talwīḥāt al-Shaykh al-Suhrawardī li-Ibn Kammūna* (cat. p. 193; Ritter I, p. 275).

Lâleli 2522, 171 ff., 24 × 16.5 cm, 25 lines to a page; containing parts I (ff. 1b–80a) and II (ff. 81a–171b). There are two undated and anonymous collation colophons (ff. 80a, 171b). The title page has: *Sharḥ al-Talwīḥāt li-l-Ḥakīm al-mudaqqiq ʿIzz al-Dawla Saʿd b. Manṣūr al-maʿrūf bi-Ibn Kammūna al-Isrāʾīlī* (cat. p. 193; Ritter I, p. 274).

Leiden Or. 137, 282 ff., 31 lines to a page, containing parts I (ff. 1b–76a), II (ff. 76b–149b) and III (ff. 150b–282a); part II completed on 6 Ṣafar 736/14 September 1335 in Mashhad (f. 149b), part III completed on 29 Ṣafar 736/17 October 1335 in Mashhad (f. 282a). It has been collated with other manuscript(s) as is indicated by several marginal collation notes (cat. 1851–77, vol. 3, pp. 349–51 no. 1496; cat. 1957, p. 362).

Majlis 1849, 227 ff., 17 × 24.5 cm, 27 lines to a page, completed on 1 Muḥarram 702/25 August 1302; containing parts II (ff. 2b–81a) and III (ff. 81b–227b). Up to f. 86a, the text of Suhrawardī is underlined in red (cat. vol. 5, pp. 314–17).

Majlis 1850, 281 ff., 26.5 × 13.5 cm, 24 lines to a page, containing parts II (ff. 1b–109b) and III (ff. 110b–291b). Copied by Muẓaffar Ḥusayn al-Kāshānī, who had completed copying part II in Ṣafar 1023/March-April 1614 (f. 109b). The text of Suhrawardī is underlined in red. The title page has: *Sharḥ al-Talwīḥāt maʿa jild-i dīgar* (cat. vol. 5, pp. 317–18).

[Microfilm copy: Majlis 2607].

Malik 934, 283 ff., 12 × 22.2 cm, 35 lines to a page. Part II completed on 14 Ṣafar 704/15 September 1304 in the Madrasa Niẓāmiyya, part III completed on 10 Rabīʿ I 704/10 October 1304. The scribe used several manuscripts, and numerous corrections and additions are noted in the margins (cf. f. 288b), and it contains a collation note. On the front folio there is a statement of ownership by Ḥājj Ḥusayn Malik: "[This] is an exceptionally fine book in respect to both the [quality of the] manuscript as well as to the author's [rank] who had been one of the great [scholars] of his age at the Niẓāmiyya Madrasa in Baghdād...I bought this book, together with 45 other volumes, from Mr. Falsafī Sharaf al-Mulk...and placed it in the library in the year 1314." (Translation provided in the introduction to Ibn Kammūna: *Tanqīḥāt*. Eds. Ziai & Alwishah, p. 47) (cat. vol. 1, pp. 390–91).

Mishkāt 273, 214 ff., 16 × 24.5 cm, 23 lines to a page; containing parts I (ff. 1b–114b) and II (ff. 115b–214a) and one page of part III (f. 214b); undated, containing an ownership statement (f. 1b) by the name of Muḥammad b. Muḥammad b. Barakāt dated 961/1553–54 (cat. 1976, p. 16; cat. 1951–85, vol. 3 i, pp. 211–13).

Nuruosmaniye 2661, 139 ff., 18.5 × 13 cm, 17 lines to a page. The text of the commentary is located in the margins whereas the text of the *Talwīḥāt* is on the main parts of the pages. At the end of part II there is a colophon dated 11 Muḥarram 713/8 May 1313 (Ritter I, p. 274).

Ragıp Paşa 852, 336 ff., 26 × 18 cm, 27 lines to a page; completed on 17 Rabīʿ I 752/14 May 1351 (see colophons at the end of parts II and III on ff. 177a and 336a). Contains parts I (ff. 1b–85a), II (ff. 85b–

177a) and III (ff. 177b–336a), together with glosses by Naṣīr al-Dīn al-Kāshānī and apparently also by some other, not identified readers (cat. p. 64; Ritter I, p. 274).

Şehid Ali Paşa 1740, 253 ff., 17.8 × 25.3 cm, 29 lines to a page. Containing parts I (ff. 1b–59a), II (ff. 59b–125a) and III (ff. 125b–253a). Copy completed on 16 Jumadā II 686/29 July 1287 in the Madrasa Niẓāmiyya (see colophon f. 253a:11–13). Part II was collated with a copy that has been read to Ibn Kammūna (see colophon f. 125a: *qūbila bi-nuskha muṣaḥḥaḥa maqrūʾa ʿalā l-muṣannif fa-ṣaḥḥa bi-qadr al-imkān wa-llāh al-mustaʿān*).

St. Petersburg RNL II Firk. Yevr.-Arab. I 857, 1 f., 5 lines (f. 1a), respectively 22 lines (f. 1b) to a page, heavily damaged fragment containing the beginning of the text: f. 1b (= ed. Mūsawī, p. 2:1–21) (cat. 1991, p. 104).
[Microfilm copy: IMHM no. 54602].

St. Petersburg RNL II Firk. Yevr.-Arab. I 3137, 18.2 × 26.2 cm, 123 ff., consisting of one quire of six leaves (ff. 1–6), four isolated leaves (ff. 7, 8, 9, 10), one quire of eight leaves (ff. 11–18), one isolated leaf (f. 19), one quire of eight leaves (ff. 20–27), one isolated leaf (f. 28), one bifolio (ff. 29–30), one isolated leaf (f. 31), one quire of ten leaves (ff. 32–41), two quires of eight leaves (ff. 42–49, 50–57), one isolated leaf (f. 58), four quires of ten leaves (ff. 59–68, 69–78, 79–88, 89–98), one isolated leaf (f. 99), one quire of eight leaves (ff. 100–107), one bifolio (ff. 108–109), one isolated leaf (f. 110), two bifolios (ff. 111–112, 113–114), two isolated leaves (ff. 115, 116), one quire of six leaves (ff. 117–122). 25 lines to a page. The fragment contains the following portions of the text: ff. 1–7 (= ed. Mūsawī, pp. 3:1–21:9), ff. 8–9 (= 23:11–28:2), ff. 10–18 (= 48:3–71:16), f. 19 (= 45:18–48:3), ff. 20–27 (= 165:7–181:2), f. 28 (= 73:16–75:23), f. 29 (= 544:8–547:6/ eds. Ziai & Alwishah, pp. 420:13–424:12), f. 30 (= 526:21–529:12/ 391:11–396:3), f. 31 (= 226:21–228:19), ff. 32–41 (= 182:23–204:20), ff. 42–49 (= 28:2–45:18), ff. 50–88 (= 75:23–163:7), ff. 89–98 (= 341:1–364:1/ 88:4–125:3), ff. 99–107 (= 204:20–226:21), ff. 108–109 (= 559:5–563:15/ 444:14–451:16), f. 110 [lower margin of the leaf heavily damaged] (= 565:14–567/ 455:5–458), f. 111 (= 556:9–559:5/ 440:9–444:14), f. 112 (= 563:15–565:14/ 451:16–455:5), f. 113 (= 181:2–182:23), f. 114 (= 163:7–165:7), f. 115 (= 524:12–

526:21/ 387:9–391:11), f. 116 (= 521:16–524:12/ 382:12–387:9), ff. 117–123a (= 529:12–543:2/ 396:3–418:13), f. 124 (= 2:1–21). The manuscript belongs to the same copy as RNL II Firk. Yevr.-Arab. I 3138 and Berlin Ms. Or. Fol. 1321. See below Appendix A: The Firkovitch/Shapira Codex of Ibn Kammūna's *Sharḥ al-Talwīḥāt*.
[Microfilm copy: IMHM no. 56523].

St. Petersburg RNL II Firk. Yevr.-Arab. I 3138, 18.2 × 25.8 cm, 9 ff., consisting of one isolated folio (f. 1) and two quires of four folios (ff. 2–5, 6–9), 25 lines to a page. The fragment contains the following portions of the text: f. 1 (= ed. Mūsawī, pp. 496:3–499:1/ eds. Ziai & Alwishah, pp. 339:6–344:2), ff. 2–5 (= 504:21–516:5/ 353:8–373:4), ff. 6–7 (= 499:1–504:21/ 344:2–353:8), ff. 8–9 (= 516:5–521:16/ 373:4–382:12). The manuscript belongs to the same copy as RNL II Firk. Yevr.-Arab. I 3137 and Berlin Ms. Or. Fol. 1321. See below Appendix A: The Firkovitch/Shapira Codex of Ibn Kammūna's *Sharḥ al-Talwīḥāt*.
[Microfilm copy: IMHM no. 56522].

Taymūriyya Ḥikma 92, containing part II (cf. GAL, vol. 1, p. 565).

Yahuda 201, ff. 1a–73a (text of Suhrawardī's *Talwīḥāt*), ff. 74b–345b (commentary by Ibn Kammūna), 28 lines to a page, completed Friday, 11 Ṣafar 702/4 October 1302 in Isfahan (see colophons ff. 141b, 345a) (cat. pp. 143–44).

Yeni Câmi 765/1, 225 ff., 32 × 21 cm, 33 lines to a page, copied for the Sulṭān Muḥammad Fātiḥ (reigned 848–850/1444–1446 and 855–886/1451–1481). It is followed by a copy of Naṣīr al-Dīn al-Ṭūsī's *Awṣāf al-ashrāf*, dated 864/1459–60, which seems to have been copied at about the same time as Ibn Kammūna's *Sharḥ*. The copy contains parts I (1b–58a), II (58b–117a) and III (ff. 117b–217b) of the commentary entitled *Sharḥ Talwīḥāt al-Shaykh Shihāb al-Dīn al-maqtūl li-l-Shaykh al-maʿrūf bi-Ibn Kammūna* (Ritter I, p. 275).

Yeni Câmi 766, 213 ff., 24 × 12.5 cm, 25 lines to a page; containing part III. Parts I and II of the same copy are contained in Ahmed III 3284 (= Topkapı 6687). On the title page is written: *al-Juzʾ al-ākhir min Sharḥ al-Talwīḥāt taʾlīf al-Ḥakīm al-Fāḍil al-Faylasūf Saʿd b. Manṣūr b. Saʿd b. al-Ḥasan b. Kammūna*. Completed between 27–30 Ṣafar 686/12–15 April 1287 by Ismāʿīl b. Ibrāhīm b. al-Zāhid for his own use (*li-nafsihī*) (cf. colophon f. 214b:21–25). On the first leaf is to be found a *waqfiyya* for the library of Mashhad ʿAlī (Najaf), dated 766/1364 (cat. p. 40; Ritter I, p. 273; *Istanbul Umumî kütüphaneleri Yazmalari*, p. 36). [Microfilm copy: Maʿhad 234 (cat. vol. 1, p. 223)].

Ẓāhiriyya 7885, 308 ff., 12.5 × 25 cm, copied in 690/1291 in Baghdad (cat. 1970, pp. 26–27; cat. 1987, p. 394).

1.1.3. *Taʿālīq ʿalā l-suʾālāt al-mūrada ʿalā l-uṣūlayn min Kitāb al-Maʿālim (Iʿtirāḍāt ʿalā Kitāb al-Maʿālim fī l-uṣūlayn li-Fakhr al-Dīn al-Rāzī/ Maqāla ʿalā mabāḥith Najm al-Dīn al-Kātibī al-Qazwīnī wa-īrāduhū ʿalā nawʿay al-ʿilm)*

Commentaries on Kātibī's critical remarks on the *Kitāb al-Maʿālim* of Fakhr al-Dīn al-Rāzī. Since Ibn Kammūna refers to them in his *Sharḥ al-Talwīḥāt*,[3] the date of completion of his *Sharḥ*, 667/1268, is the *terminus ante quem* for the *Taʿālīq*.

Introduction (Esad Efendi 1932, f. 116a):—

قال الشيخ العلامة الفاضل المفضل المحقق عز الدولة سعد بن الحسن بن هبة الله بن كمونة البغدادي رحمه الله تعالى: إني وجدت كتاب المعالم في الأصولين للإمام العلامة حجة الخلائق فخر الدين الرازي قدّس الله روحه ونوّر ضريحه مع كونه من أنفع المختصرات المصنفة في هذا الفن وأشدها تحقيقاً وتدقيقاً غير خالٍ عن مواضع غير منتقدة وأبحاث غير مهذبة. فأردت أن أنبّه على تلك المواضع إرشاداً لطالبي الحقائق إلى ما في هذا الكتاب من مواقع الأنظار ودقائق الأفكار لما شاهدته من إقبال الناس في العراق عليه وكثرة الانصراف إليه فاتفق أن وقع إلي رسالة للإمام العلامة أفضل المتأخرين وسيد المحققين نجم الدين الدبيران [الكاتبي] متّع الله أهل العلم بدوام أيامه وحراسة مدّته قد أودعها أسئلة على نوعي كتاب المعالم مشتملة على الفوائد الجمة متفردة بالفرائد المهمة حاوية بالمباحث العجيبة جامعة للدقائق الغريبة متضمنة من ذلك أشياء كثيرة ... خطرت لي عند تصفح الكتاب لكني وجدت فيها مع ذلك مواضع عدة محتملة لزيادة بحث ومفتقرة إلى معاودة نظر فاستغنيت بها عما كتب وعزمت على التنبيه عليه ودعاني ما وجدت في تلك الأسئلة من المواضع المذكورة إلى إثبات تلك المباحث والإرشاد إلى ما فيها من الفوائد مع إضافة أبحاث على مواضع من أصول الدين لم يتكلم عليه ليكون ذلك جاري مجري الحواشي على

[3] *Sharḥ al-Talwīḥāt*. Ed. Mūsawī, pp. 701:10–11 (*wa-anā fa-qad ajabtu ʿan hādhā fī jawāb al-shukūk allatī awradahū al-Imām al-ʿAllāma Najm al-Dīn al-Dabīrānī... ʿala Kitāb al-Maʿālim...*), 714:19–20 (*...wa-qultu ʿalayhi fī ajwibat mā awradahū al-Imām al-Qudwa Najm al-Dīn al-Dabīrānī... ʿalā Kitāb al-Maʿālim min al-asʾila...*).

تلك الرسالة تكثيراً لفوائدها وتثميراً لفرائدها بحيث إذا أضيف له أسئلة مذكورة وهذا التعليق إلى كتاب المعالم تم الانتفاع به ووقعت الإحاطة بحل [بعض] مشكلاته ومستصعبه على أني لم أتمكن من تفريع البحث على أصوله الكلامية لضيق الوقت بالشواغل الدنيوية فإن النظر في الدقائق المهذبة والتعمق في الأبحاث المتشعبة لا سبيل إلى إدراكه بالأفكار المشذبة ولا نؤمن فيه والحال هذه من الأنظار الغير المستصوبة. فأسأل الله تعالى العصمة والتوفيق والهداية إلى سواء الطريق إنه كريم رحيم.

End (Ayasofya 4862, f. 151a):—

أقول: البرهان المعروف ببرهان التطبيق يدل على الانتهاء إلى علة أولى وإلى معلول أخير وهو برهان آخر مشهور في الكتب الحكمية والكلامية وصاحب الكتاب ذكره في أكثر كتبه فكيف يجزم القول بأن البرهان ما قام على الانتهاء إلى معلول هو آخر المعلولات. وهذا آخر ما سنح لخاطري من التعاليق على السؤالات الموردة على الأصولين من كتاب المعالم والحمد حق حمده والصلاة على خير خلقه محمد وآله الطاهرين وصحبه الأكرمين.

### *Manuscripts*

Ayasofya 4862, ff. 141b–151a, incomplete in the beginning (cf. *Nawādir*, vol. 1, p. 167).

Esad Efendi 1932, ff. 118a–126b, 12.7 × 18.5 cm, 23 lines to a page; containing also Kātibī's remarks, ff. 99a–117b; copy completed on 11 Ramaḍān 721/4 October 1321 in Nakhjawān by Fakhr al-Dīn Hindūshāh b. Sanjar b. ʿAbd Allāh al-Ṣāḥibī (i.e. Dawlatshāh b. Sanjar al-Ṣāḥibī, fl. 724/1324) (cf. *Mukhtārāt*, p. 142 where a wrong shelfmark is mentioned).
[Microfilm copy: Ritter/Uppsala MF 15:2573 (cat. p. 169)].

### *Editions*

Eds. Reza Pourjavady & Sabine Schmidtke. Tehran: Iranian Institute of Philosophy (forthcoming).

## 1.2. *Excerpts*

### 1.2.1. *Fawāʾid min Talkhīṣ al-muḥaṣṣal (Multaqaṭ / Iltiqāṭ / Intikhāb min Talkhīṣ al-muḥaṣṣal)*

Excerpts from Naṣīr al-Dīn al-Ṭūsī's critical commentary on Fakhr al-Dīn al-Rāzī's *Muḥaṣṣal afkār al-mutaqaddimīn wa-l-mutaʾakhkhirīn*, *Talkhīṣ al-muḥaṣṣal*, completed between 1–3 Dhu l-Qaʿda 670/30 May–1 June 1272. While Ibn Kammūna paraphrases the textual selections from Ṭūsī's *Talkhīṣ* at the beginning of his excerpts, he gradually starts quoting them verbatim as the work progresses. It is noteworthy that he does not adduce anything from the section devoted to the impeccability (*ʿiṣma*) of the Prophet and selects only one comment by Ṭūsī from the chapter on the imamate, stating that most Shīʿite traditions are based on unreliable texts. The excerpts contain the following portions of the text (references are given according to British Library Add 7517):

[From *al-Rukn al-awwal: fī l-muqaddamāt:*]
Ff. 67a:5–9 (= *Talkhīṣ*, Beirut 1405/1985, p. 6:9–12), 67a:9–11 (= 7:6–11), 67a:11–67b:4 (= 7:18–23), 67b:5–7 (= 8:18–20), 67b:7–9 (= 9:8–10), 67b:9–13 (= 12:1–4), 67b:13–68a:7 (= 12:15–21), 68a:7–9 (= 14:21–22), 68a:9–68b:2 (= 23:6–10), 68b:2–6 (= 23:14–17), 68b:6–9 (= 24:3–7), 68b:9–69a:7 (= 29:13–19), 69a:7–69b:2 (= 30:7–12), 69b:2–6 (= 30:20–23), 69b:7–10 (= 31:9–11), 69b:10–70a:7 (= 31:23–32:9), 70a:8–9 (= 33:8–10), 70a:9–12 (= 34:10–12), 70a:12–70b:4 (= 35:5–9), 70b:4–5 (= 35:17–18), 70b:5–11 (= 35:23–36:4), 70b:12–71a:1 (= 36:14–16), 71a:1–4 (= 37:3–5), 71a:4–7 (= 38:3–5), 71a:8–13 (= 38:14–19), 71a:13–71b:5 (= 39:5–14), 71b:5–8 (= 49:5–7), 71b:8–10 (= 52:20–21), 71b:10–13 (= 53:4–8), 72a:1–4 (= 53:14–16), 72a:4–6 (= 55:2–5), 72a:6–9 (= 55:16–18), 72a:10–13 (= 56:8–11), 72a:13–72b:9 (= 57:8–15), 72b:9–10 (= 57:22–23), 72b:10–73a:4 (= 59:3–9), 73a:4–5 (= 61:20–21), 73a:5–6 (= 63:2–3), 73a:7–9 (= 63:20–21), 73a:9–11 (= 64:2–4), 73a:11–13 (= 67:5–6).
[From *al-Rukn al-thānī: fī taqsīm al-maʿlūmāt:*]
Ff. 73b:1–3 (= 75:11–13), 73b:3–4 (= 75:14–15), 73b:4 (= 75:18–19), 73b:4–8 (= 75:22–24), 73b:8–10 (= 77:16–18), 73b:11 (= 78:5–6), 73b:12–13 (= 81:16–18), 74a:1–2 (= 91:3–4), 74a:2–3 (= 92:13), 74a:3–6 (= 93:16–94:1), 74a:6–74b:2 (= 94:7–13), 74b:2–10 (= 95:9–

14), 74b:10–75a:2 (= 97:20–98:2), 75a:2–9 (= 98:4–8), 75a:9–12 (=98:21–99:1), 75a:12–75b:3 (=100:3–8), 75b:3–76a:6 (=100:17–101:5), 76a:6–76b:4 (= 101:18–102:5), 76b:4–12 (= 102:19–103:2), 76b:12–13 (= 103:22), 76b:13–77a:3 (= 106:10–13), 77a:3–77b:9 (= 107:12–108:4), 77b:9–11 (= 109:22–110:1), 77b:11–78a:6 (= 115:22–116:5), 78a:6–7 (= 116:10–11), 78a:7–10 (= 116:19–20), 78a:10 (= 117:3), 78a:11–78b:3 (= 118:19–22), 78b:3–4 (= 119:14), 78b:4–6 (= 120:20–22), 78b:7–13 (= 121:19–25), 78b:13–79a:7 (= 127:21–128:3), 79a:7–79b:1 (= 128:13–18), 79b:2–8 (= 130:14–19), 79b:8–12 (= 131:22–132:2), 79b:13–80a:4 (= 133:6–9), 80a:4–8 (= 133:20–23), 80a:8–80b:9 (= 134:22–135:9), 80b:9–81a:6 (= 135:15–136:2), 81a:6–9 (= 137:1–3), 81a:9–11 (= 137: 7–9), 81a:11–12 (= 137:19–20), 81a:12–81b:1 (= 139:7–8), 81b:1–2 (= 140:16–17), 81b:2 (= 146:11), 81b:3 (= 147:10), 81b:3–6 (= 155:17–19), 81b:6–7 (= 156:15–16), 81b:7–8 (= 156:21–22), 81b:8–11 (= 157: 1–3), 81b:11–82a:1 (= 157:7–9), 82a:1–3 (= 157:18–21), 82a:3–11 (= 162: 5–10), 82a:11–82b:6 (= 162:12–17), 82b:6–8 (= 164:3–5), 82b:8–10 (= 164:12–14), 82b:10–83a:4 (= 166:7–13), 83a:5–8 (= 167:8–12), 83a:8–12 (= 173:8–14), 83a:12–83b:4 (= 174:5–13), 83b:4–6 (= 175:8–9), 83b:6–7 (= 177:21–22), 83b:7–84a:11 (= 178:2–14), 84a:11–12 (= 179:6–7), 84a:12–84b:2 (= 179:10–12), 84b:2–7 (= 181:6–8), 84b:7–8 (= 181:13–14), 84b:9–11 (= 182:14–16), 84b:11–85a:6 (= 184:15–22), 85a:6–7 (= 185:11–12), 85a:7–10 (= 185:16–18), 85a:10–11 (= 185:20), 85a:11–85b:1 (= 186:18–20), 85b:1–7 (= 186:23–187:4), 85b:8–9 (= 187: 5–6), 85b:9–86a:1 (= 188:1–4), 86a:2–9 (= 188:17–22), 86a:9–11 (= 189: 12–13), 86a:12–86b:2 (= 201:12–14), 86b:3–6 (= 201:15–19), 86b:7–11 (= 202:9–12), 86b:11–13 (= 202:23–203:1), 86b:13–87a:4 (= 203:9–12), 87a:4–7 (= 203:21–204:1), 87a:8–13 (= 204:3–8), 87a:13–87b:1 (= 204:13–14), 87b:1–3 (= 204:15–21), 87b:3–6 (= 206:20–23), 87b:6–13 (= 207:13–18), 88a:1–2 (= 207:19–22), 88a:2–5 (= 207:23–208:5), 88a:5–88b:1 (= 209:8–14), 88b:1–8 (= 210:16–20), 88b:8–11 (= 211:11–13), 88b:12–89a:1 (= 212:17–18), 89a:1–8 (= 214:21–215:4), 89a:8–10 (= 215:6–8), 89a:10–89b:3 (= 216:15–19), 89b:3–12 (= 217:17–23), 89b:12–13 (= 219:19–220:1), 89b:13–90a:1 (= 222:8–9), 90a:1–6 (= 222:23–223:3), 90a:6–13 (= 223:23–224:16), 90a:13–90b:13 (= 226: 9–19).

[From *Khātima: fī aḥkām al-mawjūdāt:*]

Ff. 91a:1–2 (= 232:13–15), 91a:2–6 (= 232:16–19), 91a:6–9 (= 236:3–5), 91a:9–11 (= 238:7–9), 91a:12–13 (= 238:15–17).

[From *al-Rukn al-thālith: fī l-ilāhiyyāt, al-Qism al-awwal: fī l-dhāt:*]
Ff. 91b:1–6 (= 244:8–12), 91b:6–7 (= 245:15), 91b:8–11 (= 246:15–18), 91b:11–92a:5 (= 247:6–13), 92a:6–11 (= 252:1–4), 92a:11–12 (= 252:20–21), 92a:13–92b:2 (= 256:17–23).
[From *al-Qism al-thānī: fī l-ṣifāt:*]
Ff. 92b:2–3 (= 260:2–3), 92b:3–5 (= 261:22–262:1), 92b:5–7 (= 262:10–11), 92b:7–10 (= 265:19–266:12), 92b:10–12 (= 267:13), 92b:12–13 (= 267:20), 92b:13–93a:2 (= 269:12–19), 93a:2–5 (= 269:19–270:2), 93a:5–6 (= 275:11–12), 93a:6–8 (= 276:1–2), 93a:8–10 (= 276:7–8), 93a:10–11 (= 276:11–12), 93a:11–12 (= 276:18–20), 93a:12–14 (= 276:22), 93a:14 (= 280:2), 93a:14–93b:1 (= 280:4–5), 94a:14–94b:3 (= 293:8–11), 94b:3–4 (= 293:12–13), 94b:4–5 (= 294:5), 94b:5–7 (= 294:17–20), 94b:7–9 (= 295:17–18), 94b:9–13 (= 295:23–296:5), 94b:13–95a:6 (= 296:16–22), 95a:6–7 (= 297:20–21), 95a:7–10 (= 298:12–14), 95a:10–15 (= 299:14–18), 95a:15–95b:4 (= 300:11–14), 95b:4–6 (= 300:17–18), 95b:6–9 (= 303:23–304:2), 95b:9–10 (= 305:14–18), 95b:10–13 (= 305:19–22), 95b:13–15 (= 309:17–19), 95b:15–96a:4 (= 310:6–9), 96a:4–6 (= 310:19–20), 96a:6–7 (= 311:2), 96a:7–8 (= 313:1–2), 96a:9–13 (= 313:5–8), 96a:13–15 (= 313:10–11), 96b:1–2 (= 313:12–13), 96b:2–8 (= 315:18–22), 96b:8–9 (= 315:23–24), 96b:9–14 (= 319:18–22), 96b:14 (= 320:1–2), 96b:15–97a:2 (= 320:19–21), 97a:2–3 (= 322:11–12), 97a:3–7 (= 324:1–5).
[From *al-Qism al-thālith, fī l-afʿāl:*]
Ff. 97a:6–7 (= 327:1), 97a:7–9 (= 327:4–6), 97a:10–15 (= 327:14–17), 97a:15–97b:1 (= 327:18–19), 97b:1–3 (= 328:17–18), 97b:4–7 (= 334:19–21), 97b:7–10 (= 335:1–3), 97b:10–14 (= 335:4–6), 97b:14–98a:1 (= 335:7–8), 98a:1–5 (= 335:21–336:1), 98a:5–14 (= 338:2–8), 98a:15–98b:2 (= 338:9–11), 98b:3–6 (= 339:10–13), 98b:7–9 (= 340:20–21), 98b:9–11 (= 341:2–4), 98b:11–13 (= 341:11–13), 98b:14–15 (= 343:2), 98b:15–99a:1 (= 343:5–6), 99a:1–4 (= 344:5–8) 99a:4–8 (= 344:10–12), 99a:8–12 (= 344:16–19), 99a:12–15 (= 345:12–15), 99a:15–99b:3 (= 345:14–16).
[From *al-Qism al-rābiʿ, fī l-asmāʾ:*]
Ff. 99b:3–6 (= 347:12–13), 99b:6–11 (= 347:18).
[From *al-Rukn al-rābiʿ: fī l-samʿiyyāt:*]
[From *al-Qism al-awwal: fī l-nubuwwāt:*]
Ff. 99b:11–13 (= 351:23–352:1), 99b:13–14 (= 353:9–10), 99b:14 (= 365:14–16), 99b:15–100a:1 (= 366:2–3), 100a:1–3 (= 366:17–18).

[From *al-Qism al-rābiʿ: fī l-maʿād:*]
Ff. 100a:4–5 (= 376:1–2), 100a:5–8 (= 379:3–4), 100a:8–100b:1 (= 380:16–23), 100b:1–3 (= 381:1–2), 100b:3–6 (= 381:5–7), 100b:6–12 (= 381:23–382:4), 100b:12–13 (= 382:14–15), 100b:14–101a:4 (= 384:13–17), 101a:4–9 (= 385:14–17), 101a:9–11 (= 387:7–8), 101a:11–14 (= 387:13–17), 101a:14–101b:4 (= 387:19–22), 101b:4–9 (= 388:5–8), 101b:9–10 (= 389:18–19), 101b:10–13 (= 389:22–390:1), 101b:14–15 (= 390:1–3), 102a:1–10 (= 390:13–19), 102a:10–14 (= 391:6–9), 102a:14–102b:2 (= 391:15–17), 102b:2–4 (= 392:4–6), 102b:4–6 (= 392:7–8), 102b:6–15 (= 392:9–16), 102b:15–103a:5 (= 398:1–5), 103a:5–6 (= 399:7–11), 103a:6–14 (= 400:3), 103a:13–103b:3 (= 400:5–8), 103b:3–8 (= 400:18–22).
[From *al-Qism al-thālith: fī l-asmāʾ wa-l-aḥkām:*]
Ff. 103b:8–11 (= 401:12–16), 103b:11–12 (= 402:19–20), 103b:12–104a:2 (= 402:22–403:2), 104a:2–3 (= 403:5–6), 104a:3–6 (= 403:13–15), 104a:6–104b:1 (= 404:16–22), 104b:1–5 (= 405:5–7).
[From *al-Qism al-rābiʿ: fī l-imāma:*]
Ff. 104b:5–6 (= 421:22–422:2).

(*Dharīʿa*, vol. 2, pp. 357–58 no. 1444; vol. 22, p. 196 no. 6669; GAL, vol. 1, p. 507; GALS, vol. 1, p. 769; vol. 3, p. 668; Kholeif (*A Study on Fakhr al-Dīn al-Rāzī*, p. 198) erroneously classifies the excerpts as "a commentary on this *Talkhīṣ* by Ibn Kammūnah;" Rescher: *Development*, p. 205; Steinschneider: *Arabische Literatur*, p. 240 no. 7)

Introduction (British Library Add 7517, ff. 66b:2–67a:5):—

قال الفقير إلى الله تعالى سعد بن منصور بن سعد بن الحسن بن هبة الله بن كمونة عفا الله عنه وغفر له وأعانه على مراضيه ووفقه لطاعته: تشتمل هذه الأوراق على فوائد التقطتُها من كلام الإمام الأعظم العلامة جامع علوم الأوائل والأواخر ملك الحكماء والمحققين سيد الأفاضل قدوة الأماثل مفتي الفرق والطوائف مخترع الدقائق واللطائف المولى نصير الحق والدولة والدين فخر الإسلام والمسلمين محمد بن محمد الطوسي أعز الله تعالى أنصاره وضاعف اقتداره وأمتع أهل العلم بدوام أيامه وحراسة مدته في كتابه الذي سماه بتلخيص المحصل وينحل بها لذوى الفطانة بعد التأمل مشكلات كتاب المحصل وشكوكه ومغالطته ويتبين منها ما في أثناء مباحثه من الخلل والزلل ولعلها كافية في ذلك عند الذكى الفطن ومغنية عن استيعاب جميع ما قيل في التلخيص المذكور ولم ألتزم تمييز المباحث بعضها عن بعض ولا المحافظة على إيراد ألفاظ الكتاب والله موفق للصواب وهو الهادي وإليه المآب. المعاني تنقسم إلى نفس الإدراك . . .

End (British Library Add 7517, f. 104b:6–13):—

...وهذا آخر ما رأيت أن أذكره من كتاب تلخيص المحصل بحسب غرضي ولا بحسب غرض غيري وأسأل الله تعالى العصمة والهداية والتوفيق وحسن الخاتمة وسعادة الدارين. إنه ولي ذلك والقادرِ عليه ولا حول ولا قوة إلا بالله العلي العظيم. والحمد لله رب العالمين وصلواته على من اصطفاه من عباده من ملائكته المقربين وأنبيائه المرسلين وأوليائه المخلصين وخصوصًا على محمد وآله الطاهرين و سلامه عليهم أجمعين...

Author's colophon (Gharawiyya 572; quoted in cat. 1971, p. 25):—

وكان الفراغ منه انتخاباً ونسخاً في العشر الأوسط من ذي القعدة سنة سبعين وستمائة هجرية...

*Manuscripts*

British Library Add 7517/6, ff. 66b–104b, 13/14 lines to a page. There is a scribe's colophon following the previous item in this codex dated 9 Shaʿbān 672/18 February 1274 written in Baghdad (cf. GALS, vol. 3, p. 668; cat. 1849–71, p. 210 no. 429).

Gharawiyya 572 (cat. 1959, p. 24 no. 10; cat. 1971, p. 25 no. 15; cf. also *Dharīʿa*, vol. 22, p. 196 no. 6669; Roper: *World Survey*, vol. 2, p. 45) autograph, completed between 11–20 Dhu l-Qaʿda 670/9–18 June 1272.

1.2.2. Excerpts from *al-Shamsiyya fī l-manṭiq (Multaqaṭ / Iltiqāṭ al-Shamsiyya fī l-manṭiq)*

Excerpts from the *Risāla al-shamsiyya* of Kātibī.

(*Dharīʿa*, vol. 2, p. 286 no. 1158)

*Manuscripts*

Gharawiyya. Contained in an autograph *majmūʿa*, written in 679/1280–81 (cat. 1959, p. 24 no. 8).

### 1.2.3. Excerpts from *Kitāb al-Hayʾa (Multaqaṭ / Iltiqāṭ Kitāb al-Hayʾa)*

Excerpts from the *Kitāb al-Hayʾa* of Muʾayyad al-Dīn al-ʿUrḍī.[4]—Ibn Kammūna refers to ʿUrḍī in his *al-Jadīd fī l-ḥikma* (p. 403) praising his *Kitāb al-Hayʾa* as the first of its kind (*mā lam yuḥaqqaq min qablihī*) and paraphrasing a brief portion of the work (ibid. pp. 403–5). Ibn Kammūna was apparently the first scholar to use ʿUrḍī's work which was later on also used by Quṭb al-Dīn al-Shīrāzī.[5]

(*Dharīʿa*, vol. 2, p. 286 no. 1160)

#### *Manuscripts*

Gharawiyya. Contained in an autograph *majmūʿa*, written in 679/1280–81 (cat. 1959, p. 24 no. 7).

### 1.2.4. Excerpts from *al-Āthār al-bāqiya (Multaqaṭ / Iltiqāṭ al-Āthār al-bāqiya)*

Excerpts from the *Āthār al-bāqiya ʿan al-qurūn al-khāliya* of Abū Rayḥān al-Bīrūnī.

(*Dharīʿa*, vol. 2, p. 286 no. 1157)

#### *Manuscripts*

Gharawiyya. Contained in an autograph *majmūʿa*, written in 679/1280–81 (cat. 1959, p. 24 no. 6).

---

[4] ʿUrdī's work has been edited with an introduction by George Saliba: *The Astronomical Work of Muʾayyad al-Dīn al-ʿUrḍī. A Thirteenth Century Reform of Ptolemaic Astronomy. Kitāb al-Hayʾah*. Beirut 1990.

[5] See George Saliba: "The original source of Quṭb al-Dīn al-Shīrāzī's Planetary Model." In idem: *A History of Arabic Astronomy. Planetary Theories during the Golden Age of Islam*. New York / London 1994, pp. 119–34.

1.2.5. Excerpts from *al-Qānūn al-Mas'ūdī (Multaqaṭ / Iltiqāṭ al-Qānūn al-Mas'ūdī)*

Excerpts from the *Qānūn al-Mas'ūdī fī l-hay'a wa-l-nujūm* of Abū Rayḥān al-Bīrūnī.

(*Dharī'a*, vol. 2, p. 286 no. 1159)

*Manuscripts*

Gharawiyya. Contained in an autograph *majmū'a*, written in 679/1280–81 (cat. 1959, p. 24 no. 5).

1.2.6. Excerpts from *al-I'tirāḍāt min Kitāb Zubdat al-naqḍ wa-lubāb al-kashf (Multaqaṭ / Iltiqāṭ al-i'tirāḍāt min Kitāb Zubdat al-naqḍ wa-lubāb al-kashf)*

Excerpts from Najm al-Dīn al-Nakhjawānī's commentary on Ibn Sīnā's *al-Ishārāt wa-l-tanbīhāt*, entitled *Zubdat al-naqḍ wa-lubāb al-kashf*, written in Shawwāl 679/January–February 1281.

(*Dharī'a*, vol. 2, p. 286 no. 1161; Shabībī: *Turāthunā*, p. 101; on Nakhjawānī's commentary, cf. Mudarris Raḍawī: *Aḥwāl u āthār*, pp. 110–11; Mahdawī: *Fihrist*, p. 35)

*Manuscripts*

Gharawiyya 686/1; contained in an autograph *majmū'a*, completed in Shawwāl 679/January–February 1281 (cat. 1959, p. 23 no. 4; cat. 1971, p. 54 no. 94).

Notes from the introduction (Shabībī: *Turāthunā*, p. 101):—

...غير ملتزم بألا أخل بشيء من اعتراضاته ومناقضاته ولا أن آتي بعبارة وتركيب يفهم المقصود منها كل من يقف عليها بل ذكرت ما ذكرته بحسب غرضي لا بحسب غرض غيره ولم أعترض لشيء من شرحه للكتاب بل اقتصرت على الاعتراض والنقض فقط. وما كان منه نقضا لما يعلم قطعا أنه غير مقصود لصاحب كتاب الإشارات لم أذكره بل ربّما ذكرت مناقضه وربّما يكون مقصوداً له، وأكثر هذه الاعتراضات والمناقضات التي ذكرتها يمكن الجواب عنها وينتصر لمصنّف

كتاب الإشارات عليها وهذا إنما سطّرته بعد تلخيص لكتابه الذي سماه: لباب المنطق وخلاصة الحكمة.

End (*Dharīʿa*, vol. 2, pp. 286–87 no. 1161):—

كتبه الذي التقطه من الكتاب المذكور سعد بن منصور بن سعد بن الحسن بن هبة الله بن كمونة في شوال سنة 679.

1.2.7. *Khulāṣat mā taḍammanathu khulāṣat al-ḥikma (Talkhīṣ lubāb al-manṭiq)*

Summary of Nakhjawānī's *Lubāb al-manṭiq wa-khulāṣat al-ḥikma*, which is an abbreviation of his more extensive *Naqḍ qawāʿid al-ishārāt wa-kashf tamwīh al-shifāʾ wa-l-najāt* (cf. Shabībī: *Turāthunā*, p. 102). *Khulāṣat al-ḥikma* is the title of the second portion (*jumla*) of *Lubāb al-manṭiq*. Completed in Dhu l-Qaʿda 675/April–May 1277.

(*Dharīʿa*, vol. 4, p. 425 no. 1876; Shabībī: *Turāthunā*, p. 108)

End of the text (Shabībī: *Ibn al-Fuwaṭī*, vol. 2, pp. 124–25; see Shabībī: *Turāthunā*, p. 108 for a shorter version of the quotation):—

والواقف على الأصول يتمكن من استكناه هذه الفوائد في كتاب الله وسنة رسوله. فهذا زبدة ما تتضمنه خلاصة الحكمة المضمومة إلى لباب المنطق المشتمل عليها كتاب واحد تصنيف الإمام نجم الدين أحمد بن أبي بكر بن محمد النخجواني، ووقع الفراغ من اختيار هذا القدر منه في ذي القعدة من سنة 675 هجرية والذي اختار ذلك العبد الفقير إلى رحمة الله تعالى سعد بن منصور بن سعد بن الحسن بن هبة الله بن كمونة عفا الله عنه.

*Manuscripts*

Gharawiyya or Amīr al-Muʾminīn, contained in an autograph *majmūʿa*, completed in Shawwāl 679/January–February 1281 (cf. *Dharīʿa*, vol. 4, p. 425 no. 1876; Shabībī: *Turāthunā*, p. 108).

### 1.3. *Independent Philosophical Treatises*

#### 1.3.1. *(al-Jadīd fī l-ḥikma / al-Jadīd fī l-ḥikma wa-l-ilāhiyyāt / al-Ḥikma al-jadīda / Kitāb fī l-ḥikma / "Risāla-yi falsafa" / al-Kāshif / al-Kāshif fī l-ḥikma al-jadīda / Falsafat Ibn Kammūna / Kitāb fī l-manṭiq wa-l-ṭabīʿiyyāt wa-l-ḥikma al-jadīda / Kitāb fī l-manṭiq wa l ṭabīʿī maʿa l-ḥikma al-jadīda)*

An extensive treatise on philosophy, arranged in seven *abwāb*, each containing seven *fuṣūl*. Composed at the request of Dawlatshāh b. Sanjar al-Ṣāḥibī; completed on 19 Dhu l-Qaʿda 676/13 April 1278.[6]—The author did not give any title to this work, and it is not clear on what basis the various titles under which the work later circulated were coined.

(*Dharīʿa*, vol. 7, p. 56 no. 300 [*al-Ḥikma al-jadīda*]; vol. 16, p. 305 no. 1350 [*Falsafat Ibn Kammūna*]; GALS, vol. 1, p. 769; Rescher: *Development*, p. 205; Steinschneider: *Arabische Literatur*, p. 239 no. 1; *Kashf al-ẓunūn*, vol. 1, p. 450)

Introduction (ed. Ḥamīd Marʿīd al-Kabīsī, pp. 145–47):—

قال العبد الفقير إلى رحمة الله تعالى سعد بن منصور [بن سعد] بن الحسن بن هبة الله بن كمونة عفا الله عنه وأعانه على مراضيه: أحمد الله تعالى حمداً يقرب إلى جنابه الكريم ويوجب المزيد من فضله وإحسانه وأستغفره استغفاراً يؤمن من عقابه الأليم ويخلد في الفردوس الأعلى من جنابه وأسأله الهداية إلى صراطه المستقيم بإلهام الحق وإنارة برهانه وأن يصلي على من بالملاء الأعلى الحافين من حول العرش العظيم وعلى المصطفين لإظهار التوحيد وإعلانه. وبعد، فقد اتفق أرباب العقائد العقلية والديانات النقلية أن الإيمان بالله واليوم الآخر وعمل الصالحات هو غاية الكمالات الإنسانية وبدونه لا يفوز الإنسان بالسعادة السرمدية ولا ينجو من الشقاوة الأخروية. ومن الظاهر أن ذلك لا يتم تحصيله على الوجوه اليقينية لا الظنية وبالطرق البرهانية لا التقليدية إلا بعلم الحكمة الذي هو استكمال النفس الإنسانية بتحصيل التصورات والتصديقات، بالحقائق النظرية والعملية، على حسب الطاقة البشرية. وبهذا انقطع عذر من لم يرغب في تحصيله ولم يدأب في تمهيد قواعده وأصوله. فإن مكذب أهل الحكمة والتنزيل مع عدم المستند والدليل لا يعد من أهل العقد والتحصيل. ويجب على المتردد العمل بالأحوط وعلى المصدق أن لا يضل على علم عن سواء السبيل. ولما كان الأمير الكبير الفاضل العالم العادل عز الدين فخر الإسلام

---

[6] This date is given in Majlis 1938, p. 418. In Ayasofya 2446/1, f. 176b, only the month and the year are given: Dhu l-Qaʿda 676.

والمسلمين معتمد الدولة فخر الملك مفتخر العراقين دولة شاه بن الأمير الكبير سيف الدين سنجر الصاحبي بلغهما الله مبتغاهما وأدام أيامهما ممن اطلع على شرف هذا العلم بألمعيته الثاقبة وآرائه الصائبة التمس مني تصنيف كتاب فيه ترسمه فعملت هذا الكتاب في أثناء ما قد الجئت إليه من ملابسة الأمور الدنية والشواغل الدنيوية مشتملاً مع اختصاره على مهمات المطالب وأمهات المسائل متضمناً مع الزيادات التي من قبلي لخلاصة أفكار الأواخر ولباب حكمة الأوائل خالياً عما يقصر عن إفادة اليقين من الحجج والدلائل عارياً عن تحقيق ما لا يجدي تحقيقه بطائل. فلا تجد في هذا الكتاب إلا ما ينتفع به في العلم بالله تعالى وتوحيده وتنزيهه وصفات جلاله وعجائب مخلوقاته الدالة على كبريائه وعظمته وبيان جوده وعنايته، وفي إثبات الملائكة السماوية والنفوس الأرضية وإدراكاتها وآثارها وبقائها بعد خراب البدن وأبديتها وتزكيتها وما يعصمها من الخطأ والخذلان، وفي خصائص النبوة والولاية وحال المعاد والنشأة الثانية. وبالجملة فهو مشتمل على ما يعصم من الضلال ومزلة أقدام الجهال ويسعد النفس في المآل بما تحلت به من الكمال. وأقول مع اعترافي بتقصيري وقلة بضاعتي من العلم: إنه لا يعرف محل هذا الكتاب إلا المحقق الذي طال نظره في الكتب السابقة عليه، وقد جعلته سبعة أبواب في كل باب منها سبعة فصول. ومن الله استمد العصمة وإصابة الصواب والرحمة وجزيل الثواب إنه العفو الوهاب. الباب الأول في آلة النظر المسماة بالمنطق . . .

End (Majlis 1938, pp. 417–18):—

. . . ومن علم أن مدبر الدنيا والآخرة واحد وأنه غفور رحيم لطيف بعباده متعطف عليهم وتأمل . . . به على الإنسان من صحة البدن وسلامة الأعضاء ثم بعثة لهدايته ثم خلق الأطعمة والأشربة والأدوية لأجله وما ألهمه من التدبيرات المنجعة والدعوات المستجابة أوجب له ذلك التأمل وثوقاً تاماً لطمأنينته إلى سعة رحمة الله تعالى في الآخرة فأسئلك اللهم أن تجعلني من أهلها وأن تنفعني بما علمتني وأن تجعله يوم القيامة حجة لي لا عليّ وأن تغفر لي خطيئتي يوم الدين وأن تعصمني بنور هدايتك من ورطات المضلين وأن تبلغني درجات الصادقين المخلصين وتوفقني بجودك لما أكون به في دار الخلود من الآمنين وبسعادة الأبد من الفائزين وتدخلني زمرة عبادك الصالحين برحمتك يا أرحم الراحمين ويا أكرم الأكرمين. والحمد لله رب العالمين وصلواته على ملائكته المقرّبين وأنبيائه وأوليائه أجمعين وخصوصاً على محمد وآله الطيبين الطاهرين والحمد لله حمد الشاكرين.

Author's colophon (Ayasofya 2446/1, f. 176b):—

وقع الفراغ من تأليفه في ذي القعدة من سنة ست وسبعين وستمئة هلالية هجرية وكتبه مؤلفه العبد الفقير إلى رحمة الله تعالى سعد بن منصور بن سعد بن الحسن بن هبة الله بن كمونة عفا الله عنه وغفر له ووفقه لطاعته ومراضيه وآمنه في الدارين ولي الخيرات في الدنيا والآخرة رب اختم بخير برحمتك يا أرحم الراحمين.

*Editions*

Ed. with introduction by Ḥamīd Marʿīd al-Kabīsī. Baghdad 1403/1982 (Wizārat al-Awqāf wa-l-shuʾūn al-dīniyya. Qism al-dirāsāt wa-l-buḥūth al-islāmiyya. Iḥyāʾ al-turāth al-islāmī; 44).
[Edited on the basis of the following manuscripts: Ahmed III 3234 (= Topkapı 6757), Köprülü 1612/7].

Ed. Ḥāmid Nājī Iṣfahānī. Tehran: Iranian Institute of Philosophy (forthcoming).

*Translations*

With the exception of some chapters on logic, the entire text was translated into Persian by Quṭb al-Dīn al-Shīrāzī without any indication of the author in his *Durrat al-tāj*; see our "Quṭb al-Dīn al-Shīrāzī's (d. 710/1311) *Durrat al-Tāj* and Its Sources. (Studies on Quṭb al-Dīn al-Shīrāzī I)." *Journal Asiatique* 292 (2004), pp. 309–28.

Ömer Mahir Alper: "Özü dolayisiyla varliği zorunlu'nun kanitlanmasi hakkinda." *Istanbul Üniversitesi İlahiyat Fakültesi Dergisi Sayı* 9 (2004), pp. 139–49 [translation into Turkish of *bāb* 7. *faṣl* 1 *fī ithbāt wājib al-wujūd li-dhātihī* = ed. Ḥamīd Marʿīd al-Kabīsī, pp. 527–33].

*Manuscripts*

Ayasofya 2446/1, ff. 1b–176b, 12.5 × 18 cm. Copied by al-Ḥasan al-Murtaḍā al-Ḥusaynī al-Simnānī from the autograph, completed in

Rabīʿ II 731/January–February 1331 (cf. colophon f. 176b). The entire codex apparently consists of what were originally two different codices; ff. 1–176 is written on light, white-yellow paper, all by the hand of the person mentioned on the title page (f. 1a) in black ink only, with 21 lines to a page. Ff. 177–315 (*Kitāb fī l-ḥikma*, see below, Section 2.4) is written by (at least) one different hand, on slightly reddish paper, with black ink and red ink for chapter headings, with 13 lines to a page (cat. p. 363).

Ayasofya 2447/3, ff. 91a–319b, 24.3 × 14 cm, 17 lines to a page; copied at the end of the 9th/15th century (cat. p. 146; cf. Ritter II, p. 58).

Ahmed III 3231 (= Topkapı 6756), 147 ff., 23 × 14.5 cm, 21 lines to a page, copied in 676/1277 (cf. Ritter II, p. 58 n. 3; cat. vol. 3, p. 654).

Ahmed III 3234 (= Topkapı 6757), 155 ff., 26 × 17 cm, 23 lines to a page. Some pages of the last *bāb* are missing. Copy completed on 8 Jumādā I 732/6 February 1332 by ʿAlī b. al-Taqī (or al-Naqī ?) al-Muʾadhdhin in Mashhad; next to the copyist's colophon, there is a note by a different hand stating that the manuscript has been collated with another copy taken from the autograph (for a facsimile of the colophon, see ed. Ḥamīd Marʿīd al-Kabīsī, p. 139) (cat. vol. 3, p. 655).

[Microfilm copy: Maʿhad 87 (cat. vol. 1, p. 207)].

Esad Efendi 1930, 220 [222] ff., 12.5 × 16.6 cm, 14/15 lines to a page, copy completed on 12 Rabīʿ I 1081/29 July 1670 (cf. colophon f. 222a:16–17).

Fâtih 3141/1. The text is incomplete and the sequence of leaves is in disorder: ff. 1–158 (= ed. Ḥamīd Marʿīd al-Kabīsī, pp. 145:1–498:7), f. 159 (= 531:8–533:3), ff. 160–172 (= 498:7–531:7), ff. 173–201 (= 536:22–598:4), f. 210 (= 533:3–536:22); copied by Maḥmūd b. al-Sharīf al-Kirmānī (cf. *Nawādir*, vol. 1, pp. 165–66; *Mukhtārāt*, p. 144; in both works, the shelfmark of the manuscript is wrongly given as 1341/1).

[Microfilm copy: AUB 14 (cat. 1985, pp. 492–93)].

Köprülü 895/1, ff. 1b–137a, 26 × 17 cm, 21 lines to a page, copied by Muṣṭafā b. Sirāj al-Dīn (cf. colophon f. 137a:16–18); the title page has: *Kitāb al-Kāshif li-Ibn Kammūna al-Mūsawī*, and above by a different hand: *Kitāb al-Ḥikma*. The codex also includes a copy of *Jāmiʿ uṣūl al-hayʾa* (*Jāmiʿ al-mabādī wa-l-ghāyāt*) of Niẓām al-Dīn al-Nīsābūrī (ff. 138a–334b), written by a different hand and completed at the end of Dhu l-Qaʿda 918/February 1513 (cat. 1886, p. 57; cat. 1986, vol. 1, p. 442; cf. also *Nawādir*, vol. 1, p. 165; *Mukhtārāt*, p. 143).

Köprülü 1612/5, ff. 241–373, 24.5 × 14 cm, 27 lines to a page, copied in 691/1291–92 by ʿAbd al-ʿAziz b. Ibrāhīm b. al-Khīmī al-Māridānī in Baghdad for his own use (*li-nafsihī*) (cf. colophon f. 272a:23–26); the title page (f. 241a) has: *Kitāb al-mulaqqab bi-l-Jadīd li-l-Shaykh al-ajall al-ʿālim al-kāmil ʿIzz al-Dawla Ibn Kammūna*... *bal aslama qabla mawtihī wa-ṣannafa Kitāb Ifḥām al-yahūd bayyana fīhi faḍāʾiḥahum wa-kidhbahum wa-mukābaratahum fī l-ḥaqq wa-iftirāʾahum ʿalā llāh*... (for facsimiles of the titlepage and the colophon, see ed. Ḥamīd Marʿīd al-Kabīsī, pp. 130, 137) (cat. 1986, vol. 2, p. 361).
[Microfilm copy: Maʿhad 72 (cat. vol. 1, p. 206), Maʿhad 315 (cat. vol. 1, p. 231)].

Majlis 1938, 408 pp., 13 × 25 cm, 17 lines to a page; copied in 11th/17th or 12th/18th century (cat. vol. 5, pp. 439–44).

Majlis 1939, 217 pp., 26 × 17.5 cm, 21 lines to a page; incomplete in the end; copied in 7th/13th century. On the title page there are two indications as to title and author: *Kitāb al-Kāshif min taʾlīfāt Saʿd b. Manṣūr b. Saʿd b. al-Ḥasan b. ʿAbd Allāh b. Kammūna*, and *Kitāb fī l-ḥikma taʾlīf Ibn Kammūna*; author and title are mentioned again on f. 5b at the beginning of a table of contents of the work: *Kitāb fī l-ḥikma taṣnīf al-shaykh al-fāḍil al-ʿAllāma Saʿd b. Manṣūr al-maʿrūf bi-Ibn Kammūna*. At the beginning of the text (f. 7b) the name of the author is given as follows: *qāla al-ʿabd al-faqīr ilā raḥmat Allāh Taʿālā Saʿd b. Manṣūr b. Saʿd b. al-Ḥasan b. ʿAbd Allāh b. Kammūna*, to which the eulogy *raḥimahū llāh* is added by a different hand. Numerous statements of ownership and reader's colophons on f. 3a, among others by the renowned jurisprudent and philosopher of the Safavid period, Muḥammad Bāqir Sabzawārī (d. 1090/1679)[7] (cat. vol. 5, pp. 444–45).

Mishkāt 248, 290 pp., 20.5 × 16 cm, 13 lines to a page; copied by Mīrzā Yad Allāh Naẓar Pāk Kajūrī and completed on 3 Rabīʿ II 1351/6 August 1932. On f. 1b the title of the work is indicated as *Falsafa-yi Ibn Kammūna* (cat. 1976, p. 14; cat. 1951–85, vol. 3 i, pp. 265–66; cf. also *Dharīʿa*, vol. 16, p. 305 no. 1350).

*Studies*

Christine Fink: *Ibn Kammuna über die Grundbegriffe der Philosophie. Analyse eines Kapitels aus seinem Werk 'Das Neue in der Philosophie'*. MA thesis, Frankfurt/Main 1997.

Y. Tzvi Langermann: "Ibn Kammūna and the 'New Wisdom' of the Thirteenth Century." *Arabic Sciences and Philosophy* 15 (2005), pp. 277–327.

1.3.2. *al-Maṭālib al-muhimma min ʿilm al-ḥikma*

A short work on philosophy consisting of seven *fuṣūl* with seven *abḥāth* respectively. Completed in Shaʿbān 657/July–August 1259.

(*Dharīʿa*, vol. 21, p. 141 no. 4327)

Beginning (ed. Mūsawī, p. 66:3–11):—

الحمد لله رب العالمين وصلى الله على محمد وآله وصحبه الأجمعين. وبعد، فهذا الكتاب أوردت فيه المهم من المطالب الحكمية وما ينفع به، جامعاً في ذلك بين ما حصلته بنظري وهو الأقل وبين غيره مما تحققت صحته وهو الأكثر، وذلك على غاية من الإيجاز والاختصار لا سيما في المنطق، تاركاً لما سوى ذلك مما يطلب به المباهاة والمفاخرة دون النجاة في الآخرة. وسميتُه بالمطالب المهمة من علم الحكمة لمطابقة اسمه لمعناه، ورتبتُه على سبعة فصول تشتمل كل فصل منها على سبعة أبحاث، مقدماً على ذكر الفصول بيان إيجاب العقل لطلب العلم والسعي في تحصيله ليكون محرضاً لذوي البصائر على تحصيل ما في هذا الكتاب وما هو فيه وهو على هذا التقرير...

End (ed. Mūsawī, p. 86:29–33):—

فقد ثبت صدق المقدم وبثبوته يصدق التالي وهو أن إله العالم جل وعلا واحد. وتقرير هذا المطلوب على هذا الوجه مما لم أجده في كلام من تقدم. وقد ختمتُ الكتاب بمسألة التوحيد ثناء والأمان، يختم الله تعالى لنا به ويجعلنا في دار الخلود من الآمنين والصلاة على سيد المرسلين محمد وآله الطاهرين.

[7] On him, see Āghā Buzurg: *Ṭabaqāt*, vol. 5, pp. 71–72; Corbin: *La philosophie iranienne*, pp. 227–33.

*Edition*

Ed. Sayyid Ḥusayn Sayyid Mūsawī in *Kheradnāme-ye Ṣadrā* 8 xxxii (Tābestān 1382/ Rajab 1425/ September 2003), pp. 64–86. [Edition prepared on the basis of Āstān-i Quds 901].

*Manuscripts*

Adana 177/3 [= 181/2], ff. 226a–239a, 18.5 × 13.5 cm, 23 lines to a page, copied by Jaʿfar al-Astarābādī in Dhu l-Qaʿda 707/April–May 1308 (cf. *Mukhtārāt*, p. 144; cat. vol. 1, p. 87 no. 308).

Āstān-i Quds 901, ff. 1b–12b, 25 × 18 cm, 21 lines to a page. At the end of the manuscript (f. 15a), there is the following statement:

هذا الكتاب تصنيف الشيخ المحقق المفتي عز الدولة سعد بن الحسن بن هبة الله بن كمونة الذي صار على مذهب الإمامية آخراً ومات عليه غفر الله له.

(cat. 1926–85, vol. 4, p. 245; cat. 1990, p. 527).

[Copies: Dānishgāh microfilm 1519 (cat. 1969–84, vol. 1, pp. 382–83); Dānishgāh photography 6453/4 (cat. 1969–84, vol. 3, p. 255); Āstān-i Quds microfilm 2189].

Gharawiyya 717/2, written during the lifetime of Ibn Kammūna (eulogy following his name: *adāma Allāh ayyāmahū*) (cat. 1971, pp. 64–65 no. 125).

Muḥammad al-Samāwī, autograph completed in Shaʿbān 657/July–August 1259 (cf. *Dharīʿa*, vol. 21, p. 141 no. 4327). The current whereabouts of the manuscript is unclear.

Private Collection of Majd al-Dīn al-Nuṣayrī, dated 1080/1669–70 (cf. *Dharīʿa*, vol. 21, p. 141 no. 4327). The current whereabouts of the collection could not be verified.

St. Petersburg RNL II Firk. Yevr.-Arab. II 73, 2 ff.; written in Hebrew characters; 15 lines to a page; containing the following fragments of the treatise: f. 1 (= ed. Mūsawī, pp. 84:29–85:15), f. 2 (= 84:9–29).

1.3.3. *Taqrīb al-maḥajja wa-tahdhīb al-ḥujja (Talkhīṣ al-ʿulūm al-ḥikmiyya / Talkhīṣ al-ḥikma)*

A short treatise on philosophy, completed during the first part of Dhu l-Ḥijja 657/November 1259.

(*Dharīʿa*, vol. 26, p. 225 no. 1133)

Introduction:—

قال مولانا وسيدنا وشيخنا عز الدولة سعد بن منصور بن سعد بن الحسن بن هبة الله بن كمونة سقى الله صوبه الغمام: هذا تلخيص في الغاية القصوى من الإيجاز اقتصرت فيه من العلوم الحكمية على ما هو من المهمات عند طالبي الفوز بالسعادة الأبدية والنجاة من الشقاوة الأخروية المنوطين على ما أجمع عليه الأنبياء صلوات الله عليهم أجمعين والحكماء المحققين بالعلم بالله تعالى وصفاته وأفعاله الذي يجب على كل عاقل بمجرّد هذا الإجماع الاجتهاد في اكتسابه عملاً بالأحوط ولو كان متلادداً في تصديقهم فضلاً عن المؤمن وعنونتُه بتقريب المحجة وتهذيب الحجة فأقول: العلم منه تصور وهو الإحاطة . . .

End:—

وختمتُ به هذا المختصر الذي من اطلع على فوائده وأسراره وتأمل ما ذكروه في المطولات تثبتت له منزلته ومن الله أسأل وإليه أضرع في أن يوفّقني للهدى والرشاد ويؤمنني من عقابه في يوم المعاد ويرزقني بسعادة الأبد التي وعد بها من آمن وعمل صالحاً من العباد بمنّه ورحمته إن شاء الله تعالى.

Author's colophon (Shabībī: *Ibn al-Fuwaṭī*, vol. 2, p. 124 quoting apparently from the autograph in the Gharawiyya library):—

وقع الفراغ من نسخه في القسم الأول من ذي الحجة سنة 657 كتبه مؤلفه الفقير إلى رحمة الله تعالى وعفوه وغفرانه سعد بن منصور بن سعد بن الحسن بن هبة الله بن كمونة وفقه الله لطاعته واعانه على مراضيه إنه جواد كريم.

### *Manuscripts*

Ayasofya 2446/4, ff. 293b–308a, 12.5 × 18 cm; copy completed between 28–30 Jumādā I 727/21–23 April 1327 (cf. colophon f. 308a). The entire codex apparently consists of what were originally two different codices; ff. 1–176 (*al-Jadīd fī l-ḥikma*, see above, Section 1.3.1) is written on light, white-yellow paper, by one hand in black ink only, with 21 lines to a page. Ff. 177–315 is written by (at least) one different hand, on slightly reddish paper, with black ink and red ink for chapter headings, with 13 lines to a page (cat. p. 363).

Ayasofya 2447/2, ff. 77b–87a, 24.3 × 14 cm, 17 lines to a page; copied at the end of the 9th/15th century (cf. Ritter II, p. 58).

Esad Efendi 1933/2, ff. 49b–55a, 16.3 × 25.2 cm, 19 lines to a page; copied towards the end of the 10th/16th century (cf. Ritter II, p. 61; *Mukhtārāt*, p. 142).
[Copies: AUB microfilm 15/2 (cat. 1985, p. 493); Dānishgāh microfilm 535 (cat. 1969–84, vol. 1, p. 496); Mīnuwī photograph 294 (cat. p. 138)].

Gharawiyya; contained in an autograph *majmūʿa*, completed in 675/1276–77 (cf. *Dharīʿa*, vol. 26, p. 225 no. 1133; Shabībī: *Ibn al-Fuwaṭī*, vol. 2, p. 124).

Istanbul University al-qism al-ʿarabī 1458, ff. 192b–194b; 12th/18th century (cf. *Nawādir*, vol. 1, p. 164).

### *Edition*

## Chapter Three, Text Three

### 1.3.4. *Kalimāt wajīza mushtamila ʿalā nukat laṭīfa fī l-ʿilm wa-l-ʿamal (al-Lumʿa al-juwayniyya fī l-ḥikma al-ʿilmiyya wa-l-ʿamaliyya / Risāla fī l-ʿilm wa-l-ʿamal)*

The work was written following the order of a certain Shams al-Dīn Muḥammad al-Muʾmīn al-Qazwīnī, and is dedicated to Bahāʾ al-Dīn Muḥammad, one of the sons of *ṣāḥib al-dīwān* al-Juwaynī. The autograph of the manuscript in the Gharawiyya is dated 679/1280–81, but the work itself must have been completed before Shaʿbān 678/December

1279, when Bahāʾ al-Dīn died. The work consists of two portions (*jumla*) on *ʿilm* and *ʿamal*, each containing two *bāb*s, and each *bāb* consisting of five chapters (*fuṣūl*).

(*Dharīʿa*, vol. 18, p. 351 no. 438; GAL, vol. 1, p. 555; GALS, vol. 1, p. 769, vol. 3, p. 1232)

Beginning:—

قال العبد الفقير إلى رحمة الله تعالى سعد بن منصور بن سعد بن الحسن بن هبة الله بن كمونة عفا الله عنه في دنياه وأخراه: أحمد الله حمد مسترشد بنور هدايته، مسترفد من جوده وعنايته، وأتوسل إليه بنعوت جلاله وكمال إلاهيته، أن يهب لي توفيقاً يوجب الأمان من عقابه والخلود في جنّته، وأن يصلّي على القدّيسين من ملائكته وعلى من كرّمه بالنبوة والولاية من بريته، خصوصاً على محمد وعترته. وبعد، فإنه حيث وصل الصدر الكبير العالم الفاضل المتّقي البارع شمس الدولة والدين فخر الإسلام والمسلمين معتمد الملك محمد المؤمن القزويني أدام الله أيامه، وبلغه سؤاله ومرامه، إلى بغداد، التمس مني عند مقاربة عوده إلى مقصده أن يصحب معه شيئاً من كلامي في العقليات ليكون برسم خزانة المولى المعظم الصاحب الأعظم صاحب السيف والقلم العالم العادل المنصف الكامل شمس الحق والدين بهاء الإسلام والمسلمين ملك الوزراء في العالمين صاحب ديوان الممالك دستور الشرق والغرب مفخر الأعيان، مظهر الإحسان، محمد بن صاحب السعيد بهاء الدولة والدين محمد الجويني أعزّ الله نصره وأنفذ في المنافقين أمره، فكتبت هذه الكلمات الوجيزة على عجلة مشتملة على نكت لطيفة في العلم والعمل، ينتفع بها جماعة الطلاّب لمهمّات الحكم والآداب، ترشدهم إذا استبصروا بها وتسعدهم إذا استضاءوا بنورها وعملوا بموجبها والتكلان على واهب التوفيق.

وقد رتّبتها على جملتين يشتمل كل جملة منها على بابين وكل باب على خمسة فصول. ولعلّي بعد هذا أشرحها شرحاً يفهم من يستفسرها ويفحم من يستنكرها، وأسأل من الله تعالى أن يسهل علينا طريق الحق وأن يصرف قلوبنا جميعاً عن هذه الدار الفانية إلى الدار الباقية وأن يسعدنا في الدارين، وأن لا يجعلنا من المتكلفين، لأنه ولي ذلك والقادر عليه، ﴿وَمَا تَوْفِيقِي إِلاَّ بِاللهِ عَلَيْهِ تَوَكَّلْتُ وَإِلَيْهِ أُنِيبُ﴾.

End:—

فهذا ما رأيت أن أذكره من الفوائد على حكم ما اقتضت الحال الحاضرة وأكثره يلتقط من كلام من تقدم، إنما الأقلّ النزر هو الذي أدركته بفكري أو سنح لخاطري، وفيه مباحث غريبة كثيرة

النفع، ولولا ضعف بخاري في البصر يمنع من كثرة الكتابة وطول المطالعة وموانع أخرى لبسطت الكلام بزيادة مسائل وبراهين دقيقة ذكرتها في غير هذه الرسالة، إنما الغرض من هذه ذكر مهمات العقائد والأعمال لا غير، وفي كتب من تقدم كفاية وأنا معترف بالتقصير وقلّة البضاعة من العلم، وأسأل الله تعالى العفو والعافية، وأن يوفقنا جميعاً لإصلاح العلم والعمل، ويعصمنا من الزيغ والزلل، وأن يجعلنا لسبله من المنتهجين، وبرضائه في دار الأبد من المبتهجين، إنه لطيف كريم رؤوف رحيم.

والحمد لله رب العالمين وصلاته على ملائكته المقربين وأنبيائه المرسلين وأوليائه المخلصين وخصوصاً على محمد وآله الطاهرين. وأختم بالخير.

*Manuscripts*

Āstān-i Quds 855, 26 × 18 cm, 19 lines to a page (cat. 1926–85, vol. 4, p. 231 no. 965; cat. 1990, p. 484).
[Copies: Dānishgāh microfilm 1521 (cat. 1969–84, vol. 1, p. 373); Dānishgāh photograph 6454/5].

Fâtih 3141/5, ff. 266a–286b, 8 × 16.5 cm, 25 lines to a page; copied by Maḥmūd b. al-Sharīf al-Kirmānī in the Madrasa ʿImādiyya in Nuṣaybīn between 27–29 Rabīʿ II 696/21–23 February 1297 (see colophon f. 286b) (cf. *Nawādir*, vol. 1, p. 164; *Mukhtārāt*, p. 143).

Gharawiyya 686/2, autograph, completed between 21–30 Dhu l-Qaʿda 679/14–23 March 1281 (cat. 1959, p. 23 no. 3; cat. 1971, p. 54 no. 94).

Majlis 593/1, pp. 3–40, 24.5 × 15.5 cm, 19 lines to a page, copied in Ṣafar 1060/February 1650 from the codex written by Quṭb al-Dīn al-Shīrāzī (= Marʿashī 12868) (cf. colophon p. 40)] (cat. 2/350: "*Risāla dar ʿaqliyyāt*").
[Copy: Dānishgāh microfilm 2320 (cat. 1969–84, vol. 1, p. 673)].

Marʿashī 12868, ff. 40a–59b, 13.5 × 5 cm, 29/30 lines to a page, copied by the hand of Quṭb al-Dīn al-Shīrāzī in Konya in 685/1286 (cat. vol. 32, pp. 639–40)

Mishkāt 861/16, ff. 137b–151b, 22 lines to a page; copy completed on 21 Muḥarram 1284/26 May 1867 by Muḥammad b. Ḥusayn al-Ḥājj Ḥājjī Nād ʿAlī (cat. 1976, p. 51; cat. 1951–85, vol. 3 i, p. 335).
[Copy: Dānishgāh microfilm 6783].

*Edition*

Chapter Three, Text One

*Commentaries*

Ibn Kammūna says in the introduction to the work that he intended to write a commentary (*sharḥ*) on it later on. There is no indication that this was ever realized.

Ibn Abī Jumhūr al-Aḥsāʾī (d. after 906/1501) wrote a work entitled *al-Durra al-mustakhraja min al-Lumʿa fī l-ḥikma*, which is lost. Āghā Buzurg suggests that this was perhaps a commentary on the treatise by Ibn Kammūna (*Dharīʿa*, vol. 8, p. 89 no. 323; Schmidtke: *Theologie, Philosophie und Mystik*, p. 274 no. 22).

1.3.5. *Ithbāt al-mabdaʾ wa-ṣifātihī wa-l-ʿamal al-muqarrib ilā llāh (ilayhi) (Risāla fī l-ḥikma)*

A brief philosophical treatise which is arranged in a similar fashion to *Kalimāt wajīza* (see above, Section 1.3.4). The identical wording in many sections suggests that this treatise is a concise version of *Kalimāt wajīza*, which was written either shortly before or after the latter work. Āghā Buzurg mentions a work by Ibn Kammūna with a similar title (*Risāla fī ithbāt al-wājib wa-ithbāt mā yajibu ithbātuhū li-l-wājib wa-ithbāt al-tawḥīd wa-annahū lā shārik lahū*; cf. *Dharīʿa*, vol. 11, pp. 11–12 no. 51) which might be identical with *Ithbāt al-mabdaʾ*. He mentions that there is a manuscript of this tract in the Majlis Library as indicated in the catalogue of this library; we were unable to find any hint of such a treatise in the catalogue.—In the only manuscript available to us (Rawḍātī) there is no introduction by the author. The same apparently applies to the manuscript that was part of the Mahdī Bayānī collection.

Beginning:—

قال . . . سعد بن منصور بن سعد بن الحسن [بن] هبة الله بن كمونة عفى الله عنه في دنياه وآخره: كمال نفس الإنسان هو إما من جهة قوتها النظرية بالعلم أو من جهة قوتها العملية بالعمل وكل العلوم والأعمال على اختلافها وكثرة تشعبها ترجع إلى كمال هاتين القوتين . . .

End:—

فهذا ما كتبه ممتثلاً للمتقدم المولوي مؤلفه... ومنتهى الرغبات وصلى الله على خير خلقه محمد وآله وسلم تسليماً.

*Manuscripts*

Mahdī Bayānī; the current whereabouts of the *majmū ʿa* in which the tract appears as treatise no. 9 is not known (cat. 1969, pp. 692–93).

Rawḍātī, ff. 1b–9b, 18 lines to a page; copied by Abu l-Ḥasan b. Khalīl Allāh Mūsawī; undated.
[Microfilm copy: Dānishgāh microfilm 2712/1 (cat. 1969–84, vol. 1, p. 722)].

*Edition*

Chapter Three, Text Two

1.3.6. *Kitāb fī l-kalām wa-l-falsafa*

*Manuscripts*

Gharawiyya 717/1 (cat. 1971, p. 64 no. 125: *kitāb ṣaghīr naẓunnu annahū min muʾallafāt* Ibn Kammūna). The codex further contains a copy of Ibn Kammūna's *al-Maṭālib al-muhimma* which was copied during his lifetime (see above, Section 1.3.2).

1.3.7. *Risāla fī l-kalām*

A treatise on theology, arranged in chapters (*abwāb*), the fifth *bāb* being entitled *fī mudabbir al-ʿālam al-wājib al-wujūd li-dhātihī wa-nuʿūt jalālihī*. Completed between 21–30 Dhu l-Qaʿda 679/14–23 March 1281.

(*Dharīʿa*, vol. 18, pp. 109–10 no. 903)

*Manuscripts*

Gharawiyya; autograph (cat. 1959, p. 24 no. 9).

1.3.8. *Maqāla fī l-taṣdīq bi-anna nafs al-insān bāqiya abadan (Maqāla fī ithbāt abadiyyat nafs al-insān bi-ṭarīq burhānī / Maqāla fī ithbāt abadiyyat al-nafs / Risāla fī baqāʾ al-nafs wa-abadiyyatihā / Risāla fī abadiyyat al-nafs / Risāla [Maqāla] fī azaliyyat al-nafs wa-baqāʾihā)*

The work consists of ten chapters (*fuṣūl*): 1: *fī qawl al-shāriḥ li-maʿnā ism al-nafs*; 2: *fī ithbāt wujūd nafs al-insān*; 3: *fī l-wājib wa-l-mumkin wa-inqisām al-mawjūd ilayhimā*; 4: *fī anna wājib al-wujūd lā yuʿdamu abadan*; 5: *fī anna nafs al-insān qadīma*; 6: *fī anna li-kull mumkin qadīm ʿilla qadīma*; 7: *fī anna ʿadam al-maʿlūl lā yanfakku ʿan ʿadam ʿillatihī*; 8: *fī anna jamīʿ al-mumkināt tantahī fī silsilat al-ḥāja ilā wājib al-wujūd li-dhātihī*; 9: *fī tartīb hādhihī l-muqaddamāt maʿa muqaddamāt ukhar badīhiyya li-yuntaja anna nafs al-insān lā yuʿdamu abadan*; 10: *fī dhikr mā waqaftu ʿalayhi min ḥujaj al-qāʾilīn bi-abadiyyat al-nafs al-insāniyya li-yatabayyana anna mā salaktuhū fī hādhihī l-maqāla li-ithbāt hādhā l-maṭlūb huwa afḍal min tilka l-ḥujaj.*—Ibn Kammūna refers presumably to this treatise or to the treatise described below in Section 1.3.10 in his *al-Maṭālib al-muhimma* (completed in Shaʿban 657/July August 1259) when he states at the end of his proof for the eternity of the soul (*fī abadiyyat al-nafs*): *wa-qad afradtu lahū maqāla dhakartu fīhā mā yumkinu īrāduhū ʿalayhi wa-ʿalā muqaddamātihī fī l-iʿtirāḍāt wa-bayyantu wajh al-jawāb ʿanhā ʿalā ghāyat al-īḍāḥ* (see ed. Mūsawī, p. 85:3–4). Shaʿbān 657/July–August 1259 is therefore presumably the *terminus ante quem* for his *Maqāla fī l-taṣdīq bi-anna nafs al-insān bāqiya abadan.*

(*Dharīʿa*, vol. 11, p. 131 no. 815)

Introduction (ed. Barkhʷāh, p. 74:3–7):—

الحمد لله الذي نطقت بحكمته عجائب صنعته ووقفت أفكار بريته دون تصور ماهيته. أحمده حمد من أبصر اليقين بعين بصيرته فآثر طريق مسلوكه ومحجته وأسأله أن يفيض علينا من أنوار هدايته ما يوفقنا به للخلود في جنته بمنّه وسعة جوده ورحمته. فإن الذي يتضمنه هذه المقالة هو إثبات أبدية نفس الإنسان بطريق برهاني هو مما أدركته بفكري ولم أعرف أني سبقت إليه ويتم الغرض منها بإيراد عدة فصول. الفصل الأول في قول الشارح لمعنى اسم النفس...

End (ed. Barkhʷāh, p. 124:10–17):—

... فهذه هي الحجج المشهورة على هذا المطلوب ولم أقف على غيرها وكلها يفتقر إلى بيان أن علة النفس يمتنع عدمها وإلى مزيد تقرير لمقدماتها وإلا لم يكن برهانية. نعرف ذلك من أجال نظره في كتب المنطق وقد عورض بعضها أو نوقض بما لا يقدح فيها عند التحقيق. ومن أراد أن يقف على المعارضات التي عليها والمناقضات فعليه بالنظر في كتب المتكلمين الذين انتدوا للرد على الفلاسفة مثل كتب الإمام الغزالي وكتب الإمام فخر الدين الخطيب الرازي وغيرهما. ومن كان من أهل التحقيق ونظر إلى الحق بعين الرضا ولم يسلك طريق الهوى علم أن ما اخترعته لإثبات هذا المطلوب من الأقيسة البرهانية وأن جميع ما حكيته من الحجج على إثباته لا يفيد اليقين إلا بزيادة تقرير كما ذكرت. وهذا خاتمة الكلام في هذه المقالة والحمد لله رب العالمين وصلواته على أهل قدسه من الأنبياء والمرسلين خصوصاً على أفضل مخترعاته محمد النبي وآله الطاهرين.

*Edition*

Insīyah Barkhʷāh: *Taṣḥīḥ-i Risāla-yi Azaliyyat al-nafs wa-baqāʾihā taʾlīf-i Ibn Kammūna*. MA thesis: Madrasa-yi Murtaḍā Muṭahharī (Sipahsālār), 1379/2000, pp. 72–124.
[Edited on the basis of Majlis 3814/1].

*Manuscripts*

Majlis 3814/1, pp. 1–96, 17.5× 11 cm, 17 lines to a page, copied before Rabīʿ I 1067/December 1656–January 1657 (cat. vol. 10 iv, pp. 1832–33).

St. Petersburg II Firk. Yevr.-Arab. I 1416, 49 ff., 14 × 18 cm, consisting of one quire of nine leaves (ff. 1–9 with one leaf [f. 8 of the original ten-leaves quire] missing) and four quires of ten leaves (ff. 10–19, 20–29, 30–39, 40–49); incomplete at the beginning (f. 1a = ed. Barkhʷāh, p. 80:23 [*munqasiman*]); on the upper left margin of f. 1a is indicated: *thānī* (cat. 1987, p. 26 no. 64).
[Microfilm copy: IMHM no. 55205].

1.3.9. *Maqāla fī anna l-nafs laysat bi-mizāj al-badan wa-lā kāʾina ʿan mizāj al-badan*

Ibn Kammūna refers to this treatise in his *Maqāla fī l-taṣdīq bi-anna nafs al-insān bāqiya abadan* (see above, Section 1.3.8) (ed. Barkhʷāh, p. 91): *al-faṣl al-khāmis fī anna nafs al-insān qadīma. qad barhantu ʿalā hādhā l-maṭlab fī l-maqāla allatī bayyantu fīhā anna l-nafs laysat bi-mizāj al-badan wa-lā kāʾina ʿan mizāj al-badan*...; (ed. Barkhʷāh, p. 94): ... *qad dhakartu fī l-maqāla allatī bayyantu fīhā anna l-nafs laysat bi-mizāj*; (ed. Barkhʷāh, p. 104): ... *fī l-maqāla allatī bayyantu fīhā anna l-nafs laysat bi-mizāj*. He likewise refers to it in his *Maqāla fī anna wujūd al-nafs abadī wa-baqāʾahā sarmadī* (see below, Section 1.3.10). This suggests that this treatise predates the treatises described in Sections 1.3.8 and 1.3.10.—The treatise contains eleven chapters (*fuṣūl*) to which Ibn Kammūna did not give any specific titles. The editor, Insīyah Barkhʷāh, coined the following titles for the chapters: 1. *fī annahū lā ḥāja ilā taṣawwur māhiyyat al-nafs wa-l-mizāj wa-ḥaqīqatihī*; 2. *fī bayān sharḥ ism al-nafs wa-ithbāt wujūdihī*; 3. *fī bayān al-mizāj ʿalā l-wajh al-muttafaq ʿalayhi fīmā bayn al-ḥukamāʾ mā huwa*; 4. *fī bayān marātib al-ashyāʾ*; 5. *fī anna ʿillat al-ḥādith al-zamānī ḥāditha zamāniyya*; 6. *fī anna l-ʿilla li-wujūd al-ḥādith al-zamānī murakkaba min mawjūd dāʾim al-wujūd wa-man sabaqa ḥādith ākhar*; 7. *fī imtināʿ ṣudūr al-basīṭ ʿan al-ʿilla al-murakkaba*; 8. *fī bayān anna l nafs yastaḥīlu an takūna murakkaba*; 9. *fī tartīb hādhihī l-muqaddimāt li-yuntij anna l-nafs laysat bi-mizāj al-badan wa-lā kāʾina ʿan mizāj al-badan*; 10. *fī bayān tashkīkāt al-muʿānidīn ʿalā muqaddimāt al-burhān*; 11. *fī dhikr al-ajwiba ʿan tilka al-tashkīkāt*. The only extant manuscript of the text does not contain an introduction by the author.

(*Dharīʿa*, vol. 21, p. 406 no. 5709)

Beginning (ed. Barkhʷāh, p. 126:2–11):—

مقالة في أن النفس ليست بمزاج البدن ولا كائنة عن مزاج البدن تشتمل على أحد عشر فصلاً. فصل اعلم أنه لا بد لمن أراد أن يبين أن النفس ليست بمزاج البدن ولا كائنة عن مزاج البدن أن يبتدئ ببيان أمرين: أن المراد بقولنا: النفس، ما هو؟ والثاني أن حقيقة المزاج على الوجه المتفق عليه فيما بين الحكماء ما هو؟ وإنما يحتاج إلى ذلك لأن قولنا: النفس ليست بمزاج البدن أو ليست بكائنة عن مزاج البدن تصديق وكل تصديق فلا بد فيه من تصور عنوان المحكوم عليه والمحكوم به والحكم، والعلم بذلك من الأمور التي يشترك في العلم بها جميع العقلاء. فإن كل

أحد يعلم بالضرورة أن كل ما ليس بمعلوم يمتنع الحكم عليه أو به إلا إذا كان عنوانه معلوماً ولا يفتقر في علمه بذلك إلى اكتساب . . .

End (ed. Barkhʷāh, p. 165:9–14):—

. . . وبالجملة فليس الغرض من الحكمة تقريرها مع المعاند بل الغرض منها تنبيه ذوى الإنصاف على كيفية ما يجب أن يعتقد في حقائق الأشياء وإن لم تكن مقدمات هذا البرهان بديهية فظاهر أنها ليست حسية، استحال أن يكون القياس المركب منها برهانياً وفرضناه كذلك. هذا خلف، فالحاصل أن مقدمات هذا البرهان إما أن تكون بديهية أو لا. وعلى التقدير الأول يكون ما ذكرته من السؤالات والأجوبة مما لا حاجة إليه في تصحيحه فيكون هدراً، وعلى التقدير الثاني يلزم أن لا يتبين مما قيل في هذه المقالة.

*Manuscripts*

Gharawiyya 49 (cat. 1959, p. 27).
Majlis 3814/2, pp. 97–170, 17.5 × 11 cm, 17 lines to a page, copied in Rabīʿ I 1067/December 1656–January 1657 (cf. colophon p. 170) (cat. vol. 10 iv, pp. 1832–33).

*Edition*

Insīyah Barkhʷāh: *Taṣḥīḥ-i Risāla-yi Azaliyyat al-nafs wa-baqāʾihā taʾlīf-i Ibn Kammūna*. MA thesis: Madrasa-yi Murtaḍā Muṭahharī (Sipahsālār), 1379/2000, pp. 125–65.
[Edited on the basis of Majlis 3814/2].

1.3.10. *Maqāla fī anna wujūd al-nafs abadī wa-baqāʾahā sarmadī (Baqāʾ al-nafs / Maqāla fī azaliyyat al-nafs / Risāla fī abadiyyat al-nafs wa-baqāʾihā wa-basāṭatihā / "Treatise on the Immortality of the Soul")*

The introductory remarks by Ibn Kammūna, as well as his concluding statement, include a nonspecific reference to an earlier writing in which he says he dealt with other issues relating to the soul, consisting of questions and answers. The description fits his *Maqāla fī anna l-nafs laysat bi-mizāj al-badan wa-lā kāʾina ʿan mizāj al-badan* (see above, Section 1.3.9).

(*Dharīʿa*, vol. 11, p. 131 no. 713 where Ibn Kammūna is not mentioned as author; GALS, vol. 1, p. 769; Steinschneider: *Arabische Literatur*, p. 239 no. 3)

Introduction (ed. Barkhʷāh, pp. 167:2–168:3):—

الحمد لله عدد منه التي نعجز عن عدّها وحصرها وكفاء أنعمه التي لا نستطيع ولو استنفدنا الوسع في الشكر القيام بواجب شكرها والصلوة على ملائكته المقربين وأنبيائه وأوليائه أجمعين وخصوصاً على محمد وآله الطاهرين. وبعد، فإن نتائج الأفكار ونفائس الأنظار مما يجب إذاعتها لأجل الانتفاع بها ولا ينبغي إخفاؤها عن الراغبين وإن قلّ الراغب في طلبها لا سيما إذا كانت من أهم المطالب العلمية وأنفع المباحث اليقينية وقد أثبت في هذه الأوراق مما قدحه زناد خاطري وحصلته بفكري ونظري برهاناً شريفاً منطوياً على أن وجود النفس أبدي وبقاؤها سرمدي يندرج في ضمنه تحقيق عدة مسائل من علم النفس وغيرها. والتصرف في هذه المقالة على تقرير البرهان لأني قد ذكرت في غيرها ما أورده جماعة من الفضلاء على مقدماته من السؤالات مع الجواب عنها وما يمكن توجيهه عليها من غوامض الإشكالات مع تقرير وجه التخلص منها ومن الله تعالى استمدّ حسن التوفيق وأسأله الهداية إلى سواء الطريق إنه جواد كريم. وهذا تقرير صيغة البرهان...

End (ed. Barkhʷāh, p. 178:13–17):—

...فهذا برهان قاطع على أبدية النفس قد انطوى على أبحاث شريفة وغرائب نفيسة وهو مما لم أعرف أني سبقت إليه ولا إلى كثير من مقدماته ومن أراد أن يتحقق صحته وإفضاءه إلى المطلوب فعليه بالآلة التي يسمّى المنطق تعبيره بها. والاستقصاء فيما عليه من السؤالات والأجوبة فقد ذكرته في غير هذه المقالة فليطلب من أماكنه. والحمد لله رب العالمين وصلواته على محمد وآله أجمعين إلى يوم الدين.

Scribe's colophon with author's colophon (Landberg 510/5, f. 70b; also quoted in Nemoy: *The Arabic Treatise on the Immortality of the Soul*, p. 6):—

هذه الرسالة من تصنيف الشيخ المعظم والحكيم المقدم قدوة الحكماء اليهوديين ابن كمونة ومنقولة من خطه وقد كتب في آخرها: كتبه مصنّفه الفقير إلى الله تعالى سعد بن منصور بن سعد بن أبي الحسن بن هبة الله بن كمونة أعانه الله على مراضيه رب أختم بخير يا أرحم الراحمين.

*Manuscripts*

Ākhūnd 1187/66, ff. 286a–290a [pp. 533–41] (cat. pp. 1479–80).
Gharawiyya (cf. *Dharīʿa*, vol. 11, p. 131 no. 713).
Landberg 510/5, ff. 58–70, copied shortly after 680/1282 from the autograph (cat. p. 147 no. 1383). The second part of the *majmūʿa* of which this treatise is part was copied by the order of the circle around Rashīd al-Dīn; see Baneth: "Ibn Kammuna," p. 308 n. 1; see also the description in Nemoy: *The Arabic Treatise on the Immortality of the Soul*, pp. 1–6.
Majlis 1837/6, pp. 127–42 [ff. 69–77], 24.5 × 19.5 cm, 17 lines to a page; copied in Isfahan in 1056/1646–47 (cat. vol. 5, p. 287; vol. 9 ii, pp. 502–9).
Marʿashī 11619/1, ff. 1–7, 6 × 15 cm, 22 lines to a page, copied by Muḥammad Naṣīr b. Ḥasan Bulānakābī (?) Jīlānī in 12th/18th century (cat. vol. 29, p. 363).
Ragıp Paşa 1461, ff. 278a–281b, 12.8 × 18 cm, 21 lines to a page (cf. *Mukhtārāt*, p. 142).
Ṣadr al-Dīn Maḥallātī 17/5, ff. 39b–44b, 9.5 × 16.8 cm, copied in 1056/1646–47 [the manuscript was given to the Khānqāh-i Aḥmadiyya Library in Shiraz, the current shelfmark being no. 695] (cat. p. 273).

*Editions*

Leon Nemoy: *The Arabic Treatise on the Immortality of the Soul by Saʿd ibn Manṣūr ibn Kammūna (XIII Century). Facsimile Reproduction of the Only Known Manuscript (Cod. Landberg 510, fol. 58–70) in the Yale University Library. With a Bibliographical Note.* New Haven 1944. [Online available: www.muslimphilosophy.com/ip/kumnah.pdf].
Leon Nemoy: "Ibn Kammunah's Treatise on the Immortality of the Soul." *Ha-Rofe ha-ʿIvri. The Hebrew Medical Journal* 1 (1962), pp. 239–213. [Edition on the basis of Landberg 510 and translation into English].
Insīyah Barkhʷāh: *Taṣḥīḥ-i Risāla-yi Azaliyyat al-nafs wa-baqāʾihā taʾlīf-i Ibn Kammūna.* MA thesis: Madrasa-yi Murtaḍā Muṭahharī (Sipahsālār), 1379/2000, pp. 166–78.

[Edited on the basis of Majlis 1837/6 and Ṣadr al-Dīn Maḥallātī 17/5].

*Translations*

Leon Nemoy: "Ibn Kammūna's Treatise on the Immortality of the Soul." In *Ignace Goldziher Memorial Volume*. Eds. S. Löwinger, A. Scheiber & J. Somogyi. Jerusalem 1958, vol 2, pp. 83–99. [Translation into English].

Leon Nemoy: "Ibn Kammunah's Treatise on the Immortality of the Soul." *Ha-Rofe ha-ʿIvri. The Hebrew Medical Journal* 1 (1962), pp. 239–213. [Edition on the basis of Landberg 510 and translation into English].

Leon Nemoy: "Ibn Kammunah u-masato ʿal hishaʾarut ha-nefesh." *Ha-Rofe ha-ʿIvri. The Hebrew Medical Journal* 1 (1962), pp. 131–36. [Hebrew version of the English introduction to his edition and of the English translation in the same issue of the journal, pp. 239–213].

*Studies*

Ignaz Goldziher: "Saʿd b. Manṣūr ibn Kammūna's Abhandlung über die Unvergänglichkeit der Seele." In *Festschrift zum achtzigsten Geburtstage Moritz Steinschneider's*. Leipzig 1896, pp. 110–14.

1.4. *Comparative Writings*

1.4.1. *Maqāla fī tanqīḥ al-abḥāth li-l-milal al-thalāth (Risālat Tanqīḥ al-abḥāth li-l-milal al-thalāth)*

An overview of the polemical arguments for and against Judaism, Christianity and Islam. It consists of four chapters on prophecy in general (ch. 1), Judaism (ch. 2), Christianity (ch. 3), and Islam (ch. 4). The work was written following some discussion (*mufāwuḍāt*) and completed in Jumādā II 679/September–October 1280.

(*Dharīʿa*, vol. 4, p. 460 no. 2054; GALS, vol. 1, pp. 768–69; GCAL, vol. 2, pp. 270–71; *Kashf al-ẓunūn*, vol. 2, p. 433 no. 3672; Steinschneider: *Arabische Literatur*, p. 240 no. 8; Steinschneider: *Polemische und apologetische Literatur*, pp. 37–41 no. 19, cf. also ibid. p. 326; Wolff: *Bibliotheca Hebraea*, vol. 3, p. 643 no. 1337b; Kochius: "Observatio nova," § viii [= p. 144]; d'Herbelot: *Bibliothèque orientale*, p. 851)

Introduction (ed. Perlmann, p. 1):—

احمد الله على ما أرشد وهدى وأتوسل إليه بأسمائه الحسنى وبما أظهره على لسان أنبيائه ورسله من صفاته العلى، أن يؤمنني في الآخرة والأولى وأن يجعلني في دار الأبد من أهل السعادة العظمى وأن يصلي على من بالملأ الأعلى وعلى من اختصه بالنبوة والولاية من الورى وخصوصاً على رسوله المصطفى وعلى آله وأصحابه أولى النهى. وبعد، فإنه قد جرت مفاوضات اقتضت أن عملت هذه المقالة في تنقيح الأبحاث للملل الثلاث، أعني ملة اليهود وملة النصارى وملة المسلمين. فذكرت فيها أولاً أحوال النبوات مجملاً، ثم ذكرت ما يختص بكل ملة من هذه على ترتيب أزمنتها فابتدأت بأقدمها زماناً وهي الملة اليهودية وتلوتها بأوسطها وهي الملة النصرانية وختمت بأحدثها وهي الملة الإسلامية. وحكيت عن كل واحدة من هذه الملل أصول معتقدها سوى التفاريع لتعذر استقصائها وأردفت ذلك بحكاية أدلة أربابها على صحة نبوة النبي الآتي بها. وأوردت ما وجه من المطاعن عليها وما ذكر من الأجوبة عنها منبهاً على مواقع الأنظار فيها ومميزاً ما يصلح لأن يعول عليه من تلك الأجوبة عما لا يصلح لذلك منها. ولم أقل في شيء من ذلك مع الهوى ولا تعرضت لترجيح ملة على أخرى بل قررت مباحث كل ملة إلى غايتها القصوى ورتبت المقالة على أربعة أبوابها. ومن الله أسأل التوفيق والهدى وحسن الخاتمة والمنتهى وأن يجعلني في الآخرة في زمرة من آمن واتقى إنه ولي النعم في الدنيا وإليه منتهى الرغبات في الدار الأخرى وهو المشكور على ما أحسن وأولى وحسبي الله وكفى.

Perlmann does not mention the following introductory remarks that are to be found at the beginning of three of the extant manuscripts.

Berlin Ms. Or. Oct. 256/1, f. 1b:1–9:—

בסם אללה אלרחמן אלרחים והו רב אלעאלמין קאל אלשיך אלעאלם אלפאצל אלמויד פי מסאלבה באשרף ישחם עלאמה אלאמם משיד חגג אלאקדמין ואלמתאחרין ומרשד כאפה פרק אלבאחתין סידנא ומולאנא עז אלדולה סעד בן מנצור בן סעד בן אלחסן בן הבה אללה אלמערוף באבן כמונה ופקה אללה למא ירצי וגעל אכרתה כיר מן אלאולי ונפענא במא אפאד בה ואמלי אנה רבנא ונעם אלאולי

Marʿashī 12868, f. 59b: 30; Majlis 593, p. 69:2:—

رسالة تنقيح الأبحاث للملل الثلاث لسعد بن منصور بن سعد بن الحسن هبة [بن] الله بن كمونة

End (ed. Perlmann, p. 108:4–8):—

...لا سبيل إلى ذلك إلا بإنكار صحة تواتر اليهود والنصارى، وقد عرفت كيفية الكلام في ذلك. وقد طعن بمطاعن كثيرة وأجيب عنها. ويحتمل هذا الموضع مباحث أخرى لكن لا أرى الزيادة على قدر الذي ذكرته. ومن الله تعالى أسأل الهداية والعصمة وحسن الخاتمة والرحمة، وأن يجعلني بسعادة الأبد من الفائزين ولعقابه من الآمنين. والحمد لله رب العالمين وصلواته على أصفيائه وأنبيائه المقربين.

Author's colophon (Berlin Ms. Or. Oct 256/1, f. 136b:8–10):—

נגז תצניף דלך פי גמאדי אלאכרה מן סנה תסע וסבעין וסתמאיה הגריה הלאליה ואלסלאם.

Given in ed. Perlmann (p. 110) in Arabic transcription:—

نجز تصنيف ذلك في جمادى الآخرة من سنة تسع وسب١عين وستمائة هجرية هلالية والسلام.

*Manuscripts*

Berkeley Magnes Museum KC 24, pp. 1–118, 25 lines to a page, written in Hebrew characters in 1886.
[Microfilm copy: IMHM no. 49250]

Berlin Ms. Or. Oct. 256/1, ff. 1–137, 12 × 16.5 cm, 16 lines to a page; written in Hebrew characters. The copy is undated; it has an undated reader's colophon in Hebrew letters (f. 136b) by a certain **אברהם אלקרא אליהודי ולד מוסי אלקרא אליהודי**. Since Ms. Or. Oct. 256/2, containing Ibn Kammūna's treatise on the differences between Rabbanites and Karaites, is written by a different hand, the dated colophon (= 1341 C.E.) found at the end of the second treatise of the

codex gives no indication as to when the copy of *Tanqīḥ al-abḥāth* was completed (cat. 1878–97, vol. 1, pp. 74–76 no. 107; Steinschneider: *Polemische und apologetische Literatur*, p. 39).
[Microfilm copy: IMHM no. 1986].

Biblioteca Angelica 15; copied by a Christian scribe, dated 755/1354 (cat. pp. 64–66).

Bodleian MS Huntington 390 (Uri 361), 146 ff., written in Hebrew characters. The manuscript is undated but has an ownership statement dated 5202 = 1442 C.E. (cat. 1886, pp. 453–54 no. 1281; cat. 1787, p. 66; Steinschneider: *Polemische und apologetische Literatur*, pp. 38–39).
[Microfilm copy: IMHM no. 22095].

Chester Beatty 4965, 94 ff., 18.8 × 13.4 cm, 13/14 lines to a page; undated. At the end there is a reading colophon written by the owner, Shihāb al-Dīn Aḥmad b. Abī l-Qādir b. al-Hādī (?) (f. 94b:9–10) stating that he read the book to his Shaykh al-ʿAllāma Badr al-Dīn b. al-Ṭabbākh al-Ashʿarī at the beginning of 965/1557. Incomplete in the beginning (f. 1:1 = ed. Perlmann, p. 1:11) (cat. vol. 6, p. 157).

Fâtih 3141/2, 8 × 16.5 cm, ff. 202–263; ff. 202–209 (= ed. Perlmann, pp. 2:1–16:4), 211–263 (= 17:2–108:8, 111:9–11); copied in the Madrasa Nūriyya in Mūṣil by Maḥmūd b. al-Sharīf al-Kirmānī, dated "*khaṣūb*" (= 698/1299) according to the colophon (f. 263a:13–17):—

وقع الفراغ من تحرير هذا الكتاب يوم الخميس من ذي القعدة لسنة خصوب في المدرسة النورية من بلدة موصل حماها الله عن الآفات على يدي العبد الفقير إلى الله الغني بالله محمود بن الشريف الكرماني سعد جدّه وجدّ سعده.

This reading disagrees with Perlmann's reading of the year as *khuṣū* (= 696/1297) (cf. ed. Perlmann, pp. xii, 111).

Jewish Theological Seminary 5519 (ENA 95), 48 ff., 16 lines to a page; written in Hebrew characters. The fragment contains the following portions of the text: ed. Perlmann, pp. 73:4–93:8.
[Microfilm copy: IMHM no. 37283]

Maḥfūẓ 318 (cat. 1960, p. 56; cf. also ed. Perlmann, p. xii n. 1: "Al-Maktaba (Baghdad), Vol. III, no. 2 (1962), pp. 28 ff. mentions what is apparently a copy of this MS in the possession of Dr. Husayn Ali Mahfuz in Kazimiya; cf. Vol. III, no. 4, p. 35."). ʿAlī Naqī Munzawī, who has consulted this manuscript for his edition, states that the manuscript, which is undated, was written in the 19th century and is in fact a direct copy of Majlis 593. He explains that the copyist could apparently not read everything as there are some places left blank.[8]

Majlis 593/3, pp. 69–142 [from p. 100 onwards, the codex is also foliated, p. 100 being f. 17a], 24.5 × 15.5 cm, 19 lines to a page; incomplete at the end. The manuscript contains only one page belonging to *al-bāb al-rābiʿ fī dhikr ʿaqīdat al-muslimīn.*[9] The entire codex, including this work, was copied in 1060/1650 from a copy that Quṭb al-Dīn al-Shīrāzī had written in 685/1286 in Konya (= Marʿashī 12868) (cat. vol. 1, pp. 54–55, vol. 2, pp. 350–51).
[Microfilm copy: Dānishgāh microfilm 2320 (cat. 1969–84, vol. 1, p. 673)].

Marʿashī 12868, ff. 59b–126b, 13.5 × 5 cm, 33 lines to a page, copied by the hand of Quṭb al-Dīn al-Shīrāzī in Konya in 685/1286 (cat. vol. 32, pp. 640–41).

Shīrāz 538, 8 ff.[?], 17.5 × 24 cm, 19 lines to a page (cat. pp. 133–34).

St. Petersburg RNL II Firk. Yevr.-Arab. I 1702, 6 ff., 18 lines to a page, written in Hebrew characters. The fragment contains the following portions of the text: f. 1 (= ed. Perlmann, pp. 20:8–21:6), ff. 2–3 (= 80:22–83:11), f. 4 (= 28:2–29:4), f. 5 (= 72:13–73:16), f. 6 (= 62:20–63:22). The manuscript belongs to the same copy as RNL II Firk. Yevr.-Arab. II 490.
[Microfilm copy: IMHM no. 56251]

St. Petersburg RNL II Firk. Yevr.-Arab. II 490, 6 ff., 18 lines to a page, written in Hebrew characters. The fragment contains the following portions of the text: f. 1 (= ed. Perlmann, pp. 45:3–46:4), f. 2 (= 44:5–45:3), f. 3 (= 95:4–96:5), f. 4 (= 92:1–94:3), f. 5 (= 48:3–49:4), f. 6 (= 41:7–42:4). The manuscript belongs to the same copy as RNL II Firk. Yevr.-Arab. I 1702.
[Microfilm copy: IMHM no. 58782]

---

8 ʿAlī Naqī Munzawī: *Ibn Kammūna wa-falsafatuhū fī l-nubuwwa*, p. 56.

9 Last words on p. 142: فإن لم يكن دعوته بلغتهم أو أنها بلغتهم على وجه لا يحرّك داعية الطلب والنظر كما إذا سمعوا أن شخصا ملتبسا اسمه محمد ادعى النبوة...
Beginning of p. 143 (apparently a new text): ويتعد خلق كثيراً يحدعوا بكلامه...

*Editions*

Leo Hirschfeld: *Sa'd b. Manṣûr Ibn Kammûnah und seine polemische Schrift* تنقيح الأبحاث للملل الثلاث. Inaugural-Dissertation zur Erlangung der Doctorwürde vorgelegt der Hohen Philosophischen Fakultät der Universität Heidelberg. Leipzig 1893.
[Introduction, edition and translation of a portion of the introduction and a section of the chapter on Judaism (= ed. Perlmann, pp. 1:6–15, 24:1–29:2) on the basis of two mss: Berlin Ms. Or. Oct. 256; Bodleian MS Huntington 390].

Moshe Perlmann (ed.): *Sa'd b. Manṣūr Ibn Kammūna's Examination of the Inquiries into the Three Faiths. A Thirteenth-Century Essay in Comparative Religion.* Berkeley 1967.
[Edited on the basis of Majlis 593/3; Fâtih 3141; Biblioteca Angelica 15; Berlin Ms. Or. Oct. 256; Bodleian MS Huntington 390]. On this edition, see the reviews by Leon Nemoy (*Jewish Social Studies* 31 (1969), pp. 333–34), Rudi Paret (*Der Islam* 46 (1970), pp. 102–4), Kurt Rudolph (*Orientalistische Literaturzeitung* 67 ix/x (1972), pp. 462–65), Ernst Ludwig Dietrich (*Oriens* 21–22 (1968–69), pp. 440–41), Lenn E. Goodman (*Philosophy East and West* 22 (1972), pp. 487–88), James Kritzeck (*Journal of Near Eastern Studies* 32 (1973), pp. 506–7), H.T.N. (*Bulletin of the School of Oriental and African Studies* 36 (1973), pp. 204–5), Lawrence V. Berman (*Journal of the American Oriental Society* 91 (1971), pp. 333–34), Victor Danner (*Journal of the American Academy of Religion* 42 (1974), pp. 565–66).

ʿAlī Naqī Munzawī: *Ibn Kammūna wa-falsafatuhū fī l-nubuwwa.* PhD thesis, Université Saint Joseph, Beirut 1972. [*Al-Qism al-awwal: al-Madkhal wa-l-naṣṣ al-kāmil li-Tanqīḥ al-abḥāth; al-Qism al-thānī: al-Taʿlīqāt ʿalā Tanqīḥ al-abḥāth li-l-milal al-thalāth*]. [Edited on the basis of Majlis 593, Maḥfūẓ 318 and Perlmann's edition].

Habib [Ḥabīb] Bacha [Bāshā]: *Ḥawāshī (Notes) d'Ibn al-Maḥrūma sur le "Tanqīḥ" d'Ibn Kammūna* [*Ḥawāshī Ibn al-Maḥrūma ʿalā Kitāb Tanqīḥ al-abḥāth li-l-milal al-thalāth li-Ibn Kammūna*]. Etude et édition critique. Rome 1984. [Containing the chapters on Judaism and Christianity].

Karīmī Zanjānī Aṣl, Muḥammad (ed.): *Tanqīḥ al-abḥāth li-l-milal al-thalāth li-Ibn Kammūna.* Tehran (forthcoming). [This edition relies heavily on those by Munzawī and Perlmann.]

*English Translation*

Moshe Perlmann: *Ibn Kammūna's Examination of the Three Faiths. A thirteenth-century essay in the comparative study of religion.* Translated from the Arabic, with an introduction and notes. Berkeley / Los Angeles / London 1971.

*Christian Refutations*

Abu l-Ḥasan b. Ibrāhīm b. al-Maḥrūma: *Ḥawāshī* (cf. GCAL, vol. 2, pp. 270–71). Edition: *Ḥawāshi (Notes) d'Ibn al-Maḥrūma sur le "Tanqīḥ" d'Ibn Kammūna.* Etude et édition critique par Habib Bacha. Rome 1984. On this edition, see the review by Leon Nemoy (*The Jewish Quarterly Review* 77 i (July 1986), pp. 82–83).
On this retort, see also Moshe Perlmann: "On Maḥrūma, a Christian opponent of Ibn Kammūna." *American Academy for Jewish Research* 2 (1965) [H.A. Wolfson Jubilee Volume], pp. 641–55; Barbara Roggema: "Jewish-Christian debate in a Muslim context. Ibn al-Maḥrūma's notes on Ibn Kammūna's *Examination of the Inquiries into the Three Faiths.*" In *All Those Nations... Cultural Encounters Within and With the Near East. Studies presented to Han Drijvers at the occasion of his sixty-fifth birthday by colleagues and students.* Collected and edited by Herman J.L. Vanstiphout with the assistance of W.J. van Bekkum, G.J. van Gelder and G.J. Reinink. Groningen 1999, pp. 131–39.

*Muslim Refutations*

Muẓaffar al-Dīn Aḥmad b. ʿAlī b. al-Sāʿātī (d. 694/1294): *al-Durr al-manḍūd fī l-radd ʿalā faylasūf al-yahūd.* The work is apparently lost (cf. Steinschneider: *Polemische und apologetische Literatur*, pp. 47–48 no. 30; Abū l-Maʿālī: *Taʾrīkh ʿulamāʾ Baghdād*, p. 35; Ibn Taghrībirdī: *al-Manhal al-ṣāfī*, vol. 1, p. 421; *Kashf al-ẓunūn*, vol. 3, p. 193; *Tanqīḥ*. Ed. Perlmann, introduction, p. 8 n. 13).

Zayn al-Dīn Sarīyā b. Muḥammad al-Malaṭī (d. 788/1386): *Nuhūḍ ḥathīth al-nuhūd ilā duḥūḍ khabīth al-yahūd.* The work is apparently lost (cf. Steinschneider: *Polemische und apologetische Literatur*, p. 107 no. 86; *Tanqīḥ*. Ed. Perlmann, introduction, p. 8 n. 1; *Kashf al-ẓunūn*, vol. 6, pp. 411–12).

Anonymous: *Kitāb ithbāt al-nubuwwa* (ms. Hamidiye 1447/3). A transcription of the manuscript, together with a short analysis, was published by Moshe Perlmann: "'Proving Muhammad's Prophethood', A Muslim critique of Ibn Kammuna." [Hebrew] In *Hagut Ivrit Ba'Amerika. Studies on Jewish Themes by Contemporary American Scholars.* Eds. M. Zohori, A. Tartakover & H. Ormian. Vol. III. Tel Aviv 1974, pp. 75–97. For an improved edition, see below Appendix B: *Kitāb fī ithbāt al-nubuwwa li-Muḥammad ʿalayhi al-salām.*

## *Studies*

Habib Bacha: "Tanqīḥ al-abḥāth li-l-milal al-thalāth." *Parole de l'Orient* 11 (1971), pp. 151–62.

Geneviève Gobillot: "Ibn Kammûna. Les trois anneaux de la prophétie?" *Tsafon. Revue d'études juives du Nord* 50 (automne 2005–hiver 2006) [Hommage à Jean-Marie Delmaire], pp. 23–48.

Leo Hirschfeld: *Saʿd b. Manṣûr Ibn Kammûnah und seine polemische Schrift* تنقيح الأبحاث للملل الثلاث. Leipzig1893.

ʿAlī Naqī Munzawī: *Ibn Kammūna wa-falsafatuhū fī l-nubuwwa*. PhD thesis, Université Saint Joseph, Beirut 1972. [*Al-Qism al-awwal: al-Madkhal wa-l-naṣṣ al-kāmil li-Tanqīḥ al-abḥāth; al-Qism al-thānī: al-Taʿlīqāt ʿalā Tanqīḥ al-abḥāth li-l-milal al-thalāth*].

ʿAlī Naqī Munzawī: "Tanqīḥ al-abḥāth-i Ibn Kammūna." *Waḥīd* 6 v (1348), pp. 369–71; 6 vi (1348), pp. 503–14; 6 xii (1348), pp. 1079–83.

Friedrich Niewöhner: "Die Wahrheit ist eine Tochter der Zeit. Ibn Kammūna's historisch-kritischer Religionsvergleich aus dem Jahre 1280." In *Religionsgespräche im Mittelalter*. Eds. B. Lewis & F. Niewöhner. Wiesbaden 1992, pp. 357–69.

Moshe Perlmann: "The Medieval Polemics Between Islam and Judaism." In *Religion in a Religious Age*. Ed. S.D. Goitein. New York 1974, pp. 103–38.

Reza Pourjavady and Sabine Schmidtke: "Muslim Polemics against

Judaism and Christianity in 18th Century Iran. The Literary Sources of Āqā Muḥammad ʿAlī Bihbahānī's (1144/1732–1216/1801) *Rādd-i shubahāt al-kuffār*." *Studia Iranica* 35 (2006), pp. 69–94.
Barbara Roggema: "Epistemology as Polemics. Ibn Kammuna's Examination of the Apologetics of the Three Faiths." In *The Three Rings. Textual Studies in the Historical Trialogue of Judaism, Christianity, and Islam*. Eds. B. Roggema, M. Poorthuis & P. Valkenberg. Leuven 2005, pp. 47–70.
Simone Rosenkranz: "Judentum, Christentum und Islam in der Sicht des Ibn Kammûna." *Judaica. Beiträge zum Verstehen des Judentums* 52 (1996), pp. 4–22.
Dominique Urvoy: *Les penseurs libres dans l'Islam classique. L'interrogation sur la religion chez les penseurs arabes indépendants*. Paris 1996, pp. 202–14.

### 1.4.2. *Treatise on the Differences between the Rabbanites and Karaites*

Systematic investigation, composed at the request of *baʿḍ al-sāda al-afāḍil*, of the mutual accusations of Rabbanites and Karaites with the respective replies in three chapters (*fuṣūl*) dealing with the validity of the Rabbinic tradition (*fī ḥikāyat mā dhukira min kayfiyyat naql al-sharīʿa wa-faḍāʾil al-ḥakhamīm al-nāqilīn lahā wa-bayān ʿulūmihim wa-l-tanbīh ʿalā wujūb al-ʿamal bi-muqtaḍā nuqūlihim*), the Karaite charges against the Rabbanites (*fī dhikr jumla min al-maṭāʿin allatī yatʿanu bihā al-qarāʾiyyūn ʿalā al-ḥakhamīm wa-l-ajwiba al-dāfiʿa lahā wa-li-ashkālihā min bāqī maṭāʿinihim*) and the Rabbanite charges against the Karaites (*fī dhikr mā ṭaʿana bihī al-Rabbāniyyūn ʿalā l-Qarāʾiyyīn wa-mā yaṣlaḥu an yakūna jawāban lahum ʿanhu*). There is little indication in the text as to Ibn Kammūna's own sympathies. Whereas Steinschneider argued that the objective character of the tract and the lack of characteristic Rabbanite eulogies indicate the Karaite allegiance of the (in his view anonymous) author, Nemoy takes the fact that Ibn Kammūna "devotes twice as much space to the defense of Rabbanism as he does to that of Karaism" as an indication that "his sympathies nevertheless remained with the mother-synagogue" (ed. Nemoy, p. 109).

The authenticity of the treatise was intitially contested. Hartwig Hirschfeld designated the treatise as an "anonyme Schrift"[10] and Moritz

[10] In the preface to his edition of the *Kitāb al-Khuzarī* (Leipzig 1887); cf. Nemoy:

Steinschneider expressed doubts as to the accuracy of attributing this work to Ibn Kammūna.[11] However, the authorship of Ibn Kammūna is well attested by the typical wording of the introduction and has been uncontested since the publication of Dawid Hartwig Baneth's study on Ibn Kammūna in 1925.[12] The original work had no title. "Treatise on the Differences between the Rabbanites and Karaites" was the title given by its first editor, Hartwig Hirschfeld. The extant manuscripts are written exclusively in Hebrew characters.

---

"Ibn Kammunah's Treatise," (1968), p. 107 n. 3.—In the table of contents of his *Arabic Chrestomathy*, Hirschfeld still seems to be doubtful on the issue of authorship when he writes: "Treatise on the differences between the Rabbanites and Karaites attributed to Saʿd b. Manṣūr (cod. Berol. Oct. 256[2] Unicum)." See also Baneth: "Ibn Kammuna," p. 296: "die zweite [Schrift über Rabbaniten und Karäer] hat Hartwig Hirschfeld in seiner Arabic chrestomathy in Hebrew characters (London 1892) vollständig abgedruckt, aber auf Grund von Steinschneiders Urteil als unterschoben behandelt."

[11] Moritz Steinschneider: *Verzeichniss der Hebräischen Handschriften* 1–2. Berlin 1878–97, vol. 1, p. 76: "eine Schrift über die Differenzen zwischen den Rabbaniten und Karaiten, welche (gewissermaßen als Fortsetzung des תנקיח) untergeschoben scheint; ... Der Verf. dieser Abhandlung spricht als Jude, citirt fortwährend hebräische Schriften und gebraucht hebräische Phrasen, ganz abweichend vom Verf. des תנקיח, allerdings mit Rücksicht auf das Thema und mit dem Scheine einer ähnlichen Objectivität, hinter welcher wohl ein Karait sich versteckt, da die Rabbaniten zu einem solchen Scheine weder Veranlassung noch Neigung hatten. Auch fehlen selbst in den vorgeführten Ansichten der Rabbaniten gewisse übliche Eulogien und Formeln... wie auch unser Unicum aus karaitischen Händen kömmt und die, [sic] die Echtheit versichernde Schlußformel den Verdacht der Fälschung voraussetzt. Auch die Benutzung des Buches כוזרי ist eine halb ehrliche, nach der Art raffinirter Plagiatoren." Idem: *Arabische Literatur*, p. 240: "Die Abhandlung über die Differenzen der Rabbaniten und Karaiten (ms. Bl.) ist jedenfalls eine, wahrscheinlich karaitische Unterschiebung." Idem: *Polemische und apologetische Literatur*, p. 39: "Es folgt dann eine, dem Character nach verschiedene Abhandlung über die Differenzen der Rabbaniten und Karaiten, welches grösstentheils aus dem Kusari des Jehuda ha-Levi... schöpft, und durch eine Vorbemerkung etwas ungeschickt als eine Art Fortsetzung des vorigen Werkes eingeleitet wird. Ich vermuthe hier eine Unterschiebung."

[12] See Baneth: "Ibn Kammuna," pp. 309–11; Nemoy: "A Forgotten 13th Century Jewish Physician," p. 157.—In his study on Ibn Kammūna's *Tanqīḥ al-abḥāth*, Leo Hirschfeld followed the views of Hartwig Hirschfeld and Steinschneider; see his *Ibn Kammūnah und seine polemische Schrift*, p. 1 n.3, p. 15 n. 1.

Introduction (Berlin Ms. Or. Oct. 256/2, ff. 139b:1–140a:5):—

קאל אלסיד אלמולי אלעאלם אלפאצל סראג אלעלמא ותאג אלפצלא אוחד עצרה ופריד דהרה עלאמה אלעראקין ומפתי אלפריקין עז אלדולה ואלדין סעד אבן מנצור אבן סעד אבן אלחסן אבן הבה אללה אבן כמונה גפר אללה לה ועפא ענה ורחמה פי דניאה ואכראה. קד דכרת פי הדה אלמקאלה בעד חמד אללה תעאלי ושכר אלאיה מא אלתמסה מני בעץ אלסאדה אלאפאצל אדאם אללה מגדה ואסעד גדה והו מא ענדי פימא דכר עלי סביל אלאגמאל מן פצאיל אלחכמים אלדין יעתמד אליהוד אלרבאניון פי אמורהם אלשרעיה עלי נקולהם ואראיהם מע כיפיה נקלהם ללשריעה וביאן וגוב אלעמל במקתצאה ופי אלמטאען אלתי אורדהא פריק אלקראיין מן אליהוד עליהם תקרירא וגואבא וכדלך אלמטאען אלתי אורדהא אלרבאניון עלי אלקראיין וחררת אלביאנאת ואלאגובה מע קלה מערפתי בגזויאת אלשרעיה עלי וגוה לם אגד אכתרהא [تصحيح فوق السطر: קלילא] או כתירא מנהא פי כלאם מן סבקני ורתבת דלך פי תלתה פצול ומן אללה אסתמד חסן אלמעונה ואלתופיק...

Introduction (Arabic transcription, ed. Nemoy, lines 2–14):—

قال السيد المولى العالم الفاضل سراج العلماء وتاج الفضلاء أوحد عصره وفريد دهره علامة العراقين ومفتي الفريقين عز الدولة والدين سعد بن منصور بن سعد بن الحسن بن هبة الله ابن كمونة غفر الله له وعفا عنه ورحمه في دنياه وأخراه. قد ذكرت في هذه المقالة بعد حمد الله تعالى وشكر آلائه ما التمسه مني بعض السادة الأفاضل أدام الله مجده وأسعد جدّه وهو ما عندي فيما ذكر على سبيل الإجمال من فضائل الחכמים الذين يعتمد اليهود الربانيون في أمورهم الشرعية على نقولهم وآرائهم مع كيفية نقلهم للشريعة وبيان وجوب العمل بمقتضاه وفي المطاعن التي أوردها فريق القرائين من اليهود عليهم تقريراً وجواباً وكذلك المطاعن التي أوردها الربانيون على القرائين وحرّرتُ البيانات والأجوبة مع قلة معرفتي بجزءيات الشريعة على وجوه لم أجد أكثرها أو كثيراً منها في كلام من سبقني ورتّبتُ ذلك في ثلاثة فصول ومن الله أستمدّ حسن المعونة والتوفيق. الفصل الأول ...

End (Berlin Ms. Or. Oct. 256/2, f. 180a:4–12):—

וחיניד לא יגוז אן יפסר אלא בדלך אלוגה לא גיר והדא ואן כאן הו אלואגב אלא אן אכתר אלגמאעה גיר מלתזמין בה ודלך כטא עטים יודי אלי אן יפהם אכתר אלנאס מן כלאם אלנבואת גיר מא קצד בה אעאדנא אללה

מן אלכטא ואלזלל ווּפקנא לצואב אלקול ואלעמל וגעלנא מן אהל אלתקוי ומן אלאמנין אלואצלין פי דאר אלכלוד אלי אלסעאדה אלקצוי אנה ולי אלכיראת ומנתהי אלרגבאת

End (Arabic transcription, ed. Nemoy, lines 876–882):—

...وحينئذ لا يجوز أن يفسّر إلا بذلك الوجه لا غير وهذا وإن كان هو الواجب إلا أن أكثر الجماعة غير ملتزمين به وذلك خطأ عظيم يؤدي إلى أن يفهم أكثر الناس من كلام النبوات غير ما قصد به أعاذنا الله من الخطأ والزلل ووفّقنا لصواب القول والعمل وجعلنا من أهل التقوى ومن الآمنين الواصلين في دار الخلود إلى السعادة القصوى إنه ولي الخيرات ومنتهى الرغبات.

*Manuscripts*

Berlin Ms. Or. Oct. 256/2, ff. 139b–180, 12 × 16.5 cm, 15 lines to a page; written in Hebrew characters by a different hand from Berlin Ms. Or. Oct. 256/1 (*Tanqīḥ al-abḥāth*; see above, Section 1.4.1); copied from the autograph by a Karaite by the name of Azaryahu ha-Nasi ben Yehalalel ha-Nasi in Baghdad in 1341 C.E. by the order of a certain Kamāl al-Dawla ʿAbd al-Khāliq b. Yūnah and collated with another copy that was taken from the autograph (cat. 1878–97, vol. 1, pp. 76–77 no. 107; Baneth: "Ibn Kammuna," pp. 309–11). [Microfilm copy: IMHM no. 1986].

St. Petersburg RNL II Firk. Yevr.-Arab. I 913, 7 ff.; written in Hebrew characters; consisting of two bifolios (ff. 1–2, 3–4), one isolated folio (f. 5), quire of four leaves (ff. 6–9, ff. 8–9 being empty), 13.6 × 18.2 cm, 17/19 lines to a page; containing the following fragments of the treatise: ff. 1a–2b (= ed. Nemoy, lines 644–703), ff. 3a–b (= 559–588), ff. 4a–5b (= 760–808), ff. 6a–7b (= 837–882). The fragment may belong to the same copy of the treatise as RNL II Firk. Yevr.-Arab. I 914. [Microfilm copy: IMHM no. 54504].

St. Petersburg RNL I Firk. Yevr.-Arab. I 914, 6 ff.; written in Hebrew characters; consisting of isolated folios (ff. 1, 2, 5, 6) and one bifolio (ff. 3–4), 14.2 × 18.5 cm, 19 lines to a page; containing the following fragments of the treatise: ff. 1a–2b (= ed. Nemoy, lines 364–424), f. 3a–b (= 336–363), f. 4a–b (= 424–451), f. 5b [preceding f. 5a]

(= 309–323), f. 5a (= 323–336), f. 6a–b (= 451–478). The fragment may belong to the same copy of the treatise as RNL II Firk. Yevr.-Arab. I 913.
[Microfilm copy: IMHM no. 54505].

St. Petersburg RNL II Firk. Yevr.-Arab. I 1703, 2 ff. (bifolio); written in Hebrew characters by a different hand than Yevr.-Arab. I 913 and 914; 13.5 × 18.6 cm, 15/16 lines to a page; containing the following fragments of the treatise: f. 1b (= ed. Nemoy, lines 1–9), f. 2 (= 135–154) (cat. 1991, p. 110).
[Microfilm copy: IMHM no. 56285].

*Editions and Translations*

Hartwig Hirschfeld (ed.): *Arabic Chrestomathy in Hebrew Characters. With a Glossary*. London 1892, pp. 69–103: "Treatise on the differences between the Rabbanites and Karaites attributed to Saʿd b. Manṣūr (cod. Berol. Oct. $256^2$ Unicum)." On the edition see the emendations suggested by Leon Nemoy: "Remarks on Ibn Kammūna's Treatise on the Differences between the Rabbanites and the Karaites." [Hebrew] *Tarbiẓ* 24 (1955), pp. 343–53, vi [English Summary] and in *Sefer yovel mugash le-khevod Yisrael Elfenbeyn*. Ed. Y.L. Maimon. Jerusalem 1962, pp. 201–8.

Leon Nemoy: "Ibn Kammunah's Treatise on the Differences between the Rabbanites and the Karaites." *Proceedings of the American Academy for Jewish Research* 36 (1968), pp. 107–65 [Edition of the text transcribed into Arabic characters on the basis of Berlin Ms. Or. Oct. 256/2, summarizing translation, and introduction].

Leon Nemoy: "Ibn Kammunah's Treatise on the Differences between the Rabbanites and the Karaites." *Jewish Quarterly Review* 13 (1972–73), pp. 97–135, 222–46 [Introduction and English translation].

### 1.5. *Correspondence*

#### 1.5.1. *Correspondence with Naṣīr al-Dīn al-Ṭūsī I*

This correspondence was initiated by Ibn Kammūna, who asked Ṭūsī about a *mughālaṭa* formulated by Kātibī on the createdness of the world (*ḥudūth al-ʿālam*). On the subsequent reply by Ṭūsī, Ibn Kammūna sent another letter.—The problem as formulated by Ibn Kammūna in the correspondence runs as follows: "the [temporal] originatedness of the world does not necessitate the negation of reality, and whatever does not necessitate the negation of the reality is itself a reality that resulted from the originatedness of the real world" (*ḥudūth al-ʿālam lā yastalzimu irtifāʿ al-wāqiʿ wa-kull mā lā yastalzimu irtifāʿ al-wāqiʿ fa-huwa wāqiʿ yuntiju ann ḥudūth al-ʿālam wāqiʿ*).

(Mudarris Raḍawī: *Aḥwāl u āthār*, p. 523 no. 78; Mudarrisī Zanjānī: *Zarguzasht*, pp. 211–12)

Beginning of Ibn Kammūna's initial query (Mudarris Raḍawī: *Aḥwāl u āthār*, p. 523):—

كتب المولى العالم المحقق عز الدولة سعد بن منصور بن كمونة عن مغالطة للحكيم الفاضل نجم الملة والدين الكاتبي القزويني رحمه الله. حدوث العالم لا يستلزم ارتفاع الواقع وكل ما لا يستلزم ارتفاع الواقع فهو واقع ينتج أن حدوث العالم واقع . . .

Beginning of Ṭūsī's reply (Mudarris Raḍawī: *Aḥwāl u āthār*, p. 523):—

خطر ببال الداعي المخلص على الحجة المذكورة على إثبات الجواز في صورة النزاع هو هذا . . .

Beginning of Ibn Kammūna's reply (Mudarris Raḍawī: *Aḥwāl u āthār*, p. 523):—

هذه النكات صنفها مولانا ملك الحكماء أفضل المتأخرين سلطان المحققين نصير الملة والدين . . .

### *Editions*

ʿAbd Allāh Nūrānī: "Muṭāraḥāt manṭiqiyya bayn al-Kātibī wa-l-Ṭūsī." In *Collected Texts and Papers on Logic and Language*. Eds. M. Mohaghegh & T. Izutsu. Tehran 1974, pp. 285–86 (containing Ṭūsī's reply only; apparently due to a technical error, the text of p. 286 is not part of Ṭūsī's reply, as becomes apparent from a comparison with the manuscripts of the text).

Chapter Three, Text Five (Ṭūsī's reply only)

*Manuscripts*

Ākhūnd 1187/76, ff. 267a–b [pp. 495–96] (containing Ṭūsī's reply only).

Mahdī Bayānī (containing all three letters of the correspondence; cf. Mudarris Raḍawī: *Aḥwāl u āthār*, p. 523 no. 78). The current whereabouts of the manuscript is unknown.

Ragıp Paşa 1461/76, ff. 258b–259a, 12.8 × 18 cm, 21 lines to a page (containing Ṭūsī's reply only).

### 1.5.2. *Correspondence with Naṣīr al-Dīn al-Ṭūsī II*

Seven questions addressed to Ṭūsī, together with the latter's replies. The questioner does not introduce himself. The only indication of his identity appears at the beginning of Ṭūsī's response, where the latter addresses the questioner by name. However, in Majlis-i Sinā 286 there is a blank space in place of the name and only the second part of the name is extant, which can be read as *al-mūla* (see infra). The editor of the correspondence, ʿAbd Allāh Nūrānī, who used only this manuscript, does not mention this in his recent edition of the text; he has simply corrected the name to Ibn Kammūna.

(Mudarris Raḍawī: *Aḥwāl u āthār*, pp. 512–13 no. 67)

Beginning of Ibn Kammūna's questions (ed. Nūrānī, p. 25 [variant readings of Majlis-i Sinā 286, f. 74a:9–22 in square brackets]):—

آخر كتب إليه دام علوه: المسؤول من عميم فضل مولانا ملك الحكماء وقدوة المحققين، أمتع الله العلم وأهله بحراسة مدته ودوام [ودوام: + ملكه] وتثبت دولته، إيضاح ما قد سأله العبد من المسائل التي لم يزل منذ اطلاعه على شيء من العلوم الحقيقية والمباحث الحكمية يمعن النظر فيها وفي ما كان من الإشكال لها نظراً فرجع طرف فكره بعد كونه حاسراً عن ساعد الاجتهاد في ذلك حسيراً، وأن يقال [يقال: يقام] عذره أو يصفح عن الله في تهجمه على الجناب المولوي، حرس الله دولته وأدام إقباله، بما عساه يقع بسببه الاشتغال عن بعض المهام. فإن علمه بما جبلت عليه الأخلاق المولوية، أجلها الله تعالى، من السخاء والكرم هو الذي أقدمه على ذلك. وليست هذه المسائل كل ما هو عند العبد مشكل ولا هي أيضاً أهم من كل ما سواها عنده إنما هي حسبما سنح كتابة هذه الضراعة ووجد الحال غير مقتضية للزيادة إلا بعد الإذن المولوي، ظاهر الله جلاله، والأمر أعلى. مسألة . . .

Beginning of al-Ṭūsī's replies (ed. Nūrānī, pp. 28:20–29:3 [variant readings of Majlis-i Sinā 286, f. 76a:2–8 in square brackets]):—

الجواب من خطه، قدس الله روحه: طالعت الأسرار التي أوردها الحكيم الفاضل أكيس الأقران وأوحد الزمان ابن كمونة [ابن كمونة: الموله]، حرس الله [الله: كلمة غير مقروءة] فضله، على بعض مباحث الحكمة والتمس الجواب عنها فحررت ما سمح به الخاطر في الوقت وساسح الزمان مع كثرة الشواغل والعوائق من أجوبتها فإن اشتمل على سهو أو خطأ فالعذر ظاهر وما توفيقي إلا بالله عليه توكلت وإليه أنيب. أما الجواب عن المسألة الأولى فأقول . . .

*Edition*

ʿAbd Allāh Nūrānī (ed.): *Ajwibat al-masāʾil al-Naṣīriyya, mushtamil bar 20 risāla*. Tehran 1384/2005, pp. 25–34.

*Manuscripts*

Amīr al-Muʾminīn 1515; copied by al-Khaṭṭāṭ Ismāʿīl al-Marāghī, 13th/19th century (cat. 2003–4, p. 251 no. 732).

Majlis-i Sinā 286, ff. 74a–78b, incomplete at the end; part of a *majmūʿa* written in 738/1337–38 by ʿAlī b. al-Ḥasan b. al-Raḍī al-Ḥāfiẓ at the request of Shams al-Dīn Āwī in Sulṭāniyya (cat. vol. 1, p. 146). [Microfilm copy: Dānishgāh microfilm 1436 (cat. vol. 1, pp. 600–1)].

Mudarris Raḍawī evidently used a different manuscript of the text, the identity of which still needs to be established, as he indicates the beginning of the text as follows (*Aḥwāl u āthār*, p. 512):—

هذه مسائل سبع سأل عنها عز الدولة أبو الرضا سعد بن كمونة مولانا الأعظم أفضل المتقدمين أستاذ الدنيا نصير الملة والدين محمد بن محمد بن الحسن الطوسي برد الله مضجعه. المسألة الأولى . . .

### 1.5.3. *Correspondence with Najm al-Dīn al-Kātibī I*[13]

This correspondence, which consists of three epistles by Ibn Kammūna to Kātibī, together with the latter's replies to all three epistles, gives evidence that Ibn Kammūna had participated in one of Kātibī's *majālis*, where the discussion on the logical problem dealt with in the correspondence started.

Ibn Kammūna refers to this correspondence in his *al-Jadīd fī l-ḥikma* where he gives a summary of it (pp. 203–5).

(GALS, vol. 1, p. 769)

Beginning:—

استفاد العبد المخلص سعد بن كمونة من مجلس مولانا علامة العالم ملك العلماء قدوة الفضلاء مفتي الفرق جامع علوم المتقدمين والمتأخرين نجم الملة والدين أدام الله أيامه وحرس مدّته وأمتع أهل العلم بطول حياته: إن ثبوت الإمكان لا يلزم منه إمكان الثبوت فلا يلزم من صدق بعض «ج» «ب» بالإمكان العام إمكان صدق بعض «ج» «ب» بالفعل، لأن الأول حكم بثبوت الإمكان والثاني حكم بإمكان الثبوت، ومستند المنع من اللزوم أن الحادث ثبت إمكان وجوده في الأزل ولا يمكن ثبوت وجوده في الأزل، ففي هذه المادة قد ثبت الإمكان ولم يمكن الثبوت. وهذا المنع في غاية الدقة والمستند في غاية الحسن.

End:—

وكيف ما كان يحصل المطلوب على ما هو غير خاف اللهمّ إلا أن يعني بالضرورة مصطلح المحققين من أرباب فن المنطق وحينئذ يكون النزاع لفظياً لا فائدة فيه، و العبد يسأل أن ينظر في هذا وأن يقام عذره في كثرة الإضجار والتثقيل فإن الذي يحمله على هذا كون فوائد مولانا أعز الله أنصاره مغتنمة ودقائق أنظاره ولطائف أفكاره يتنافس على استفادتها ورأيه الكريم المولوي أعلى والسلام على أهل الحق ومن اتّبع الهدى.

---

[13] Perlmann erroneously assumed that the addressee of Ibn Kammūna's epistle was his son Najm al-Dawla (see the introduction to his edition of the *Tanqīḥ*, p. xii); this was corrected by Bacha (*Ḥawāshī*, p. xxii).

*Manuscripts*

Adana 177/4 [= 181.2], ff. 239b–241b, 18.5 × 13.5 cm, 23 lines to a page (cat. vol. 1, p. 88 no. 312) [given the scanty information provided in the catalogue it is not certain whether the manuscript indeed contains this correspondence].

Ākhūnd 1187/67, ff. 290b–292b (cat. p. 1480).

Ayasofya 4862, ff. 269b–271b, copied in the 8th/14th century (cf. *Nawādir*, vol. 1, p. 166) .

Majlis 630/2, pp. 9–11 (cat. vol. 2, pp. 388–89).

Majlis-i Sinā 286/5, ff. 39b–41b, part of a *majmūʿa* written in 738/ 1337–38 by ʿAlī b. al-Ḥasan b. al-Raḍī al-Ḥāfiẓ upon the request of Shams al-Dīn Āwī in Sulṭāniyya (incomplete at the end) (cat. vol. 1, p. 146, where Naṣīr al-Dīn al-Ṭūsī is erroneously mentioned as correspondent instead of Kātibī).
[Microfilm copy: Dānishgāh microfilm 1436].

Pertev Paşa 617, ff. 173a–176b, copied in the 8th/14th century (cf. *Nawādir*, vol. 1, p. 166).

Ragıp Paşa 1461, ff. 276b–278b, 12.8 × 18 cm, 21 lines to a page, copied in the 8th/14th century (cf. *Nawādir*, vol. 1, p. 166).

*Edition*

Chapter Three, Text Six

1.5.4. *Correspondence with Najm al-Dīn al-Kātibī II*

Jalāl al-Dīn al-Dawānī mentions in his second supercommentary on Qūshchī's commentary on Naṣīr al-Dīn al-Ṭūsī's *Tajrīd al-ʿaqāʾid* (Āstān-i Quds 19827, ff. 195b:21–196a:8) and later on again in his *Nihāyat al-kalām fī ḥall shubhat 'kull kalāmī kādhib'* that Ibn Kammūna corresponded with Kātibī on the liar paradox and quotes from the correspondence (ed. A. Farāmarz Qarāmalikī in *Nāme-ye Mufīd* 2 i (1375/1996), pp. 113:21–114:6). Nothing of this correspondence is preserved apart from Dawānī's quotation. In the following, the entire quote from Dawānī's supercommentary is given. Variants of the more concise quotation in his *Nihāyat al-kalām* are indicated in square brackets:

ومنها ما كتب [كتبه] ابن كمونة في جواب الكاتبي حين كتب استفساراً عن هذا الإشكال [استفساراً عن هذا الإشكال: استفسر ذلك وهو قوله]: أقول: لا نسلم أنه إما أن يكون كلامه في هذه الساعة كاذباً أو صادقاً والحصر ممنوع. فإن قيل: هذا خبر وكل خبر لا يخلو عنها، إذ بذلك يمتاز التركيب الخبري عن باقي [سائر] التراكيب، قلنا: لا نسلم أن امتيازه عن غيره بذلك، بل [بل: + امتيازه] بأنه يحتمل أن يحكم عليه بأنه صادق أو كاذب، والحكم بذلك لا ينافي أن لا يكون صادقاً ولا كاذباً. وليس من شرط الحكم المذكور أن يكون صادقاً أو كاذباً. فالحاصل أن احتمال الحكم بأنه صادق أو كاذب لا ينافي أن لا يكون في نفسه أحدهما. هذا ما سنح لي وأسأل أن ينظر فيه مولانا حرسه الله تعالى.

واعترض عليه بأن ما استدل به على [واعترض... على: واعترض عليه بأن حصر الموضوع الموجود في الملكة وعدمها أمر مشهور عند الجمهور. فلما سلم أن هذا الكلام خبر فمنع حصره في الصادق والكاذب ما لم يتبين وجه ذلك غير مسموع. وما ذكره في بين] خلو هذا الخبر عن الصدق والكذب يجري في سائر الأخبار بأن يقال: احتمال الحكم بأنه صادق أو كاذب لا ينافي أن لا يكون في نفسه أحدهما، فلو اقتضى ذلك أن لا يكون هذا الخبر صادقاً ولا كاذباً، لاقتضى أن لا يكون شيء من الأخبار صادقاً ولا كاذباً لاستواء الكل في ذلك. وأيضاً لو استدل بعدم خلو الخبر عن الصدق والكذب بما ذكره المجيب وعليه ما أورده وللمستدل أن يقول عليه بأن طرف هذا الخبر الموجود المحقق أمران معقولان حكم بالاتحاد بينهما فإن تحقق بينهما الاتحاد في نفس الأمر كان الحكم صادقاً وإلا كان كاذباً ولا يمكن الخلو عن الاتحاد وعدم فإنهما نقيضان معه لا يتمشي الجواب المذكور كما لا يخفى [وأيضاً... اليخفي: -].

### 1.5.5. *Correspondence with Kamāl al-Dīn al-Baḥrānī*

The addressee in this piece of correspondence is Kamāl al-Dīn al-Baḥrānī, who is presumably to be identified as Maytham b. Maytham al-Baḥrānī (d. 699/1300). In three manuscripts, this correspondence is the first of a number of *fawāʾid*, mostly on logic, formulated by various scholars who cannot be identified in every case. Reference is made to Kātibī, Shams al-Dīn al-Samarqandī (d. 702/1303), Nakhjawānī, and Sirāj al-Dīn al-Urmawī (d. 682/1283–84). This is followed by a section on *mughālaṭāt*, some of which are ascribed to Kātibī and contain a reference to the ʿAllāma al-Ḥillī. The eulogy following the latter's name indicates that the text was written after his death in 726/1325. Āghā Buzurg, who has seen Majlis 3850 [current shelfmark: Majlis 1830] containing

Ibn Kammūna's exchange with Baḥrānī, together with the other *fawāʾid* and *mughālaṭāt*, has no doubt that all are by Ibn Kammūna: "*al-Fawāʾid li-ʿIzz al-Dawla Ibn Kammuna, fawāʾid mutafarriqa wa-fīhi al-mughālaṭāt lahū*" (*Dharīʿa*, vol. 16, p. 318 no. 1478). Mīrzā Muḥammad Ṭāhir Tunkābunī (d. 1320/1941–42), the previous owner of this codex (see cat. vol. 5, pp. 270–73), attributed thc last part of the text to Kātibī. It is, however, not clear where he suggested "the last part" should start.

The manuscripts do not support the conclusion that Ibn Kammūna collected all these notes together, or even wrote them, and it is certain only that the first epistle addressed to Baḥrānī was composed by him.

(*Dharīʿa*, vol. 16, p. 318 no. 1478)

Beginning of *fāʾidat Ibn Kammūna*:—

فائدة لعز الدولة ابن كمونة: العلة التامة للمجموع المركب من أفرادٍ كل واحد منها ممكن، لا بد وأن تكون علة لكل واحد من تلك الأفراد . . .

Beginning of Baḥrānī's reply:—

أجاب عنه العلامة المحقق كمال الدين البحراني رحمه الله بأن الذي يلزم من جواز اجتماع وجود العلة التامة مع وجود ذلك الغير المستلزم لإمكان وجوده المستلزم لإمكان عدمه . . .

*Manuscripts*

Ākhūnd 1187/76, ff. 348b:8–349a:1 (cat. pp. 1481–82).

Majlis 1830/13 pp. 56–57; copied apparently in 1058/1648–49 by Muḥammad b. Ḥusayn al-ʿAqīlī al-Astarābādī (cat. vol. 5, pp. 271–72).

Ragıp Paşa 1461, ff. 346b:7–347a:2, 12.8 × 18 cm, 21 lines to a page.

Ṣadr al-Dīn al-Maḥallātī 17/6, f. 45a, 9.5 × 16.8 cm, copied in 1056/1646–47. The manuscript contains only the *fāʾida* of Ibn Kammūna, followed by the reply (*jawāb*) of Baḥrānī, but none of the other *fawāʾid* that follow thereupon in all other three manuscripts. The codex was given to the Khānqāh-ī Aḥmadiyya Library in Shiraz, the current shelfmark being no. 695 (cat. p. 273).

*Edition*

Chapter Three, Text Seven

1.5.6. *Correspondence with Fakhr al-Dīn al-Kāshī*

Extant in a single manuscript is Ibn Kammūna's correspondence with Fakhr al-Dīn al-Kāshī.

Beginning (Esad Efendi 3642/8, f. 168a:3–14):—

استدلت الحكماء على إثبات العقل الفعال بالذهول والنسيان العارضين للنفس حالتي إمكان ملاحظتها للمعقولات التي غائب عنها من غيرٍ تجسم كسب وافتقار ملاحظتها باكتساب جديد كما في الأول لاحتياجها في الملاحظة شيئاً غيرها حافظاً لتلك الصور التي تكون موجودة فيه بالفعل حالة الذهول، غير موجودة حالة النسيان لامتناع اتحادهما. وإلا لزم عدم الذهول أو الاحتياج بتجدد الكسب واستحالة تجزئتها بجزئين أحدهما حافظ والآخر مدرك كما قالوه، لحصل منهما موجود مفارق يرتسم بجميع الصور الموجودة بالفعل غير جسم وجسماني ونفس وهو المسمى بالعقل الفعال. وإذا ثبت هذا فنقول: لا شك أن النفس...

End (Esad Efendi 3642/8, f. 169a:15–18):—

والأعم منهما هو ما في نفس الأمر، والأخصِ هو الذي في الخارج على هذا الاصطلاح. وهاهنا احتمالات أخرى يندفع بها التشكك أيضاً لا حاجة إلى ذكرها. والله الهادي بمنّه وجوده ورحمته.

*Manuscripts*

Esad Efendi 3642/8, ff. 168a–169a, 11.5 × 19.4 cm, 18 lines to a page [*Su'āl sa'alahū Fakhr al-Dīn al-Kāshī min al-'Allāma Ibn Kammūna*]. The copy of the tract itself is not dated, but some other writings in the codex written by a single hand have dated colophons (f. 134b: 28 Rabī' I 752/25 May 1351; f. 153a: 3 Sha'bān 753/13 September 1352 *bi-baldat* Ḥilla) (cat. 1885, p. 265).

*Edition*

Chapter Three, Text Eight

### 1.5.7. *Correspondence with Ibn al-Fuwaṭī*

Ibn al-Fuwaṭī writes in his *Majmaʿ al-ādāb fī muʿjam al-alqāb* (ed. M. Kāẓim, vol. 1, pp. 190–91; transl. Nemoy: "New Data," p. 508): "...He was sought by many people who wished to profit by his instruction. It so happened that I myself was prevented from meeting him personally by an illness which overtook me, but I wrote him requesting him to let me have a token of his instruction, so that I might embellish my work with it. He wrote in reply, through my colleague and friend Shams al-Dīn Muḥammad b. Abī al-Rabīʿ, the calculator, known as al-Ḥashaf, in the year 683:

Keep knowledge always away from ignorant people,
And do not entrust it to him who is unfit for it.
He will only use it for violent, hateful, and evil purposes,
And his lack of intelligence will turn it into ignorance.
Always guard it zealously from him,
And seek not excellence from him who has none to start with..."

*Editions*

Ibn al-Fuwaṭī: *Talkhīṣ majmaʿ al-ādāb fī muʿjam al-alqāb*. Ed. Muṣṭafā Jawād. Damascus 1962, vol. 4 i, pp. 159–61 no. 189.

Ibn al-Fuwaṭī: *Majmaʿ al-ādāb fī muʿjam al-alqāb* 1–6. Ed. Muḥammad al-Kāẓim. Tehran 1416/1995–96, vol. 1, pp. 190–91.

## 1.6. *Poetry*

### 1.6.1. *Poetry I*

*Manuscripts*

Majlis 630/3, p. 20. This piece of poetry is not mentioned in the catalogue (cat. vol. 2, pp. 388–89), yet was identified by Dānishpazhūh in the preface (*dībāchah*) to his edition of Naṣīr al-Dīn al-Ṭūsī's *Akhlāq-i muḥtashamī* (Tehran 1377/1998, p. *bīst u nuh*).

Mahdawī 591 *bayāḍ*. Facsimile edited as *Jung-i Mahdawī. Tārīkh-i kitābat az 753 qamarī be ba'd. Chāp-i 'aksī az rū-yi nuskha-yi khaṭṭī-yi Kitābkhāna-yi Duktur Aṣghar Mahdawī*. Tehran: Iran University Press, 1380/2002, p. 193.

*Edition*

Chapter Three, Text Nine

1.6.2. *Poetry II*

Another piece of poetry is contained in Ibn Kammūna's reply to Ibn al-Fuwaṭī, see above, Section 1.5.7.

## 2. *Works of Uncertain Authenticity*

### 2.1. *Ḥall ishkālāt al-Ishārāt*

The treatise which is extant in a single manuscript (Vazīrī 2377/5, ff. 49–77) was copied in Shaʿbān 956/August–September 1549 and consists of two unconnected parts. The unconventional beginning of each part suggests that both are extracts from two different texts. The text itself contains no indication either of its author or title. The only indication that Ibn Kammūna is its author is the table of contents in the codex (f. 1b), which also gives the title which clearly refers to the second part only. That the title is in Persian (*Ḥall-i ishkālāt-i Ishārāt*) suggests that it was added later on.

Part One (ff. 49b–52b:5) contains two sections; the first, which seems incomplete in the beginning, deals with the Aristotelian proof of the existence of God on the basis of cause and effect.

Beginning of Part One (f. 49b:1–5):—

بسم الله الرحمن الرحيم كل جملة مركبة من علل ومعلولات، فلا بد وأن تنتهي إلى طرف هو علة ليس بمعلول، لأن تلك الجملة إما أن تكون متناهية أو غير متناهية، والقسم الآخر قد أبطل في الطبيعيات حيث ذكر أن كل مقدار أو عدد ذي ترتيب بالطبع أو الوضع موجود معاً فلا بد وأن يكون متناهياً وفرض الخط المعلم عليه وأطبق بالوهم خط على جزء من ذلك الخط على ما هو مذكور في موضعه. والقول في الأعداد هكذا . . .

The second section (*faṣl*) deals with God's knowledge; it starts as follows (f. 50b:12–13):—

فصل من كلامه أيضاً اعلم أن المعلوم هو ليس الصورة الموجودة من خارج وجوداً عينياً لأنه لو كان كذلك في علمه تعالى لكان كل موجود وجوداً عينياً معلوماً لنا وهذا التالي محال . . .

End of Part One (f. 52b:2–5):—

الهمّ أنت المرجوع إليه والمعول عليه في تيسير هذا الأمر العظيم والإنزال في هذا المنزل المبارك الكريم وأمر العاملين من عبادك إلى محل الشوق إلى مثل ذلك العالم والمشتاقين منهم إلى مرتبة العشق إنك أنت الرحيم الرؤف الكريم وصلى الله على محمد وآله أجمعين.

Part Two (ff. 52b:6–77a) contains commentaries on twelve remarks from Ibn Sīnā's *Ishārāt*, and one on a remark from his *al-Mabdaʾ wa-l-maʿād*. The author refers to Ibn Sīnā as *ṣāḥib Kitāb al-Ishārāt* or *ṣāḥib al-kitāb*. Each commentary is introduced with a complete or partial quote of the respective words of Ibn Sīnā. None of the material contained in Part Two is taken from Ibn Kammūna's commentary on Ibn Sīnā's *Ishārāt* (see above, Section 1.1.1), and the overall style of the entire text is different from Ibn Kammūna's usual style.

The twelve remarks from the *Ishārāt* commented on are taken exclusively from the second part of *Ishārāt* on philosophy, to the exclusion of the first part on logic. Moreover, none of the remarks of *namaṭ* three (psychology) and *namaṭ* nine and ten (mysticism) are commented on. With the exception of the first, the remarks are introduced as *masʾala*. The text seems complete at the end, as the last remark is entitled *al-masʾala al-akhīra* (f. 73b). The remarks are arranged, as the author himself asserts (f. 64a:12–15), in an order that is different from the order that they appear in the *Ishārāt*.

The following remarks of Ibn Sīnā are commented on in Part Two of the text (reference is given to Mujtabā al-Zāraʿī's edition of Ibn Sīnā's *Ishārāt* (Qum 1381/2002) with the respective *namaṭ* and *ishāra / tanbīh* indicated in square brackets:

[1] (52ب:6–15) متن كلامه في وحدانية الإله الواجب الوجود جل وعلا اعلم أن الكثرة منفية عن ذات واجب الوجود على وجهين أحدهما أن يكون للذات مثل أن يوضع الصفات أجزاء الذات فيكون الذات ... من أجزاء ومثل ما أن يوضع الوجود عاماً ثم يقترن به الوجوب اقتران الفصل الجنس فيكون الذات مركبة من الجنس والفصل وأشار ذلك ما هو مناف لوجوب الوجود فإن المركب من الأجزاء هو معلول الأجزاء والمركب من الأجزاء هو معلول الأجزاء من المركب من الجنس والفصل هذا حاله وقد بين هذا في موضعه بياناً شافياً والوجه الثاني أن يكون ذاتان كل واحد منهما واجب الوجود وهذا الوجه الثاني هو الذي أراد صاحب كتاب الإشارات ... في هذا الموضع فقال: واجب الوجود المتعين إن كان تعينه ذلك ... لأنه واجب الوجود فلا واجب وجود غيره (ص 270 [18/4]) ...

[2] (57أ:4) مسألة أخرى: إن الناس يشيرون إلى جهات لا تتبدل (ص 213 [1/2]) اعلم أن هذا البرهان على إثبات الجسم ...

[3] (59أ:13–14) مسألة أخرى: وأنت تعلم أن تبدّل النسبة عند المتحرك قد يكون للساكن والمتحرك فيجب أن يكون عند ساكن (ص 221 [14/2]) لما علمت أن الجهات المختلفة ...

[4] (60ب:4–5) مسألة أخرى: قوله: الحادث بعد ما لم يكن له قبل لم يكن فيه (ص 282 [4/5]) اعلم أن هذه الإشارة فيها مسألتان أحدهما ...

[5] (61ب:9–10) مسألة أخرى: ولأن التجدّد لا يمكن إلا مع تغير حال إلى آخره (ص 283 [5/5]) هو إثبات الحركة بواسطة الزمان وإثبات المتحرك الذي هو الجسم الموضوع للحركة بواسطة الحركة ...

[6] (64أ:15–64ب:1) المسألة: قوله: الجسم ينتهي ببسيطه وهو قطعه إلى آخر الإشارة (ص 204 [28/1]) اعلم أنه تبين من هذه الإشارة أن الطول والعرض والعمق التي هي الأبعاد الثلاثة المحسوسة ليست مأخوذة في حد الجسم ...

[7] (66أ:4–6) المسألة الثانية: قوله: لما كانت الجهة مما تقع نحوها الحركة لم يكن من المعقولات التي لا وضع لها فيجب أن يكون الجهات لوضعها تتناول الإشارة (ص 207 [33/1]) اعلم أنه قد بين وجود ...

[8] (66أ:11–12) المسألة الثالثة: وأحكامها أنه لما كانت الجهة ذات وضع إلى آخر الإشارة (ص 208 [34/1]) لما تبين حقيقة الأمر فيه وهو كونها حد الحركة وهذا أيضاً . . .

[9] (67أ:10–11) المسألة الرابعة: قوله: لو كان كل موجود بحيث يدخل في الحس والوهم إلى آخرها (ص 264 [3/4]) هذه المسألة واضحة المعنى وقد تبين منه أن الموجود ينقسم إلى محسوس ومعقول . . .

[10] (68أ:11–68ب:2) المسألة الخامسة: كل أشياء تختلف بأعيانها وتتفق في أمر مقوم لها فإما أن يكون ما تتفق فيه لازماً من لوازم ما تختلف به فيكون للمختلفات لازم واحد وهذا غير منكر، وإما أن يكون ما تختلف به لازماً لما تتفق فيه فيكون الذي يلزم الواحد مختلفاً متقابلاً. وهذا منكر، وإما أن يكون ما تتفق فيه عارضاً عرض لما تختلف به. وهذا غير منكر، وإما أن يكون ما تختلف به عارضاً عرض لما تتفق فيه. وهذا أيضاً غير منكر (ص 269–270 [16/4]) اعلم أن منها أربعة أقسام من حقائق الأشياء ولوازمها وعوارضها والنسب التي يحدث منها أحدها . . .

[11] (69ب:4–6) المسألة السادسة يجب علينا أن نحلل معنى قولنا فَعل وصَنع وأَوجد إلى الأجزاء البسيطة من مفهومها إلى آخره (ص 280 [2/5]) هذا الكلام بيان منه أن المعلوم يحتاج إلى علته في وجوده لا في حدوثه . . .

[12] (71ب:7–10) المسألة التي في آخر النمط فهو مذاهب الفرق في البدؤ وأما مذاهب الفرق فلا حاجة إليها . . . ذكره المصنف في الكتاب عنه إلا أنا نورد شبه المذهب الأخير ... فإن صاحب الكتاب شرح من جملة ذلك المذهب . . . (ص 288 [12/5])

[13] (73ب:3–) المسألة الأخيرة التي من غير هذا الكتاب وهي من كتاب المبدأ والمعاد هو أن التحرك لا بد له من محرك والمحرك غير والمتحرك غير

End of Part Two (f. 77a:4–8):—

. . . فتبين من جميع ما مضى أن الجسمية في الوجود يلحقها العلل المحركة من خارج فلهذا صار الجسم متحركاً دون الآخر ولم يكن هكذا بل لو كان المحرك من الداخلات في الجسمية والمقومات لها لكان كل جسم متحركاً وما جعلناه ملتحقاً بالجنس الذي هو البسائط فهو أمر يعرضه العقل من خارج فافهم هذا الفصل جيداً فهو أصل عظيم الجدوى والصلاة والسلام على محمد وآله. تم.

Colophon (f. 77a:9–13):—

قيل: لما كان صاحب شرح هذه المسائل في حالة الذي نُزع منه ثوب الحياة سأله بعض من حضر عن المكان الذي يدفن فيه، فقال: إذا خرجت أنا من البين فلا بد لمن في داره جيفة أن ينقلها عن داره. تم في شعبان ٩٦٥.

*Manuscript*

Vazīrī 2377/5, ff. 49–77, 13 × 19 cm, 16 lines to a page; copied in Shaʿbān 965/May–June 1558 (cat. vol. 4, p. 1269).
[Microfilm copy: Āstān-i Quds microfilm 14473].

2.2. *Min taʿālīq al-Ḥakīm ʿIzz al-Dawla Abi l-Riḍā Ibn Kammūna (Taʿrīfāt)*

The text, extant in a single manuscript (Esad Efendi 3642/7), lacks both conventional beginning and end and may consist of notes from a larger text—there is no intelligible context to connect the respective terms and examples discussed. The only indication that Ibn Kammūna is the author is to be found in the heading (f. 166a): *min taʿālīq al-Ḥakīm ʿIzz al-Dawla Abi l-Riḍā Ibn Kammūna*. An additional heading *taʿrīfāt* was added on the upper margin of the text (f. 166a) by a different, certainly later hand; the same hand has added titles to various writings throughout the codex; some of them are inaccurate, which suggests that the title given to this tract should not be relied upon. The title "*taʿrīfāt Ibn Kammūna*" is also to be found on the title page of the codex (f. Ia) by a third, different hand. The copy of the tract itself is not dated, but some other writings in the codex which are written in a single hand have dated colophons (f. 134b: 28 Rabīʿ I 752/25 May 1351; f. 153a: 3 Shaʿbān 753/13 September 1352 *bi-baldat* Ḥilla). It is possible that these notes were taken from some of Ibn Kammūna's glosses; at least, there are similarities in style with his *Taqrīb al-maḥajja*.

Beginning (Esad Efendi 3642/7, f. 166a:3–5):—

وجه تقسيم الألفاظ اللفظ إما أن يكون متكثراً أو متحداً والأول إن تكثر مفهومه فهو المتباين كالسماء والأرض . . .

End (Esad Efendi 3642/7, f. 167b:10–13):—

...والصدر لغةً مقدم المجلسِ واصطلاحاً المرجوع إليه في الأمور المستاذن فيها النظام لغةً ما ينظم فيه وما يُنظمُ اصطلاحاً الإنسان الذي ينتظم به أمور الملك.

*Manuscript*

Esad Efendi 3642/7, ff. 166a–167b, 11.5 × 19.4 cm, 18 lines to a page (cat. 1885, p. 265).

*Edition*

Chapter Three, Text Four

2.3. *Masāʾil mutafarriqa wa-fawāʾid mushattata*

This collection of various *suʾālāt*, *nukat* and *fawāʾid* on various logical problems, extant in a single manuscript (Ragıp Paşa 1481, ff. 283b–290a), was written, according to Ramazan Şeşen, by Ibn Kammūna (*Mukhtārāt*, p. 144). The manuscript itself contains no evidence that would support this. Şeşen may have been led to his conclusion because the preceding item in the same codex (ff. 278a–290a) is Ibn Kammūna's *Maqāla fī anna wujūd al-nafs abadī wa-baqāʾahā sarmadī* (see above, Section 1.3.10).

Beginning (Ragıp Paşa 1481, ff. 283b:21–284a:1):—

مسائل متفرقة وفوائد مشتتة سؤال منطقي قبل التقابل بين اللفظ المفرد والمركب تقابل العدم والملكة وذلك باطل لأنه يلزم منه أن لا يركب المركب من المفرد وغيره وإلا لزم وجود الكل مع عدم الجزء...

End (Ragıp Paşa 1481, ff. 289b:20–290a:2):—

لا يقال: إنا لو فرضنا جميع الأذهان الإنسانية معدومة أو ذاهلة (؟) عن ذلك الشيء المتحرك والطرفين فإنا نعلم بالضرورة تحقيق التوسط لذلك الشيء فلو كان أمراً اعتبارياً لا يعدم بانعدام الاعتبار لأنا نقول: إن من الأمور ما يختص محققه بنفس الأمر كاا[...] والقضاياء النفيسة (؟) وليست بموجودة في الخارج ولا أمور اعتبارية فلم لا يكون التوسط من هذا السبيل فائدة.

*Manuscript*

Ragıp Paşa 1481, ff. 283b–290a, copied in the 8th/14th century (cf. *Mukhtārāt*, p. 144; *Nawādir*, vol. 1, p. 166).

## 2.4. *Kitāb fī l-manṭiq wa-l-ṭabīʿa wa-mā baʿdahā (Kitāb fī l-ḥikma)*

An extensive treatise on philosophy, composed following the request by *baʿḍ al-ṭalaba al-mustarshidīn*. The authorship of the work as one by Ibn Kammūna is questionable. Neither introduction, colophon nor the text itself contain any indication that it is by Ibn Kammūna. All three extant copies of the work are contained in codices with other works by Ibn Kammūna. The earliest of the three known manuscript copies (completed in 720/1320) is Ayasofya 2446, which also contains Ibn Kammūna's *al-Jadīd fī l-ḥikma* (ff. 1–176b) and his *Taqrīb al-maḥajja* (ff. 293b–308a). On the leaf prior to f. 1a in reference to the contents of the entire *majmūʿa* appears: *kitābān li-Ibn Kammūna fī l-ḥikmiyyāt al-falsafiyya*...). On f. 1b is indicated by a later hand: *al-Kāshif*...*li-Ibn Kammūna*, which is correct, as this treatise indeed starts on the same folio. On f. 177a is written in the same later hand: *al-Ḥikma al-jadīda li-Ibn Kammūna*, which is correct if it is supposed to refer back to the treatise ending on f. 176b (*al-Kāshif* and *al-Ḥikma al-jadīda* being alternative titles for *al-Jadīd fī l-ḥikma*). The manuscript contains no indication of the title or the author of the work under consideration.—The other two extant manuscripts of the work, which Ritter tentatively dates to the 9th/15th (Ayasofya 2447) and 11th/17th century (Esad Efendi 1933) respectively, have title pages: in Ayasofya 2447 on f. 3a: *Kitāb fī l-ḥikma taʾlīf al-faylasūf al-kāmil al-fāḍil ʿIzz al-Dawla Saʿd b. Manṣūr Ibn Kammūna*...and in Esad Efendi 1933 on f. 1a: *hādhā l-Kitāb fī l-ḥikma taʾlīf al-ḥakīm al-fāḍil wa-l-faylasūf al-kāmil ʿIzz al-Dawla Saʿd b. Manṣūr b. Kammūna*.... In Esad Efendi 1933, the treatise is also mentioned with Ibn Kammūna as author on the table of contents prior to f. 1a.

On the basis of a mistake that is to be found in all three manuscripts (cf. Ayasofya 2446/2, f. 214a, Ayasofya 2447/1, f. 25b, Esad Efendi 1933/1, f. 17a where it says *al-qāʿida al-rābiʿa* instead of *al-qāʿida al-sābiʿa*, as would be required by the context), it seems safe to assume that all three copies belong to the same textual transmission, although it is unlikely that Esad Efendi 1933/1 and Ayasofya 2447/1 were copied from Ayasofya

2446/2: two diagrams, included in Esad Efendi 1933/1 and Ayasofya 2447/1, are missing in Ayasofya 2446/2 (ff. 233b, 254b); moreover, each of the three sections of the text—logic, physics, metaphysics—is preceded in Ayasofya 2446/2 by the *basmallah*—something that is not found in the two other manuscripts. It is thus more likely that all three manuscripts were either copied from the same source or go back to the same archetype. The fact that the earliest of the manuscripts lacks any indication of the title that is found in the later copies, suggests that this was also the case with their respective source or archetype.

There are other formal indications arguing against authorship by Ibn Kammūna. One of the characteristics of his systematic independent philosophical treatises is that they are arranged in a strictly even manner. *Al-Jadīd fī l-ḥikma* consists of seven *abwāb* each containing seven *fuṣūl*; *al-Maṭālib al-muhimma* consists of seven *fuṣūl* each containing seven *abḥāth*; *Kalimāt wajīza* consists of two *jumal*, each containing two *abwāb* consisting of five *fuṣūl* each; in none of these treatises does he arrange the material according to the tripartite scheme logic-physics-metaphysics. The text structure of the *Kitāb fī l-ḥikma* is strikingly different; it comprises the three disciplines of knowledge, viz. logic, physics and metaphysics, and even within these the work does not reveal a comparably even division. *Al-ʿilm al-awwal fī l-manṭiq* is subdivided into 6 *maqālāt*; *maqāla* 1 comprises 10 *qāʿida*, *maqāla* 2: 2 *qāʿida*, *maqāla* 3: 2 *qānūn* (*qānūn* 1 comprising 8 *qāʿida*, *qānūn* 2: 4 *qāʿida*), *maqāla* 4: 14 *qāʿida*, *maqāla* 5: no further subdivision, *maqāla* 6: 2 *fann* (*fann* 1 comprising 5 *qāʿida*, *fann* 2: no further subdivision). *Al-ʿilm al-thānī wa-huwa al-ʿilm al-ṭabīʿī* is subdivided into 2 *maqālāt*; *maqāla* 1: no further subdivision, *maqāla* 2 comprises 5 *fann*, *fann* 1: 6 *qāʿida*, *fann* 2 & 3: no further subdivision, *fann* 4: 3 *qāʿida*, *fann* 5: 3 *qāʿida*. *Al-ʿilm al-ilāhī* is subdivided into 5 *maqālāt*, *maqāla* 1: 13 *qāʿida*, *maqāla* 2: 3 *qāʿida*, *maqāla* 3: 7 *qāʿida*, *maqāla* 4 & 5: no further subdivision.

Moreover, there are doctrinal reasons for assuming that the work is not one of Ibn Kammūna's writings. In his discussion on the soul, the author of the work maintains that the soul has no existence prior to the body; he moreover rejects the notion of *tanāsukh* (Ayasofya 2446, ff. 261bff; Ayasofya 2447, ff. 58bff; Esad Efendi 1933, ff. 36bff). Both doctrines contradict the doctrinal position of Ibn Kammūna, who upheld the pre-eternity of the soul in all his authenticated writings and had a more favourable view of metempsychosis.

Mīnuwī, who inspected one manuscript of the work (Esad Efendi 1933) in 1956, also seriously doubted the authenticity of this treatise as a work by Ibn Kammūna. He specifically pointed out that the anonymous author refers by way of example to the river Nile, which is a highly unlikely thing for a writer from Baghdad to do. He assumes that the work was written in the 4th/10th century (Mujtabā Mīnuwī: "Az khazāʾin-i Turkiyah." In *Mīnuwī bar gustara-yi adabiyyāt-i fārsī. Majmūʿa-yi maqālāt*. Ed. Māhmunīr Mīnuwī. Tehran 1381/200, pp. 231–33].

Beginning (Ayasofya 2446/2, f. 177b:1–11):—

أما بعد فالحمد لواجب الوحدة والوجود المنزه عن الأبعاض والجهات والحدود والصلاة على صاحب الفضل المشهور وسلم وبعد فقد سألني بعض الطلبة المسترشدين أن أجمع لهم كتاباً حاوياً على العلوم الثلاثة أعني المنطق والطبيعة وما بعده بحيث لا يكون اختصاره مفضياً إلى الإهمال ولا تطويله مؤدياً إلى الإملال فأجبتهم إلى ما طلبوه وأسعفتهم بما أملوه راجياً في ذلك حسن الفوز في المعاد والغبطة عند قيام الاشهاد وهو المسؤول أن يعصمنا فيما تحاوله منه بالرفه أو عاد وأن يبلغنا غاية المنية والمراد في العلم الأول المنطق يشتمل على تمهيد وستة مقالات . . .

End (Ayasofya 2446/2, ff. 293b:8–294a:3):—

. . . ثم المعاد وما يشتمل عليه من الشقاوة والسعادة على حسب ما يفهمونه ويقرب لهم ذلك بضرب الأمثال التي يعرفونها والأشياء التي ألفوها ولا يأتيهم بما يوقعهم في الحيرة والاختلاف والأعظم الفساد وكثر العناد ثم يبين لهم بعد ذلك ما يتعلق بالطاعات والعبادات الدينية ثم ما يتعلق بالمعاملات ويأمرهم بالمعروف وينهيهم عن المنكر وليس لهم ما يوجب لهم التذكار على الدوام لأن لا يتناسبوا وأن يبين لهم أن ذلك من عند الله لأن يتلقوه بالقبول وأن يفعل ما يتمم به الخيرات وتكمل لهم السعادات عند خالق البريات وفيما ذكرنا من التنبيهات غنية وكفاية لمن أيده الله بالهداية.

*Manuscripts*

Ayasofya 2446/2, ff. 177b–294a, 12.5 × 18 cm, 13 lines to a page; copy completed on 20 Jumādā I 720/28 June 1320 in Ḥalab (cat. p. 363).

Ayasofya 2447/1, ff. 1–77b, 14 × 24.3 cm, 17 lines to a page. The title page has *Kitāb fī l-ḥikma ta'līf al-faylasūf al-kāmil al-fāḍil 'Izz al-Dawla Sa'd b. Manṣūr Ibn Kammūna*. Copied in 9th/15th century (Ritter II, p. 58).

Esad Efendi 1933/1, ff. 1–49a, 16.3 × 25.2 cm, 19 lines to a page. The title page has *Kitāb fī l-ḥikma ta'līf al-faylasūf al-kamil al-fāḍil 'Izz al-Dawla Sa'd b. Manṣūr Ibn Kammūna*. Above the title on f. 1a is written: *Majmū'a min rasā'il fī l-ḥikmiyyāt fī awwalihā risālatān li-Ibn Kammūna*. Copied in 11th/17th century (Ritter II, p. 61; cat. Ritter/Uppsala, p. 65).

[Microfilm copies: AUB 15/1 (cat. 1985, p. 493); Dānishgāh microfilm 535 (cat. 1969–84, vol. 1, p. 496); fotograph copy: Mīnuwī 294 (cat. p. 138)].

### 2.5. *al-Kāfī al-kabīr fī l-kuḥl*

A work on ophthalmology ascribed to Ibn Kammūna by Ṣadaqa b. Ibrāhīm al-Miṣrī al-Ḥanafī al-Shādhilī (8th/14th or 9th/15th century) in his *al-'Umda al-kuḥliyya fī l-amrād al-baṣariyya*. Shādhilī refers to Ibn Kammūna as "Sa'd Ibn Kammūna *ṣāḥib al-Kāfī al-kabīr*", "Ibn Kammūna *jāmi' al-Kitāb al-Kāfī*", "*ṣāḥib al-Kāfī al-kabīr*", "*ṣāḥib Kitāb al-Kāfī*", and "Sa'd Ibn Kammūna". His quotes from the *Kitāb al-Kāfī*, which is apparently not extant, show that Ibn Kammūna treated a number of cases in Ḥalab.[14] Whether the author of the *Kitāb al-Kāfī* was Sa'd b. Manṣūr Ibn Kammūna cannot be conclusively decided on the sole basis of Shādhilī's testimony. However, since there is strong evidence that Ibn Kammūna spent most of his professional life in Baghdad, it is unlikely that he treated patients in Ḥalab at any stage of his life. Moreover, there is no other evidence confirming that Ibn Kammūna was indeed

[14] The possibility that the *Kitāb al-Kāfī* is extant yet so far has not been identified should not be ruled out. The numerous quotations by Shādhilī from this work might help to identify it. See Ṣadaqa b. Ibrāhīm al-Shādhilī: *al-'Umda al-kuḥliyya fī l-amrāḍ al-baṣariyya*. Ms. Munich 834, ff. 34a:16ff, 55b:10ff, 55b:17ff, 59b:2ff, 75a:21ff, 77b:16ff, 80b:10ff, 141b:4ff, 148b:4. For the respective references in the manuscript copy of the text in St. Petersburg, see Victor Rosen: *Les manuscrits arabes de l'Institut des Langues Orientales*. St. Petersburg 1877 (reprinted Amsterdam 1971), pp. 101–2.

a physician. On the other hand, there are other likely candidates for the "Ibn Kammūna" or "Saʿd Ibn Kammūna" mentioned by Shādhilī. One possibility would be Ibn Kammūna's grandfather who was also called "Saʿd". Moreover, we know of a contemporary of Ibn Kammūna by the name of Abu l-Qāsim Saʿīd b. Maʿālī b. Futūḥ Ibn Kammūna[15] who might also qualify.

(Rosen: *Manuscrits arabes*, p. 101; Steinschneider: *Arabische Literatur*, p. 239 no. 2; GALS, vol. 1, p. 769; Ullmann: *Medizin*, pp. 213–14)

### 2.6. *al-Tadhkira fī l-Kīmiyā*

The title is included by modern scholars among the works of Ibn Kammūna on the basis of the information provided by Ḥājjī Khalīfa. However, the authenticity of the work as one by Ibn Kammūna is doubtful. One of Ḥājjī Khalīfa's sources presumably was (directly or indirectly) Shihāb al-Dīn al-Qalqashandī's (d. 821/1418) *Ṣubḥ al-aʿshā fī ṣināʿat al-inshāʾ*, completed in 814/1412. Qalqashandī lists the work under the heading *amthal kutub al-islāmiyyīn* within the discipline of *kīmiyā* and does not offer any specifics as to the name of its author, "Ibn Kammūna".[16] Since it is well known that the names "Kammūna" and "Ibn Kammūna" were relatively frequent also among Muslims (see above, Section 2.5), it seems likely on the basis of the evidence of Qalqashandī that here a different, Muslim, Ibn Kammūna is meant.

(*Kashf al-ẓunūn*, vol. 1, p. 393; MS, vol. 1, p. 346; Steinschneider: *Arabische Literatur*, p. 240 no. 6).

[15] See Nājī Maʿrūf: *Tārīkh ʿulamāʾ al-Mustanṣiriyya* 1–2. Baghdad 1379/1959, vol. 2, p. 339.—For other possible candidates, see Muḥsin Amīn: *Aʿyān al-shīʿa*, vol. 7, pp. 442–3 (Āl Kammūna).

[16] Shihāb al-Dīn al-Qalqashandī: *Ṣubḥ al-aʿshā fī ṣināʿat al-inshāʾ* 1–14. Cairo 1913–16, vol. 1, p. 475.

## CHAPTER THREE

# EDITIONS OF SELECTED WRITINGS OF IBN KAMMŪNA

## *Text 1*

*Kalimāt wajīza mushtamila ʿalā nukat laṭīfa fī l-ʿilm wa-l-ʿamal*

The following edition is based on the following manuscripts:

أ = Āstān-i Quds 855
ج = Majlis 593/1, pp. 3–40s
ف = Fâtih 3141/5, ff. 266a–286b
م = Mishkāt 861/16
ي = Marʿashī 12868, ff. 40a–59b

بسم الله الرحمن الرحيم[1]

قال العبد الفقير إلى رحمة الله تعالى[2] سعد بن منصور بن سعد بن الحسن بن هبة الله بن[3] كمونة عفا الله عنه في[4] دنياه وأخراه:

أحمد الله حمد[5] مسترشد بنور هدايته، مسترفد من جوده وعنايته، وأتوسل إليه[6] بنعوت جلاله وكمال إلاهيته، أن يهب لي توفيقاً يوجب[7] الأمان[8] من عقابه والخلود في جنته[9]، وأن يصلي على

1 الرحيم: + وبه نستعين، أ.
2 العبد الفقير . . . تعالى: الشيخ العالم الفاضل الكامل المتبحر العلامة عز الدولة، أ م.
3 بن: –، م.
4 عفا الله عنه في: بياض في ج لانهدام الورقة.
5 حمد: بياض في ج لانهدام الورقة.
6 وعنايته واتوسل إليه: بياض في ج لانهدام الورقة.
7 يوجب: لوجب، ج ف.
8 الأمان: الإيمان، أ.
9 عقابه . . . جنته: بياض في ج لانهدام الورقة.

القدّيسين من[10] ملائكته وعلى من كرمه بالنبوة والولاية من بريته[11]، خصوصاً على محمد وعترته[12]. وبعد[13]، فإنه حيث وصل الصدر الكبير العالم الفاضل المتقي البارع شمس الدولة والدين فخر[14] الإسلام والمسلمين معتمد الملك[15] محمد المؤمن القزويني أدام الله[16] أيامه، وبلغه سؤاله ومرامه[17]، إلى بغداد، التمس[18] مني عند مقاربة عوده إلى مقصده أن يصحب معه شيئاً من كلامي[19] في العقليات ليكون[20] برسم خزانة المولى[21] المعظم الصاحب الأعظم صاحب السيف[22] والقلم العالم العادل المنصف الكامل شمس الحق والدين بهاء الإسلام والمسلمين ملك الوزراء[23] في العالمين[24] صاحب ديوان الممالك دستور الشرق والغرب مفخر الأعيان، مظهر الإحسان[25]، محمد بن الصاحب السعيد بهاء الدولة والدين محمد الجويني أعزّ الله نصره وأنفذ في المنافقين[26] أمره[27]، فكتبت هذه الكلمات الوجيزة على عجلة مشتملة على نكت لطيفة في العلم والعمل[28]، ينتفع بها جماعة الطلّاب لمهمات الحكم والآداب، ترشدهم إذا استبصروا بها وتسعدهم[29] إذا استضاءوا بنورها وعملوا بموجبها والتكلان على التوفيق[30].
وقد رتبتها على جملتين[31] يشتمل كل جملة منها على بابين وكل باب على خمسة فصول. ولعلّي بعد[32] هذا[33] أشرحها شرحاً يفهم من يستفسرها[34] ويفحم من يستنكرها، وأسأل من الله تعالى

[10] على القدّيسين من: بياض في ج لانهدام الورقة.
[11] بريته: قربته، ج.
[12] والولاية... وعترته: بياض في ج لانهدام الورقة.
[13] وبعد: -، م.
[14] العالم... فخر: بياض في ج لانهدام الورقة.
[15] الملك: الملوك، أ م.
[16] الله: -، أ ج م.
[17] المؤمن... ومرامه: بياض في ج لانهدام الورقة.
[18] التمس: والتمس، ج.
[19] مقاربة... كلامي: بياض في ج لانهدام الورقة.
[20] ليكون: كمون، أ م.
[21] المولى: الملك، م.
[22] الملك... السيف: بياض في ج لانهدام الورقة.
[23] شمس... الوزراء: بياض في ج لانهدام الورقة.
[24] العالمين: العالم، ج.
[25] الشرق... الإحسان: بياض في ج لانهدام الورقة.
[26] المنافقين: الخافقين، أ ج ف.
[27] محمد الجويني... أمره: بياض في ج لانهدام الورقة.
[28] عجلة... والعمل: بياض في ج لانهدام الورقة.
[29] والأدب... وتسعدهم: بياض في ج لانهدام الورقة.
[30] والتكلان على التوفيق: بياض في ج لانهدام الورقة.
[31] وقد... جملتين: بياض في ج لانهدام الورقة.
[32] خمس... بعد: بياض في ج لانهدام الورقة.
[33] هذا: + إن شاء الله، أ ي.
[34] يفهم من يستفسرها: بياض في ج لانهدام الورقة.

أن يسهل علينا طريق الحق وأن يصرف قلوبنا[35] جميعاً عن هذه الدار الفانية إلى الدار الباقية وأن يسعدنا في الدارين، وأن لا يجعلنا من[36] المتكلفين، لأنه[37] ولي ذلك والقادر عليه، ﴿وَمَا تَوْفِيقِي إِلاَّ بِاللهِ عَلَيْهِ تَوَكَّلْتُ وَإِلَيْهِ أُنِيبُ﴾[38].

## الجملة الأولى في العلم

## الباب الأول منها في إثبات مدبر العالم الواجب وجوده ووحدانيته وبيان جملة من صفات جلاله وعنايته

### الفصل الأول[39]

تمهيد: قد تبين في العلوم الحكمية أن للنفس الإنسانية قوتين نظرية وعملية، وكمال النظرية بالعلم[40] وكمال العملية بالعمل. والعلوم والأعمال كثيرة ومتشعبة بعجز القدرة البشرية عن استقصاء فن[41] واحد منها فكيف عن جميعها، وقد جوز[42] الأوائل صنوف العلم وتقسيم قسمها[43] وتجزئة أجزائها وتوضيح سبلها وتبيين مآخذها، وكذلك فعلوا[44] في وجوه الأدب وضروب[45] الأخلاق والسياسات، فمن أراد الاطلاع على ذلك أجمع فعلية بالإمعان في كتب من تقدم منهم ومن تأخر، فإنهم لم يبقوا[46] في جليل من الأمر لقائل من بعدهم مقالاً ولا أخلوا للزيادة مجالاً، فأما ما يقال[47] من لطائف الأمور[48] إن لم يكونوا سبقوا إليه بعينه فهو مستفاد

35 وأسأل... قلوبنا: بياض في ج لانهدام الورقة.
36 وأن... من: بياض في ج لانهدام الورقة.
37 لأنه: إنه، أ م.
38 سورة هود (11): 88.
39 الفصل الأول: أ، ج ف ي.
40 بالعلم: بالعلوم، ج ي.
41 فن: في كل، أ م.
42 جوّز: قرر، ج ي.
43 قسمها: مقسمها، م.
44 فعلوا: جعلوا، ف.
45 وضروب: بياض في ج لانهدام الورقة.
46 لم يبقوا: -، م.
47 يقال: تعالى، أ م.
48 الأمور: كلمة غير مقروءة في م.

مما أسسوه لنا من الأصول التي لم نجدهم غادروا شيئاً من أمور الدنيا والآخرة إلا قرروه[49] فيها وفتحوا[50] أبواب علمه منها. وإذ لا سبيل[51] إلى[52] الإحاطة بجميع ما قيل في العلميات والعمليات، فالواجب على طالب الكمال أن يجرد عنايته لما هو الأفضل والأهم في أمر دينه ودنياه منها. وقد اتفق أرباب الديانات النقلية[53] والعقائد العقلية أن النجاة وسعادة الأبد منوطتان[54] بالإيمان بالله واليوم الآخر وعمل[55] الصالحات، ويندرج في ذلك[56] العلم بصفات الله تعالى وتنزيهه وتوحيده وبيان جوده وعنايته والعلم بنفس الإنسان، وبقاءها[57] بعد خراب البدن وأبديتها وتزكيتها[58] وحال معادها ونشأتها[59] الثانية وما السبب الموجب لكمالها وسعادتها والموجب لخطائها وخذلانها. فواجب على العاقل أن يجتهد في علم ذلك أجمع لأنه إن كذب أهل الحكمة والتنزيل من غير أن يستند في ذلك إلى برهان ودليل فليس معدوداً من أهل العقل والتنزيل[60] والتحصيل، وإن آمن بما اتفقوا عليه وصدقهم فيه، وذلك هو الواجب عليه فوجوب اجتهاده[61] في العلم بذلك ظاهر، وإن تردد بين التصديق والتكذيب وكان شاكاً في قولهم[62] قبل تحققه صدقهم فالأحوط له ذلك الاجتهاد. أيضاً كما[63] لو حذر من سلوك طريق فيها أسد يفترس وكان شاكاً في صدق من أخبره بذلك فإن العقل يقتضي أن[64] لا يسلك تلك الطريق احتياطاً إذ لا ضرر في ذلك إذا[65] لم يكن هناك أسد. وينبغي أن تعلم أن العلم هو القطب الذي عليه المدار وهو الأصل المتبوع والعمل هو الفرع التابع، وليس ما أكبّ أكثر الناس عليه وسموه علماً هو علم الحقيقة، بل العلم الحقيقي بعد الذي ذكرته هو ما يزيد في بصيرة العبد بعيوب نفسه وفي معرفته بعبادة[66]

---

49 قرروه: قد رووه، م؛ قرره، ج.
50 وفتحوا: وفتحة، ج.
51 سبيل: يتقبل، أ م.
52 إلى: على، أ م.
53 النقلية: الفعلية، ف.
54 منوطتان: منوطان، ج ي؛ منوطين، أ م.
55 وعمل: -، أ م.
56 ذلك: تلك، ج.
57 وبقاءها: ويقابلها، أ م.
58 وتزكيتها: -، ج.
59 ونشأتها: ويشابهها، أ م.
60 والتنزيل: -، أ م ي.
61 فوجوب اجتهاده: -، م؛ فوجب اجتهاده، أ.
62 قولهم: قوله، أ م.
63 أيضاً كما: وأيضاً، ج.
64 أن: -، أ م.
65 إذا: إن، أ م.
66 بعبادة: بعباد، أ م.

ربه ويقلل من رغبته في الدنيا ويزيد من رغبته في الآخرة، وهذا أيضاً إذا لم يقصد بجهة[67] الله تعالى[68] ولم ينو بتعلمه بذلك[69] وإن كان في نفسه صالحاً له[70] فهو أيضاً في حق متعلمه ذاك من علوم الدنيا لا من علوم الآخرة. والعلم الحقيقي إنما يراد به ليتوجه به إلى الجناب[71] الأعلى وإلى الحضرة القدسية، فمن[72] عدل به عن تلك الجهة منحطاً إلى أن يحصل[73] به من[74] حطام الدنيا والتقرب بها[75] إلى آبنائها فقد أمعن في الضلال وانضم إلى زمرة الأغبياء والجهال. فنعوذ بالله أن ينافي قولنا فعلنا ويباين ظاهرنا باطننا.

وعلم الحقيقة أيضاً بحر واسع، وطريق شاسع، فمن لم يتمكن من التغلغل في بحار حقائقه والتوغل في[76] سبله ومضائقه، فلا أقلّ من أن يقتصر على ما أنا ذاكره في هذه الأوراق.

## الفصل الثاني[77]

## في إثبات وجود مدبّر العالم الواجب الوجود لذاته وبيان أنه قادر عالم مريد حكيم[78] وأن له عناية بمخلوقاته

أول ما يجب على طالب النجاة والكمال هو أن يتحقق وجود موجود مخرج لهذا العالم إلى الوجود، هو الواجب لذاته وهو الإله[79] عزّ وعلا. والطرق إلى تحقيق[80] ذلك كثيرة ومشهورة لكن أسهلها مأخذاً وأحضرها مسلكاً وأتمّها بياناً وأقربها فهماً هو الذي من جهة الاعتبار بما هو[81] في موجودات العالم من الإحكام والإتقان وحسن التأليف المؤدي إلى الغاية المقصودة من النظام، فإن ذلك كما يدل على وجود المؤثر فهو يدل أيضاً على علمه وقدرته وإرادته وعنايته.

---

67 بجهة: برحمة، أ م؛ به جهةٍ، ي.

68 تعالى: -، ف ي.

69 بذلك: ذلك، ج ي.

70 له: -، أ م.

71 الجناب: جناب، م.

72 فمن: وإن، أ ج م.

73 إلى أن يحصل: مكرر في م.

74 من: -، أ ج ف ي.

75 بها: -، ي.

76 في: به، ج.

77 الفصل الثاني: ب، ج ف ي.

78 مريد حكيم: فريد، ج ي.

79 الإله: إلا، أ م.

80 تحقيق: تحقق، أ م.

81 هو: -، ف ي.

والعاقل لا يشك إذا أبصر كتابة حسنة الهيئة والترتيب منتظمة الكلام المقروء منها[82] أنها[83] من فعل كاتب حاذق لا واقعة بالاتفاق من جريان الحبر على الكاغذ مثلاً، فكيف إذا نظر إلى هيئة الأفلاك[84] وكواكبها وأوضاعها وحركاتها ومقادير أجرامها وأبعادها وتأثيراتها في العالم السفلى، وترتيب العناصر وتفاعلها بصورها وكيفياتها وحصول الأمزجة التي يتبعها تكوّن المركبات المعدنية والنباتية والحيوانية، وعرف القوى والنفوس السماوية والأرضية ومبادئها وخواصها وتفكر في الأعضاء المفردة كالعظام والعضلات والعروق والأعصاب، وفي الأعضاء المركبة كالرأس واليد والمعدة والعينين والأذنين وكيف جعل بعضها[85] رئيساً وبعضها مرؤوساً، وكيف ارتبطت النفوس بالأبدان وسرّت قواها فيها[86] وكيف ألهمت الحيوانات مصالحها وأعدّ لها[87] من[88] الآلات والجوارح ما يناسب المركوز[89] في طباعها كآلات المفترس والسبايح وكخرطوم البقة الذي خلق فيه مع لينه[90] قوة يتمكن بها من الغوص في البشرة الجاسية[91] لامتصاص الدم الموافق لها، وغير ذلك من العجائب والحكم التي لا سبيل لنا على استقصائها، وما أظن عاقلاً يعتبر[92] هذا كله إلا ويدله دلالة قطعية على علم المدبّر بما عليه الوجود في نظام الخير وأنه مريد للخير والكمال على الوجه الأبلغ بحسب الإمكان.

وإذ قد[93] تقرر هذا فأقول: إن هذا المدبّر إن كان واجب الوجود لذاته فهو إله سبحانه وتعالى[94] وإلا فليس بممتنع[95] الوجود بعد[96] ما فرض موجوداً، فلو لم[97] يكن واجباً لكان ممكناً لا يوجد إلا من غيره، ويعود الكلام في ذلك الغير، ولا بد من الانتهاء إلى الواجب[98] الأول، وإلا لكان كل موجود قبل الذي فرض أنه مدبّر ممكناً لا يوجد من[99] ذاته فكان حكمه حكم الواسطة

---

82 منها: عنها، أ م.
83 أنها: + كانت، أ م.
84 الأفلاك: الفلك، ج ي.
85 بعضها: منها، م.
86 فيها: -، م.
87 وأعدّ لها: وأخذ لها، أ م.
88 من: -، أ ج م.
89 المركوز: الركون، أ م.
90 لينه: -، أ م.
91 الجاسية: الماشية، أ ج م ي.
92 يعتبر: معتبر، ج ف ي.
93 قد: -، أ م.
94 وتعالى: -، أ م.
95 بممتنع: يمنع، أ م.
96 بعد: وبعد، أ م.
97 فلو لم: فلم، أ م.
98 إلى الواجب: مكرر في أ.
99 من: -، أ م.

سواء كان متناهياً أو غير متناه، وذاك لأن المعلولات وكل واحد واحد[100] منها يفتقر إلى موجد لا يكون منها وإلا لدخل في حكمها، وإذا علم من وجود كل واحد وجود ما قبله، ولو كثرت الوسائط، فآخر ما يعلم وجوده من المعلولات يدل[101] كذلك على وجود أول العلل وتحقق ذلك وجوب مصاحبة العلة لوجوب المعلول، إذ الحوادث التي سبق[102] بعضها بعضاً بالزمان لا مانع في العقل من عدم تناهيها في الأزل، ويدل على هذه المصاحبة أنه لا معنى لوجوب المعلول بعلته إلا صدوره عنها، ولا يؤثر[103] فيه حال عدمه، وإلاّ لكان[104] ذلك جمعاً بين وجوده وعدمه[105]، فهي إذن تؤثر حال وجوده وحال عدم العلة لا يصدر وجود المعلول، فالموجِد والموجَد يتصاحبان[106] في الوجود، فالعلل التامة ومعلولاتها كائنة ما كانت هي موجودة معاً، فلها أول هو الواجب لذاته، وكما أنه لولا[107] وجود العلم[108] البديهي الذي لا يفتقر إلى الاكتساب لما حصل لنا شيء من العلوم المكتسبة، كذلك لولا وجود الواجب المستغني عن الموجد لما وجد الممكن المحتاج إليه، ولما وجب في موجد الأثر المحكم أن يكون أبلغ في الأحكام من أثره وجب الانتهاء إلى الموجد المحيط علمه الكاملة قدرته البالغة حكمته[109]، وهو الإله[110] تعالى.

## الفصل الثالث[111]

## في بيان ثبوت جملة مما[112] يجب ثبوته للواجب ونفي جملة مما[113] يجب[114] تنزيهه عنه

وعلم الواجب فعلي، نشأت عنه[115] التغيرات لإحاطته بها وبأزمنتها على الدوام، فلهذا لا يتغير بتغير[116] معلوماته وإنما يلزم الجهل والتغير لو كان علمه انفعالياً يتبع وجود الشيء وزواله،

---

[100] واحد واحد: واحد، أ.
[101] يدل: إضافة في هامش ي.
[102] سبق: يسبق، أ م ي.
[103] يؤثر: مؤثر، ج ي.
[104] لكان: كان، ج.
[105] وعدمه: + فهي إذن تؤثر حال عدمه وإلا لكان ذلك جميعاً من وجوده وعدمه، أ م.
[106] يتصاحبان: يصاحبان، أ ج م.
[107] لولا: لا، أ.
[108] العلم: -، ف.
[109] الحكمته: -، أ م ف.
[110] الإله: الآلاء، أ م.
[111] الفصل الثالث: ج، ج ف ي.
[112] بيان ثبوت جملة مما: ما، ج؛ بيان ثبوت جملة ما، ي.
[113] مما: ما، ج.
[114] يجب: + مما، ف.
[115] نشأت عنه: عنه نشأت، أ م.
[116] بتغير: به تغير، ج.

والواجب واجب الأزلية لاحتياج[117] كل حادث إلى الموجد، وواجب[118] الأبدية إذ لو جاز عليه العدم لافتقر إلى مرجح يرجح جانب وجوده على جانب عدمه فكان ممكناً ﴿هُوَ الْأَوَّلُ وَالْآخِرُ وَالظَّاهِرُ وَالْبَاطِنُ﴾[119]، وإذا قد ثبت علمه بمخلوقاته الدال على علمه بذاته بالضرورة فواجب[120] أن لا يكون مركباً، لأن العلم بالمركب يتوقف[121] على العلم بكل واحد من أجزائه، فيتقدم العلم بالجزء على العلم بالكل تقدماً بالذات، لكن لو علم الكل جزؤه الذي هو غيره لكان عالماً بأنه هو[122] الذي علمه، وذلك يتضمن علمه بذاته، فيتقدم العلم بالكل على العلم بالجزء، وقد كان الأمر بخلاف هذا، وذلك محال.

وإذ ليس بمركب فليس بجسم لأن كل جسم مركب ذو أبعاض، وإذا[123] ليس بجسم فليس بمتحيز[124]، لأن المتحيز[125] الذي لا ينقسم إما نقطة في محل وإما جوهر فرد لا يتجزأ بوجه إن[126] جاز وجوده، وكلاهما يقتضي أن يكون موجد هذا العالم في غاية الصغر[127] والحقارة، تعالى الله عن ذلك علواً كبيراً، ويخص النقطة افتقارها في وجودها إلى غيرها لتحلّ فيه، والافتقار في الوجود إلى الغير ينافي وجوب الوجود كيف كان. فالواجب منزّه عن أن يحل في محل، وحقيقته المقدسة لا تساوي حقيقة شيء من الممكنات وإلا لكانت[128] مقتضية للإمكان لا للوجوب لوجوب تساوي الأمثال في اللوازم. ولو حصل من نوع الواجب شخصان لامتاز كل واحد منهما عن الآخر بتعيّنه، وما به الاشتراك غير ما به الامتياز، فيتركب الواجب منهما، ولأن[129] تعين[130] كل واحد إن كان لأنه واجب الوجود فلا واجب إلا هو، وإن كان لأمر آخر فهو معلول في تعينه لغيره، وإن كان غير معلل البتة كان اختصاص كل واحد[131] بتعينه لا لمخصص وهو باطل.

---

117 لاحتياج: لاحتاج، أ ج م.
118 وواجب: واجب، أ م.
119 سورة الحديد (57): 3.
120 فواجب: وجب، م.
121 يتوقف: متوقف، ي.
122 هو: + العلم، م؛ -، ج ي.
123 وإذا: وإذ، ي.
124 بمتحيز: بمتجزء، م.
125 بمتحيز: بمتجزء، م.
126 إن: + كان، ج.
127 الصغر: الضجر، أ م.
128 لكانت: كانت، م.
129 ولأن: مكرر في م.
130 تعين: تغير: أ م.
131 واحد: -، ج ف ي.

وإذ لا يوجد ما يشاركه في الحقيقة فلا صاحبة له ولا ولد. ثم كيف يتصور انفصال شيء[132] مما لا تركيب فيه فيصير مساوياً له في النوع أو الجنس حتى يكون ولداً[133] له؟ ولو ساواه غيره من مخلوقاته في تمام حقيقته[134] لما كان أحدهما بكونه مخلوقاً والآخر بكونه خالقاً أولى من العكس[135]، وإذ لا يحلّ الواجب في محل فلا ضد[136] له يعاقبه على محله ولا يجامعه فيه، وأيضاً لا ضد له بمعنى ما يساويه في قوته ويمانعه، ولا يتعلق ببدن كالنفس، فإن قدرته تعالى أوسع وأفعاله أعمّ وأكثر من أن تتخصص ببدن تصدر[137] عنه، ولا يتّحد بغيره بمعنى الاتصال[138] والامتزاج فإن ذلك لا يتصور إلا للمتحيز[139]، ولا بمعنى أن يصير أحد الشيئين هو الآخر بعينه من غير اثنينية، فإنهما إن بقيا حال الاتحاد فهما اثنان لا واحد، وإن عدما أو أحدهما فقط لم يتحد المعدوم بالمعدوم ولا بالموجود.

وهو الجواد على الحقيقة واسم الجود على غيره مجاز، لأنه أفاد الخير والأنعام وأعطى كل شيء ما يحتاج إليه من الهداية والكمال والزينة من غير غرض وفائدة يرجع إليه.

وهو الغني التام لكونه غير متعلق بشيء[140] خارج عنه لا في ذاته ولا في صفاته، سواء كانت تلك الصفات زائدة على ذاته أو لم تكن، وسواء عرض لها إضافة إلى الغير أو لم تعرض. ولا يحتمل هذا المختصر تحقيق أن الحق من ذلك أي شيء هو.

وهو الملك الحق، فإن ذات كل شيء[141] من جميع الوجوه[142] له، لأنه منه أو مما[143] هو منه[144] وجوده وكماله ولا يفتقر هو إلى شيء، وإذا لحظ كل ما سواه[145] باطلاً في نفسه وحقاً به فهو حق في نفسه ومحقق لحقيقة[146] كل ما عداه.

---

132 شيء: الشيء، أ م.
133 ولداً: والد، أ م.
134 حقيقته: حقيقة، أ م.
135 العكس: الممكن، أ م.
136 ضد: صنف، أ م.
137 تصدر: يصدر، ي.
138 الاتصال: كلمة غير مقروءة في م.
139 للمتحيز: للمحصر، أ م.
140 بشيء: كشيء، أ م.
141 ذات كل شيء: ذلك أت كله، م.
142 الوجوه: + هي، ج ي.
143 مما: ما، أ م.
144 منه: به، أ م.
145 سواه: + كان، أ ج م.
146 لحقيقة: بحقيقة، أ م.

وكل كمال فهو معشوق، وإذا كان الواجب يدرك ذاته على ما هي عليه من الجمال والبهاء والذي[147] هو مبدأ كل جمال وبهاء ومنبع كل حسن ونظام[148]، فهو أجلّ مبتهج بذاته لأنه أشرف مدرك لأشرف مدرَك بأتمّ إدراك، وقياس ابتهاجه بذاته[149] إلى ابتهاجنا بذاتنا[150] كقياس كماله إلى كَمالنا، ولا يدرك[151] كماله كما هو[152] إلا هو، ونحن فلا[153] نقدر على فهم شيء من نعوت الواجب لذاته إلا بالمقايسة إلى ما نعرفه من أنفسنا، لكن نعلم أن ذلك أكمل وأشرف وأعلى مما نفهمه في حق أنفسنا، ولا نفهم[154] حقيقة تلك الزيادة لأن مثلها لا يوجد في حقنا، وكيف يقاس المتناهي إلى ما لا يتناهى؟ لا في الشدة ولا في الكمية، وكل نعت في الواجب الأول لا نظير له فينا، فلا سبيل لنا إلى فهمه، إنما نذكر ما في[155] وسعنا أن نعلمه لا ما يستحقه هو لذاته.

## الفصل الرابع[156]
## في أن مدبّر العالم واحد لا شريك له

ويدل على وحدانيته أن الإنسان إذا أدرك ذاته وأشار إليها فلا يجد في ذاته إلا أمراً يدرك ذاته، وكل ما عدا ذلك مما هو مجهول سواء كان وجودياً أم سلبياً فهو خارج عن ذاته، ولا يدرك ذاته بصورة زائدة على ذاته، لأن كل صورة زائدة على الذات فهي مشار إليها من جهته هو، وهو مشير[157] إلى[158] ذاته بأنا، فلا ينفصل إدراكه عن[159] ذاته، والحقائق الإدراكية لا تختلف إلا بالكمال والنقص وبأمور خارجية. وإذا ثبت ذلك فلو صح إلهان[160] واجبان كل منهما يدرك ذاته

147 والذي،: الذي، ج ف ي.
148 حسن ونظام: حي نظام، م؛ نظام حسن ، أ.
149 بذاته: -، أ ج م.
150 لأنه . . . بذاتنا: -، م.
151 ولا يدرك: ويدرك، ف.
152 كما هو: -، ج.
153 فلا: + هو، م.
154 نفهم: نريهم، أ م.
155 في: -، م.
156 الفصل الرابع: د، ج ف ي.
157 مشير: + لا، أ م.
158 إلى: مكرر في م.
159 عن: على، ف ي.
160 إلهان: الأمان، م؛ الأعيان، أ.

لما زاد إدراكه لذاته على ذاته، فإن كان مخالفاً للآخر بالكمال والنقص فالكامل هو الإله دون الناقص، وإن تساويا في ذلك فهما من نوع واحد فلا بد وأن يكون أحدهما ممتازاً عن الآخر بعد مشاركته له في الحقيقة[161] الإدراكية بأمر ما فكان مركباً فلم يكن إلهاً.

ويدل أيضاً على امتناع وجود إلهين لهذا العالم شدة ارتباط أجزائه بعضها ببعض[162] كافتقار[163] أعراضه إلى جواهره باعتبار وجواهره إلى أعراضه باعتبار آخر، وكذا متحيزاته ومجرداته وعنصرياته وفلكياته وحيوانه ونباته وبسائطه ومركباته، وكذا احتياج بعض الحيوانات إلى بعض واحتياج بعض أعضاء الشخص الواحد منها وأخلاطه وأرواحه إلى بعض، فقد أعان بعض العالم بعضاً إعانة صيرته كشخص واحد مركب، ولهذا سمّاه بعض المحققين بالإنسان الكبير، فلو استقل[164] أحد الإلهين بتدبيره لم يكن للآخر تأثير فيه لئلا يجتمع مؤثران على مؤثر[165] واحد فيستغني بكل منهما عن كل منهما. ولو استقل ببعضه فقط والآخر البعض الآخر لم يكن الارتباط والتعاون بين[166] البعضين، ثم الذي يفعل شيئاً فيتبع وجوده وجود آخر[167] وينتفع به آخر فقد استقل[168] بإيجاد الشيئين أو بتدبيرهما لا[169] بواحد منهما فحسب، وامتناع أن يستقل واحد بتدبير الكل في زمان والآخر بتدبيره في زمان آخر ظاهر عند أهل الحقائق.

ولو أمكن[170] وجود عالم آخر لارتباط له بهذا العالم لكان جود الواجب ورحمته وقدرته التي لا تتناهى تقتضي إيجاده[171]، إذ كماله الذي لا يتناهى يقتضي أن لا يبقى شيء من الممكنات ولا وجه من وجوه حسن الترتيب والنظام إلا يصدر[172] عن ذاته المقدسة ويقتضيه عنايته وحكمته البالغة.

---

161 في الحقيقة: بالحقيقة، أ م.
162 بعضها ببعض: ببعضها، أ م.
163 كافتقار: لافتقار، أ م.
164 استقل: اشتغل، أ م.
165 مؤثر: مؤثران، م.
166 بين: من، أ م.
167 آخر: أخرا، ف.
168 استقل: اشتغل، ج.
169 لا: -، م.
170 أمكن: أنكر، أ م.
171 إيجاده: الجادة، م.
172 يصدر: ويصدر، ي.

## الفصل الخامس[173]

## في كيفية فعل الواجب المدبّر وصدور الممكنات[174] عنه وإثبات وجود ملائكته

ولا يوجد ممكن[175] أخسّ إلا والممكن الأشرف قد وجد قبله، فإنه لا يمكن وجود ما هو أفضل من المعلول الأول، فإن الواجب اقتضاه بجهته الوحدانية فلم تبق جهة تقتضي[176] ما هو أشرف منه، ولو فرض وجود ما هو أشرف منه لاستدعي جهة تقتضيه هي أشرف[177] مما عليه الواجب وذلك محال. والواحد الحقيقي لو صدر عنه بجهته الواحدة شيئان لاختلفا بمميز ما وإلا لم[178] يكونا شيئين، والمعلولات إذا تساوت نسبتها إلى مقيد وجودها لم يكن لأحدها منه ما ليس للآخر، وكان ما هو أكثر من واحد واحداً وهو ظاهر البطلان. ولهذا يظهر وجود الملائكة المقربين التي يسمّيها بعضهم[179] عقولاً، إذ لو[180] لا وجود ممكنات هي أشرف من النفوس المجردة لما وجدت النفس البتة على الأصل الذي قد عرفت من أن الواحد من كل الوجوه لا يصدر عنه الكثرة إلا بوسائط، وكل[181] من استعدّ لحظ ونصيب من المعرفة فجود الخالق لا يقتضي إهمال[182] إيجاده[183]، فإن كان ذلك الاستعداد في أول الفطرة من غير واسطة فهو عالم العقول والملائكة المقربين، وإن توقف على اكتساب ما لم يقتض بواسطة المقربين أو بغيرها فهو عالم ملائكة السماوات وهي[184] النفوس السماوية، وبعد ذلك مراتب سائر النفوس.

فلا خير في الإمكان إلا ويتعلق به علم الخالق وتقتضي قدرته[185] وجوده إيجاده، ولا شر في الإمكان إلا ويتعلق به علمه، ورحمته تقتضي دفعه، فالخير مقضي بالذات والشر مقضي بالعرض، والبارئ[186] فلا يترك ما كثر خيره لأجل شر قليل فيه إذ في تركه تفويت للخير الكثير، ولا

---

[173] الفصل الخامس: ٥، ح ف ي.
[174] الممكنات: الكتاب، م.
[175] ممكن: -، أ م.
[176] تقتضي: -، ج.
[177] منه ... أشرف: -، م.
[178] لم: لا، م؛ ما، أ.
[179] بعضهم: بعضا، م؛ بعضها، أ.
[180] لو: -، أ م.
[181] وكل: فكل، أ م.
[182] إهمال: الإهمال، أ م.
[183] إيجاده: بإيجاده، أ م.
[184] وهي: وبين، أ م.
[185] قدرته: قدرة، أ م.
[186] والبارئ: + كلمتان غير مقروئتان، ف.

يطلب رشح الخير ليستكمل به بل[187] رشح الخير على الغير من كماله وكون طلب الخير وإرادته أولى به[188]، وهو[189] مثل كون الوجود أولى له من العدم، ولا يقتضي شيء من ذلك أن يكون كاملاً بغيره، وإنما[190] يلزم ذلك إن لو كان كل واحد من الإرادة والمراد أولى به. أما إذا كانت إرادة نفع الغير فقط هي الأولى من غير أن يستكمل بالمراد فلا. فهذا ما اقتضت الحال أن أذكره في[191] حال المبدأ، وها أنا أتلو ذلك بحال المعاد بعون الله وحسن توفيقه.

## الباب الثاني من الجملة الأولى
## في إثبات جملة من أحوال النفس المجردة من المادة[192] لا سيما في ما يتعلق بتحقيق تجردها ونقلها بعد الموت وامتناع عدمها وكيفية سعادتها وكمالها

### الفصل الأول[193]
### في كيفية تعلق النفس بالبدن وارتباطها به وتصرفها به[194] وتصرفها في قواه

استيعاب[195] الكلام في مباحث النفس وتقسيم قواها وخواصها وأفعالها وكيفية ارتباطها بالبدن طويل، وهو مشهور في كتب من تقدم، وأهمّ ما ذكر من ذلك وأقربه إلى الفهم ما أنا مقرر له في هذه اللمعة. فأول ذلك هو[196] أن يتحقق أن النفس هي التي يشير إليها كل أحد بقوله أنا حين يقول أنا أدركت كذا أو فعلت كذا وكل القوى من توابعها وهي آلات لها، وإلا لما ارتبط فعل بعضها بفعل بعض، فما كان لك[197] أن تقول أحسست فغضبت وأدركت فحركت، فجميع قوى البدن تحت تصرفها، إذ القوة المحركة ليست إلا لجلب النافع أو دفع الضارّ الراجعين إليها

187 بل: -، م.
188 به: -، م.
189 وهو: هو، ج ف ي.
190 وإنما: + كان، ف ي.
191 في: من، أ ج م.
192 المادة: الإنسان، أ ج م؛ -، ي.
193 الفصل الأول: أ، ج ف ي.
194 وتصرفها به: -، ي.
195 استيعاب: واستيعاب، أ ج م.
196 هو: -، ي.
197 فما كان لك: فله، م؛ مما كان، أ.

بوجه ما، والمدركة ليست إلا كالجواسيس التي تفيض[198] الأخبار وتؤديها إليها، والمصورة[199] والحافظة هي لحفظ ذلك فحكمها حكم الخازن[200] لها. وعلى هذا حال جميع القوى البدنية إذا اعتبرتها، وكذا كل عضو من البدن فإنه إنما أعدّ لغرض يرجع إلى النفس، وجاز أن يتعلق بالبدن الواحد أكثر من نفس واحدة، لكنها تكون متفاوتة في رتبة الاستكمال وينتهي الترتيب[201] إلى نفس واحدة هي الرئيسة على الكل. ولعل[202] هذه النفوس هي القوى المطيعة لها لكن لا يجوز تعلق نفسين[203] ببدن واحد[204] تعلقاً هو كتعلق هذه الرئيسة[205] وإلا لحصل في البدن الواحد المتقابلان معاً كالحركة والسكون والنوم واليقظة، وجميع إدراكاتنا[206] وتحريكاتنا[207] الإرادية هي لنفس[208] واحدة[209] مدركة بجميع أصناف الإدراكات بجميع أصناف المدركات[210] ولولا ذلك لما حكمت ببعض المدركات على بعض[211]، فإن الحاكم[212] على شيء بشيء يجب أن يكون مدركاً لهما معاً، وإن كان[213] إدراكه بعضه بآلة بدنية وبعضه بغير آلة وهي الموصوفة بالشهوة والنفرة واللذّة والألم والإرادة والقدرة والفعل، ولو لم يكن الأمر كذا ما[214] لزم من إدراكها حصول هذه الأشياء لها ولما ارتبطت هذه الأشياء بذلك الإدراك هذا الارتباط الذي نجده من أنفسنا وهذا فطري وعند العقل إذا تأمل[215].

---

198 تفيض: تقتضي، ف ي.
199 والمصورة: والمتصورة، ج ف ي.
200 الخازن: الجازب، ج.
201 الترتيب: كلمة غير مقروءة في أ م.
202 ولعل: ، أ م.
203 نفسين: نفس، ج ي.
204 واحد: واحدة، م.
205 الرئيسة: الرتبة، ج ي.
206 إدراكاتنا: إدراكاتها، أ م.
207 وتحريكاتنا: وتحريكاتها، أ م.
208 لنفس: النفس، أ م.
209 واحدة: الواحدة، أ م.
210 بجميع أصناف المدركات: -، أ م.
211 بعض: البعض، أ م.
212 الحاكم: + بشيء، ج ف ي.
213 كان: كانت، ج.
214 ما: مما، ي.
215 إذا تأمل: -، ج؛ إذا تعقل، ف.

## الفصل الثاني[216]

## في أن النفس[217] ليست بجسم ولا حالة فيه وأنها بسيطة لا تركيب فيها في الخارج وأنها جوهر قائم بذاته

وهذه النفس الإنسانية ليست هذه البنية الفانية ولا مزاج[218] البدن ولا تناسب أركانه، فإن هذه الأشياء دائماً[219] في التبدل والتحلل، مع أن الإنسان يجد من نفسه أنه[220] هو الذي كان منذ سنين عديدة، فليست هي هذه الجملة، وليست[221] هي أيضاً عضواً مخصوصاً أو أعضاء مخصوصة لا يتطرق إليها التخلخل[222] في مدة العمر[223] لأنه[224] ما من جزء في البدن إلا وقد يغفل الإنسان عنه حال ما يكون مدركاً لذاته، وكيف يجتمع[225] في شيء أن يكون مدركاً وغير مدرك في حالة واحدة؟ وقد عرفت أن كلما يدرك ذاته فلا يزيد إدراكه لذاته على ذاته وكلما هو كذا فما دامت ذاته موجودة فهو مدرك لها، ويلزم من هذا أن كلما[226] يغفل عنه الإنسان في حالة ما فهو مغاير لذاته وليس بداخل فيها سواء كان جسماً أو جسمانياً أو لا جسماً ولا جسمانياً[227].

وقد سبق أن كل ما يعلم ذاته[228] فهو بسيط لا تركيب فيه في خارج الذهن، بل إن اعتبر فيه تركيب فهو بحسب الاعتبار الذهني لا في الوجود الخارجي، فواجب أن تكون نفس الإنسان كذلك ولا يجوز أن تكون جسماً أو حالّة فيه حلول سريان وإلا لتجزأت بالاتصال. ولو كانت متجزئة بالتجزئة الاتصالية[229] لما حصلت لها الملكات التي لا تتجزأ[230] كذلك، بل يكون لها اشتداد وضعف فقط كملكة العلم والفطنة والشجاعة والجبن والتهور، وإلا فلو حصلت هذه

---

216 الفصل الثاني: ب، ج ف ي.
217 في أن النفس: -، أ م.
218 مزاج: خراج، أ م.
219 دائماً: واهماً، ج.
220 أنه: أن، أ ج م.
221 وليست: وليس، أ.
222 التخلخل: التخلل، ي.
223 التخلخل في مدة العمر: الخلل، م.
224 لأنه: لأن، أ م.
225 يجتمع: تجتمع، ي.
226 كلما: كان ما، أ؛ كان ما كان، م.
227 أو لا جسماً ولا جسمانياً: -، ج.
228 ذاته: بذاته، م.
229 بالتجزئة الاتصالية: بتجزة الاتصال، م.
230 تتجزأ: تجزى، ي.

لها وأمثالها لانقسمت بانقسام ما حلّت فيه وكانت[231] متجزئة بالتجزئة الاتصالية[232] وهي[233] على خلاف ما فرض، والنفس ليست بحالة في شيء سواء كانت مجرداً أو غير مجرد، وإلا لكان ذلك الشيء قائماً بذاته أو ينتهي إلى القائم بذاته[234]، وكان[235] هو الشيء[236] الذي تنسب الأفعال إليه بما فيها[237] من العرض لا إلى ذلك العرض بها[238]. ألست ترى أنه يقال: فعلت النفس بالقدرة أو بالإرادة، ولا يقال: فعلت الإرادة والقدرة بالنفس، وهذا واضح عند كل ذي[239] بصيرة.

## الفصل الثالث[240]
## في أن النفس لا تموت بموت البدن ولا[241] تعدم البتة

وإذ[242] لا تركيب في النفس بوجه، ولا هي حالة في محل، فهي لا تعدم، فإن كل ما هو بالفعل وتعدم[243] فقوة وجوده وعدمه في غيره، وذلك الغير هو حامله، فإن الشيء من حيث هو بالفعل لا يكون ذاته[244] بالقوة لذاته وإن جاز أن يكون بالقوة لأمر آخر، كقوة الإبصار الذي هو في العين التي هي غير الإبصار، ولو كانت تلك القوة في الإبصار لكان الشيء بالقوة والفعل معاً. فلو قبل البسيط الذي لا يحلّ فيه[245] غيره العدم لكان موجوداً في حال ما فرض معدوماً بوجوب[246] وجود القابل[247] عند حصول المقبول، وكل ما أدى ثبوته إلى نفيه فهو ممتنع، فعدم النفس ممتنع. ولهذا إذا استقرأت[248] الموجودات بأسرها لم تجد المنعدم إلا التركيب وما لا يقوم

231 وكانت: فكانت، ي.
232 بالتجزئة الاتصالية: بتجزة الاتصال، م.
233 وهي: وهو، ج ي.
234 أو ينتهي إلى القائم بذاته: -، ج ف ي.
235 وكان: فكان، ج ف ي.
236 الشيء: النفس، ي.
237 فيها: فيه، ي.
238 بها: به، ف ي.
239 ذي: -، أ م.
240 الفصل الثالث: ج، ج ف ي.
241 ولا: وإياها لا، ف؛ وأنها لا، ي.
242 وإذ: وإن، أ م.
243 وتعدم: ويعدم، ي.
244 ذاته: بذاته، م.
245 فيه: في، ي.
246 بوجوب: لوجوب، ي.
247 القابل: التقابل، أ ج؛ التقابل بل، م.
248 استقرأت: استقرت، ي.

بنفسه، فأما البسائط القائمة بأنفسها فينحلّ التركيب إليها ولا تنعدم البتة، وغير القائمة بأنفسها فستعرف الحال فيها، وبدن الإنسان لما استعدّ لأن[249] يكون له مدبّر هو النفس فإذا زال ذلك الاستعداد بسبب من الأسباب المقتضية للموت صار البدن بحيث لا يقبل هذا المدبّر فتنقطع علاقته عنه[250]، ولا يقتضي ذلك عدم المدبّر، وحكم فساد جملة[251] البدن كحكم فساد عضو واحد منه وليس الموت إلا زمانة عامة لكل الأعضاء[252]، وفساد الآلة لا[253] يقتضي عدم مستعملها، والبسيط من جهة بساطته لا يصدر عن المركّب باعتبار تركبه لأنه إن استبدّ واحد من أجزائه بالعلية فلا تأثير للباقي وإن لم يستبدّ، فإن أثر في شيء من المعلول فالمعلول مركب وإن لم يؤثر البتة فلا مدخل له في العلية. وصورة الاجتماع[254] أحد أجزاء المركب فلها حكمها، فكل ما يصدر عن مركب ففيه اثنينية ما لا يمنع من كون ذاته الأصلية واحدة، فذات النفس البسيطة لما كانت ممكنة فهي معلولة لذات بسيطة لا تنقص في مرتبة[255] الوجود عن النفس لأن علم النفس بذاتها لا يزيد على ذاتها، فهو[256] كمال للذات من حيث هي والعلة[257] الفاعلية لما له هذا الكمال الذاتي يمتنع أن تكون قاصرة عنه فيه، والذي من الفاعل التام الفاعلية، فهو كظلّ له فلا يكون أتمّ وأكمل منه، والكمال التابع[258] للذات يجوز أن يكتسب من غير فاعلها بخلاف الذي هو في نفس الذات، فقد ثبت إذن وجود مجردات بسيطة أكمل من النفوس الإنسانية في[259] العلم والحياة هي العلل الفاعلة للنفوس، سواء سمّيتها ملائكة أو عقولاً أو ما شئت من الأسماء. ولا يمكن عدم النفس إلا إذا عدمت علتها البسيطة. والكلام في عدم ما هو علتها كالكلام في عدمها، فيؤدي ذلك إلى عدم الواجب الأول[260] تعالى عن ذلك علواً كبيراً.

فالنفس[261] وفواعلها القريبة والبعيدة لا[262] تعدم، وكيف لا وهي لا تعدم ذاتها، ولا يعدمها وجود ما لارتفاعه مدخل في وجودها وهو الضد إذ لا محل لها فيمانعها عليه ضد، ولا يعدمها[263]

---

249 لأن: أن، أ م.
250 علاقته عنه: من فيه، م؛ على فيه عنه، أ؛ فيه بجبه، ج.
251 جملة: الجملة إلى، ج؛ + من، ف ي.
252 الاعضاء: -، ج.
253 لا: -، ف ي.
254 الاجتماع: -، ج.
255 مرتبة: قرينة، ف ي.
256 فهو: من، م.
257 والعلة: في العلة، م.
258 التابع: المتابع، ف ي.
259 في: هي، ج ف ي.
260 الواجب الأول: الواجبة الأولى، م؛ الواجب الأولى، أ.
261 فالنفس: والنفس، أ م.
262 لا: ولا، م.
263 يعدمها: عدم، أ م؛ يعدم، ج ف ي.

ما لعدمه[264] مدخل في عدمها وهو الشرط، إذ مع بقاء العلة المقتضية لوجودها لا تأثير لعدم المباين[265] في عدمها ولا ما[266] يباينها، كالأعراض المتضادة[267] لكمالها لا ينقض إدراكها لذاتها الذي لا يزيد على ذاتها. فتغير أعراضها واختلافها بكونها كمالاً لها[268] أو نقصاً لا يؤثر في إعدامها، فبقاء[269] ما أفاد وجودها يوجب أن لا ترتفع عن الوجود ولا تعدم، وعدم المركبات إنما هو بمعنى انحلال تركيبها، والانحلال حركة ما يستند[270] إلى الحركات[271] السماوية كما هو مقرر في الكتب المشهورة، وإذا قيل[272] الهيئة أو العرض قد عدم فالمنعدم هو[273] تعلقه بمحل ما يظهر[274] له، وأما تعلقه بفاعله فلم ينعدم، بل جاز أن يظهره بمحل آخر. والانتقال الممتنع على الأعراض إنما هو الانتقال المستلزم لاستقلالها بالوجود أو بالجهات أو بالحركة المكانية أو ما يجري مجرى هذه.

وأما انتقالها بمعنى أن فاعلها يظهرها للحسّ أو لغيره من محلّ ثم يظهرها كذلك من محل غير ذلك المحل كالصورة التي[275] تظهر بالمرآة وليست فيها على الحقيقة فليس بممتنع ولعلّ[276] كل ما تعرفه[277] من الأعراض إنما تكون محالّها مظهرة لها على هذا الوجه، والانعدام عن المحل على هذا الوجه ليس بعدم على الحقيقة، فكلما اقتضى جود الواجب فيضانه عنه فهو لا يعدم بمعنى ارتفاع حقيقته عن الوجود[278]، بل عسى أنه[279] ينعدم بمعنى انعدامه عن الحسّ أو عن محل يظهر به أو عن شيء كان ملتئماً معه فتباعد عنه بحركة ما، كما في انحلال المركبات أو بمعنى آخر يكون جارياً هذا المجرى.

ومن نظر في حكمة خلق الإنسان وتحقق ما بيّن وأوضح من عناية البارئ لمخلوقاته[280] ورحمته بهم[281] تبين أن الغاية المطلوبة من خلقه ليس أمراً زائلاً، لأنه[282] كلما كان أحلى كان خوف

264 لعدمه: لعدم، م.
265 لا تأثير لعدم المباين: لا تأثير لعدم التباين لا تأثير لعدم التباين، أ م؛ + لها، ف ي.
266 ولا ما: وما لا، أ؛ ولا، م.
267 كالأعراض المتضادة: والأعراض المضادة، أ م.
268 لها: -، ج.
269 فبقاء: ببقاء، ج.
270 يستند: يسند، ف.
271 الحركات: الحركة، م؛ حركات، ي.
272 قيل: قال، أ م ج.
273 هو: -، ف ي.
274 يظهر: مظهر، ف.
275 يظهرها للحسّ... التي: -، م.
276 ولعل: داخل، أ م.
277 تعرفه: نعرفه، ي.
278 الوجود: الجود، ج.
279 أنه: أن، أ م.
280 لمخلوقاته: المخلوقاته، أ؛ بمخلوقاته، ج ي.
281 بهم: لهم، ج ي.
282 لأنه: وأنه، أ م.

زواله[283] أمرّ، فلا يصلح غاية لمثل هذا الحكيم عزّ سلطانه، بل الغاية نعيم دائم، فلا بد للإنسان من نفس باقية هي غير هذه البنية الفانية، بل إنما أحكمت[284] هذه البنية وخلقت فيها هذه العجائب لتكون آلة لها في الاستكمال، وهذه حجة حدسية قد لا ينتفع بها من لا يجد ذلك الحدس من نفسه، والبرهان المعول عليه هو ما سبق.

## الفصل الرابع[285]
## في كمال النفس ونقصها وكيفية اتصالها بالعالم الأعلى واكتساب المعارف منه وذكر جملة في خواصها وآثارها

والنفس الإنسانية ليس لها من الحياة إذا اعتبرنا ذاتها فقط إلا إدراك ذاتها. وأما إدراك غيرها وأفاعيلها فبالقوى[286] البدنية وقوتها العقلية، وتحتاج في ذلك إلى مخرج يخرجها من القوة إلى الفعل. فإذن حياتها من دون هذه الأشياء حياة ناقصة، والشعور بالوجود والسعادة[287] والموجودات متفاضلة، فأشرفها هو الوجود الأول، والسعادات تتفاوت بحسب ما تنال[288] من هذه الموجودات[289]، ومحال أن يخلو ذات عن وقوع ظل الأول عليها، فهو النور الحقيقي وكل ما عداه عكس نوره، وتتفاوت المجردات في إدراكه كتفاوتها في إشراق نوره عليها، وتتفاوت لذاتها بإدراكه كتفاوتها في ذلك الإدراك والقرب من ذات المدرك على حسب شدة الإدراك له[290]، فالمجردات المفارقة تتفاوت في اللذة بحسب تفاوت قربها وبعدها من الواجب، وبهذا تختلف مراتب النفوس[291] ودرجاتها في الشرف والانحطاط.

واشتغال النفس بالقوى البدنية يصدّها عن الالتفات إلى الجناب[292] الأعلى وقبول المعارف منه[293]. يتضح لك ذلك بتأمل أن الأمور التي قد يقع في النوم عليها اطلاع ليس إلا لقلة الشواغل بترك النفس استعمال الحواسّ الظاهرة، وليس ذلك الاطلاع عن فكره إذ الإنسان

283 زواله: ونوال، أ م.
284 أحكمت: حكمت، أ م.
285 الفصل الرابع: د، ج ف ي.
286 فالقوى: فبالقوة، أ م ج.
287 والسعادة: السعادة، أ؛ سعادة، ف ي.
288 تنال: ينال، ي.
289 الموجودات: الوجودات، ج ي.
290 له: -، ج.
291 النفوس: الأنفس، ج ف ي.
292 الجناب: الحياة، أ م.
293 منه: منها، م.

في حال اليقظة أمكن من الفكر، فليس إلا اتصال النفس بسبب قلّة شواغلها البدنية بالمبادئ العالية المفيدة لها ذلك الغيب. وقد تتأكد وصلة النفس بالجناب الأعلى وتطلع[294] على المغيبات في حال اليقظة كما تطلع عليها في حال[295] النوم لا سيما إذا كانت النفس قوية وافية بالجوانب المتجاذبة لا يشغلها البدن عن الاتصال بالمبادئ المذكورة والإخبارات النبوية من هذا السبب، ومن لا تكون نفسه قوية فقد يستجلب ذلك بما يدهش حسّه ويحير خياله، وربما كان حسّه وخياله في أصل جبلته كذلك[296]، وهذا نقص[297] ليس بمحمود، وإذا اجتمع ضعف العائق البدني وقوة النفس سواء كان ذلك أصلياً أو مستجلباً كما تقوى النفس بالتدريب[298] أو بالرياضة المنسوبة إلى العارفين فقد يتأكد الاتصال تأكداً يحصل[299] به للنفس مغيبات كثيرة ومعارف غزيرة وملاذ إشراقية لا يدرك كنهها إلا من وجدها من نفسه، ومن لم[300] يجدها فينبغي أن يؤمن بها بقيام أدلّتها، ولا أقل من أن لا ينكرها بمجرد عدم وجدانه لها، وقد تقوى النفس فتؤثر في غير بدنها كإنزال غيوث[301] وإزالة أمراض وإحداث زلازل وما جرى هذا المجرى، ولا يستنكر ذلك إلا من يظن أن النفس إنما أثرت في البدن لانطباعها فيه.

والإصابة بالعين من هذا القبيل وهي مجربة، وسببها خاصية لبعض النفوس تحدث نهكاً في[302] من تعجبت منه، والتضرع[303] والإنابة إلى الله قد يكسبان النفس استعداداً لقبول الهداية إلى الصواب كما تعد الفكرة لقبول الفيض[304] الفاعل للمعرفة، والسحر من التأثيرات النفسانية، وغرائب النفوس كثيرة تعرف بالتجارب والتواتر وبضرب من القياس[305]. ومقامات العارفين من الأنبياء وغيرهم كثيرة يتعذر علينا حصرها وتصورها، ومن كان شريراً غير مزكٍ[306] لنفسه فلا يحمد[307] قوة نفسه لأنه يستعين بها على تأثير الشر دون الخير، والأخلاق الذميمة حجاب عظيم للنفس عن الاتصال بالعالم الأعلى واكتساب السعادة[308] العظمى لأنها تُميل النفس إلى

[294] وتطلع: فتطلع، ف ي.
[295] حال: حالة، أ م ي.
[296] كذلك: ذلك، أ م.
[297] نقص: نقض، ف.
[298] بالتدريب: بالتطريب، ي.
[299] يحصل: يصل، م.
[300] ومن لم: ولم، أ م ج.
[301] غيوث: الموت، أ م.
[302] في: -، أ م.
[303] والتضرع: والتعجب، م.
[304] الفيض: البعض، م.
[305] من القياس: بالقياس، م.
[306] مزكٍ: مزكي، أ م ف.
[307] يحمد: يجد، م.
[308] السعادة: المعاني، م.

جانب البدن فتمنع من الاتصال بمحلّ السعادة ما دامت النفس متلطخة بها سواء كانت متعلقة بالبدن أو متخلية عنه، وإنما كانت النفس مع هذا البدن مصدودة عن حقائق الأشياء لا[309] لكونها منطبعة فيه، بل لاشتغالها به ونزوعها[310] إلى شهواته وشوقها إلى مقتضياته، وهذا النزوع والشوق هيئة[311] للنفس ترسخ فيها وتتمكن منها، فإن كان هذا التمكن حاصلاً بعد المفارقة كان[312] حالها بعد المفارقة كحالها قبلها، إلا أن تزول تلك الهيئات منها بمزيل لا نعرفه. وإذا فارقت النفس البدن ولم تتعلق بغيره كان شعورها بذاتها غير مخلوط بالشعور بالبدن فكان أتمّ. وكذلك تكون معقولاتها أشدّ تجرداً لأنا لا نعقل[313] شيئاً ونحن بدنيون إلا ويقترن به خيال أو ما يشبه[314] الخيال، فإذا[315] زال هذا الشوب كان التذاذ النفس بحياتها أكمل.

## الفصل الخامس[316]
## في اللذة والألم وبيان أن العقليين منهما[317] أقوى من الحسّيين

والإدراك العقلي أتمّ من الحسي لأنه أفضل وأتم وأدوم وأكثر وأوصل إلى المدرك، فاللذة به أبلغ. والكمال الخاص بالنفس الناطقة من جهة قوتها العقلية هو أن يصير عالماً عقلياً مرتسماً فيه صورة الكل والنظام المعقول فيه والخير الفائض إليه، فيكون حينئذ موازياً للعالم الموجود كله مشاهداً لما هو الحسن والخير مطلقان، وهذا أفضل وأتم من كمالات القوى الأخرى[318] كالشهوة والغضب والوهم والحفظ وغيرها، بل لا نسبة له إليه فضيلةً وتماماً وكثرة وسائر ما يتم به التذاذ[319] المدركات[320]. ثم كيف يقاس دوام الأبدي بدوام الفاسد المتغير[321]؟ وكيف يكون حال ما وصوله بملاقات[322] السطوح إلى ما هو[323] متغلغل في كنه المدرك؟ وكيف يقاس كمال

---

309 لا: إلا، أ ج م.
310 ونزوعها: ونزعها، أ م.
311 هيئة: جنة، أ م.
312 كان: مكرر في ف.
313 لأن لا نعقل: بأن يعقل، أ؛ بالعقل، م.
314 يشبه: نسميه، م.
315 فإذا: وإذا، أ م.
316 الفصل الخامس: ٥، ج ف ي.
317 منهما: منها، أ م.
318 الأخرى: الأخر، ف.
319 التزاز: الزا، ج.
320 المدركات: والمدركات، أ م ج.
321 المتغير: التغير، أ.
322 بملاقات: بملاقاة، ي.
323 هو: -، م.

الإدراك إلى الإدراك والمدرك إلى المدرك، فإن العقل أكثر عدد مدركات من الحس وأشد تقضياً للمدرك وتجريداً له عن الزوائد وخصوصاً[324] في باطنه وظاهره. وشواغل البدن هي المانعة من الاشتياق إلى ذلك الكمال اشتياقاً يناسب مبلغه، فإن اشتغال النفس بالمحسوسات يمنعها من الالتفات إلى المعقولات فلا تجد منها ذوقاً فلم يحصل لها إليها شوق[325]، كالعنين الذي لا يشتاق إلى الجماع[326] والأصم الذي لا يشتاق إلى سماع الألحان. وتأمل أن كريم النفس إذا أكبّ على عويص يهمه وعرض عليه شهوة وخيّر بين الطرفين[327] استخف الشهوة. والأنفس العامية أيضاً قد تؤثر الغرامات والآلام القادحة بسبب افتضاح أو شوق إلى أمر عقلي، واللذات الباطنة وإن لم تكن عقلية هي أقوى من الحسّية الظاهرة. واعتبر ذلك بإيثار الحيوان ولده على نفسه في المأكول وغيره.

وأقول: إن لذة الجماع إنما بلغت في القوة هذا المبلغ لا لكونها حسية[328] فقط، بل لمعنى روحاني تدركه النفس من المجامع، ولهذا لا يلتذ بدن[329] بإتيان[330] الميت على تقدير أن لا تتغير صورته[331] المبصرة. ولهذا كان الالتذاذ بمجامعة[332] البعض أشد من الالتذاذ بمجامعة البعض[333] مع تساوي قضاء[334] الشهوة فيها[335]، فالجماع جمع بين لذّتين حسية وروحانية. ونصيب الروحانية من اللذة أقوى وأشد من الحسية، فلا يرجح اللذة الحسية على العقلية إلا من هو في غاية دناءة النفس وسقوطها أو من لا يفرق بين العقلية وغيرها. وإذا تعلقت النفس بعد الموت ببدن كما في المعاد المقبول[336] من الشرائع الحقة أو بجرم ما سماوي أو أرضي فلا يلزمها[337] أن يمنعها ذاك من أن تحصل لها لذات وآلام عقلية مع اللذات والآلام الحسية الحاصلة بسبب التعلق بالجسم. وهذا القدر من أحوال النفس ومعادها لا أرى أن أذكر أكثر منه فإنه هو الأهم والأوجب بحسب

324 وخصوصاً: خوصا، أ م.
325 شوق: شوقا، م.
326 الجماع: جماع، م.
327 بين الطرفين: من الطفرين، ج أ م؛ بين الطفرين، ي.
328 حسية: -، م.
329 بدن: -، ي.
330 بإتيان: -، أ م.
331 صورته: صورة، أ م.
332 بمجامعة: يجامعه، أ م.
333 أشد... البعض: -، أ م.
334 قضاء: قضية، ف.
335 فيها: فيهما، م ي.
336 المقبول: والمقبول، م.
337 يلزمها: يلزم، أ م.

غرض[338] هذا المختصر، وما عداه من مباحثها كثير وليس ذكر بعضه وترك باقيه بأولى من العكس بمقتضى ما نحن فيه، وإذ لا تتم السعادة الحقيقية إلا بإصلاح القوة العملية[339] فلأشرع بعد ذلك في ذكر أصول نافعة فيه، ومن الله العصمة والعون.

## الجملة الثانية
## في العمل
## الباب الأول منها
## في ذكر جملة من إصلاح حال النفس وسياستها وتعديل أخلاقها

### الفصل الأول[340]
### في[341] التوبة والعبادة والزهد وما يتعلق بذلك

اعلم أن مبدأ طريقة السالكين ومفتاح سعادة المريدين هو التوبة، وهي عبارة عن تألم النفس على ما ارتكبت من الرذائل مع جزم القصد إلى تركها وتدارك الفائت بحسب الطاقة[342] منها، وهي رجوع من طريق البعد عن الله تعالى[343] إلى طريق القرب إليه، وقبولها هو أن يحصل في القلب استعداد القبول لتجلي أنوار المعرفة، وقد يرتفع من أحوال الجوارح أنوار إلى القلب كما ينحدر من معارف القلب أنوار[344] إلى الجوارح، وهذا هو سرّ انتفاع الأنفس بالعبادات البدنية. وروح العبادة الإخلاص وحضور القلب. والعبادات البدنية[345] جارية مجرى رأس المال الذي هو[346] أصل التجارة، والنوافل بعد ذلك جارية مجرى الربح الذي به الفوز بالدرجات، فيجب القيام بوظائف الواجبات وأن لا يقع الإخلال بما يمكن[347] فعله من النافلات. والنية والعمل بهما[348] تمام

338 غرض: عن، م.
339 العملية: العلمية، أ م.
340 الفصل الأول: أ، ج ف ي.
341 في: -، أ م.
342 التوبة ... الطاقة: راجع السهروردي: كلمة التصوف، ص 6:133–7:133؛ مقامات الصوفية، ص 9:68–10:68
343 تعالى: -، ي.
344 أنوار: آثار، ج ف ي.
345 البدنية: الواجبة، أ م.
346 هو: به، أ م ف ي.
347 يمكن: يتمكن، أ م.
348 بهما: لهما، أ م.

العبادة، لكن النية خير الجزئين[349] لأنها تجري مجرى الروح والعمل يجري مجرى الجسد. والنية ميل القلب إلى الخير، فاستكثِرْ منها، وثواب العمل على قدر النية، فالنوم[350] بنية عود النشاط إلى فعل الخير خير من العبادة على الملال، وحبّ الدنيا رأس كل خطيئة، وإنما يدفع بالتجرد عنها والزهد منها، والزهد[351] هو الإمساك عن الاشتغال بملاذ البدن وقواه إلا بحسب ضرورة تامة، وهو[352] يزيد عن القناعة بترك كثير من الكفاية العرفية[353].

والدنيا عند ذوي البصائر كلا شيء، إذ الدنيا متناهية[354] والآخرة لا نهاية لها. فمن اعتدّ بالزهد[355] فيها والإعراض عنها فقد عظمها[356] ومال[357] إليها من حيث يظن أنه[358] معرض عنها، ولهذا كان زهد المحققين[359] استحقاراً لها وزهد غيرهم معاملة يطلبون بها الاستكثار منها في الدار الآخرة، أعني من ملاذها الحسّية التي هي حقيرة جداً بالنسبة إلى الملاذ الروحانية كما عرفت.

والنفس واحدة إذا اشتغلت بشيء انصدّت عن ضده، ولهذا مثلت الدنيا والآخرة بالضدين[360]، إن أرضيت إحداهما أسخطت الأخرى. ومما يهون الزهد فيها النظر إلى قلة غنائها وشدة عنائها[361] وسرعة فنائها، وكل ما لا يبقى بعد الموت فهو من الدنيا. وإنما خلقت للتزوّد[362] منها إلى الآخرة، فمن اقتصر منها على قدر الضرورة فقد نجح، ومن اشتغل بلذاتها ونسي المقصد منها هلك، ومن ظن أنه يلابسها ببدنه ويخلو عنها بقلبه فهو مغرور.

وفي العزلة عن الخلق وقطع الطمع عنهم فوائد شتى، وكذلك في مخالطتهم. ويترجح أحد الأمرين على الآخر بحسب الشخص والحال، وإنما يجب أن تكون المصاحبة والأخوة لله لا للدنيا[363]، والعزلة للعالِم أنفع. وأما الجاهل فيجب عليه الخلطة للتعلم ولاشتغال[364] المريدين بالذكر الدائم

---

349 الجزئين: المحزون، أ م.
350 فالنوم: والنوم، أ م.
351 منها والزهد: -، أ م.
352 هو: -، أ م.
353 والزهد هو ... العرفية: راجع السهروردي: كلمة التصوف، ص 133:14–133:15؛ مقامات الصوفية، ص 69:8–69:9.
354 متناهية: شيء متناه، ي.
355 اعتدّ بالزهد: اعتدنا، م.
356 عظمها: -، ج؛ عظمتها، أ.
357 ومال: ونال، ج؛ وحال، ف.
358 أنه: -، م.
359 المحققين: الحفر، م؛ المتحقراً، أ.
360 بالضدين: بالغيرين، أ؛ بالقربين، م؛ بضدتين، ف.
361 غنائها وشدة عنائها: عنايتها، م؛ غنائها وكثرة عنايتها، ج ي.
362 للتزوّد: لتزود، ج؛ للمرد، أ م.
363 لله لا للدنيا: للدنيا، م؛ لله للدنيا، أ.
364 ولاشتغال: واشتغال، ي.

وترك الأحساس والحركات وقطع الخواطر التي تجرّ إلى هذا العالم مما يعينهم في مقصدهم إعانة لا ريب فيها.

ويجب التحرز من مكايد الشيطان وهي متفننة. فمن مكايده أن يدعو إلى الخير وقصده[365] في ذلك الشر[366]، كما يدعو إلى المفضول ليمنع من الفاضل، وكما يدعو الإنسان إلى خير ليجرّه[367] إلى ذنب عظيم لا يفي خيره بذلك الشر، وكم من يغضب مثلاً لغير الله[368] وهو يظن أنه يغضب لله. وفي البدن الواحد قوى كثيرة[369] هي في كفالة النفس ورعايتها، وكيفية سياسة النفس لهذه القوى وتسلطها عليها والاحتراز من الانخداع بتسويلات بعضها أمر صعب لا يتيسر إلا للمطلعين على علاج أدواء النفس وخواطرها[370].

والخاطر هو ما يرد على النفس من السوانح الداعية إلى أمر ما كان متعلقاً بالحسية[371] العالية أو السافلة.

ومن خواطر الشيطان خاطر الوهم المجرد، وهو معارضة الوهم للعقل في أمور غير محسوسة، كإنكاره لموجود[372] لا في جهة،[373] فاجتهد في التمييز بين خاطر الخير وخاطر الشر.

## الفصل الثاني[374]
## في التقوى وحفظ الجوارح الظاهرة

وينبغي أن تلجم النفس بلجام التقوى وهو تنزيه القلب عن الذنوب. وترك المناهي أشد على الإنسان من فعل الطاعات، وإنما تقضي بجوارحك[375] وقواك التي هي نعمة من الله عليك، فاحذر أن تستعين بنعمته[376] على عصيانه، فلا تنظر بعينك[377] إلى محرّم ولا إلى شيء من مخلوقات

---

365 وقصده: وإلى قصده، أ م؛ ويصده، ج.
366 الشر: -، م.
367 ليجرّه: لجره، أ م.
368 لغير الله: -، م.
369 كثيرة: -، أ م؛ قصيرة، ج.
370 وخواطرها: -، م.
371 بالحسية: بالجنبة، ف؛ بالحس، م.
372 لموجود: الموجودة، أ م؛ الموجود، ي.
373 والخاطر هو ...لا في جهة: راجع السهروردي: كلمة التصوف، ص 132:7–132:10؛ مقامات الصوفية، ص 67:1–67:4.
374 الفصل الثاني: ب، ج ف ي.
375 بجوارحك: بخواطرك، ف ي.
376 بنعمته: بنعمه، أ ج.
377 بعينك: نفسك، أ م.

الله تعالى بعين الاحتقار، وإذا رأيت الفاسق فقل لنفسك: لعل له سريرة مع الله تمحو[378] ذنوبه وتربي[379] على كل ما أفعله أنا من العبادات. ولا تصغِ بأذنك إلى الغيبة[380] والفحش أو الخوض في الباطل وذكر مساوئ الناس.

واحفظ لسانك من الكذب ولو هزلاً، فإن النطق من جملة ما شرف به الإنسان على غيره من الحيوان، وإنما يراد[381] الكلام ليعرف به الإنسان ما في نفس صاحبه، فالكاذب قد عكس هذا الغرض وصار من هذه الجهة أخسّ[382] من الحيوان[383] الأعجم. وإذا أردت أن تعرف قبح الكذب فانظر إلى استحقارك لصاحبه واستقبحْ مِن نفسك ما تستقبحه من غيرك. ولا تخلف الوعد وحافظ على الوفاء بالعهد. ولا تذكر إنساناً بما يكرهه لو سمعه، لا تصريحاً ولا تعريضاً. واطرح المراء والمناقشة والجدال[384] في الكلام ولو كنت محقّاً ومماريك مبطلاً. ولا تزكِ نفسك وتمدحها، فذلك نقص[385] من[386] قدرك عند الناس، ولا تلعن شيئاً مما خلق الله، إنساناً كان أو غيره، ولا تدخل بين[387] العِباد وبين الله، فلا إثم في ترك لعن إبليس، فكيف غيره؟ ولا تَدعُ على أحد، وإن ظلمك وَكِلْ أمرهُ إلى اللهِ. واترك السخرية والاستهزاء وكثرة المزاح. واحفظ بطنك[388] من دخول طعام أو شراب محرّم عليك، وتحفَّظ من تناول الحرام والشبهة واحرص على أن لا تتناول إلا الحلال، واقتصر على ما دون الشبع، فإن الشبع يثقل الأعضاء وينصر جنود الشيطان.

وفرجك فاحفظه من[389] كل ما حرم الله، واستعن[390] على ذلك بترك الشبع وبأن لا تنظر إلى محرم ولا تتفكر فيه. واحفظ يديك من أن تؤذي بهما أحداً ورجليك من أن تسعى[391] بهما إلى حرام أو تمشي بهما إلى ظالم، فهذا ما ينبغي أن تحفظ به جوارحك الظاهرة.

---

378 تمحو: تسحا، أ م.
379 وتربي: وتزكى، أ.
380 إلى الغيبة: للغيبة، أ م.
381 يراد: يردد، ج ف.
382 أخسّ: احسن، أ م.
383 الحيوان: + ان، أ م.
384 المراء والمناقشة والجدال: والمراء والمناقشة والجدال، ج؛ والمراء والمناقشة، م؛ والجدال والمناقشة، أ.
385 نقص: بعض، أ م.
386 من: -، م.
387 بين: من، أ.
388 بطنك: فطنك، م.
389 من: عن، ي.
390 واستعن: واسغني، م.
391 تسعي: تستعين، ج ف ي.

## الفصل الثالث[392]
## في صيانة القلب من الآفات المختصة به

وإنما القلب عين[393] النفس لا العضو المعروف، فهو ملكٌ مطاع[394] ورئيس متّبع، والأعضاء كلها له تبع. وإذا أصلح[395] المتبوع صلح التابع، وإذا استقام الملك استقامت الرعية. وهو خزانة كل جوهر للعبد نفيس وكل معنى خطير كالعقل ومعرفة الله تعالى والنية الخالصة وأنواع العلوم والحكم وسائر الأخلاق الشريفة، وحق لمثل هذه الخزانة أن يصان[396] عن الأدناس والآفات، ويحرس من السرّاق والقطاع ويكرم[397] بضروب الكرامات لئلا يلحق تلك الجواهر دنس ويظفر[398] بها عدو.

ومن[399] الآفات القلبية طول الأمل، فإنه العائق عن كل خير وطاعة والجالب لكل شر وفتنة. والعاقل لا يغفل عن ذكر الموت. ويجب على الإنسان أن لا يخلي أوقاته من أن يتفكر في من مضى من القرون الخالية والأقران، وما كانت عليه حال واحد واحد منهم من جاه وعزّ ومال وأعوان. ويحدّث نفسه بأنه إذا أمسى فقد لا يصبح وإنه إن[400] لم يمت فجأة فقد يمرض مرض الموت فجأة[401].

ومن آفات القلب الحسد، والحاسد لا يظفر[402] البتة. وكيف يظفر بمراده[403] ومراده زوال نعمة الله تعالى عن عباده، وكفى بالحسد رذيلة كونه يضرّ بصاحبه إضراراً[404] أكثر من استضرار الحسود به، بل ربما لا يستضرّ به الحسود البتة. والخلق كلهم عباد الله وعياله فيجب محبّتهم والنظر في مصالحهم ومن أحبّه[405] أحبّ عياله وعباده.

---

392 الفصل الثالث: ج، ج ف ي.
393 عين: أعني، ي.
394 مطاع: مطلع، أ م.
395 أصلح: صلح، م.
396 أن يصان: يصان، أ؛ -، م.
397 ويكرم: ويلزم، أ م.
398 ويظفر: أو يظفر، أ م.
399 ومن: من، أ م.
400 وإنه إن: بأن، أ؛ وإن، م.
401 فقد يمرض . . . فجأة: -، أ م.
402 يظفر: + مراده، أ م.
403 بمراده: بإرادته، أ م.
404 إضراراً: إضرار، ف.
405 أحبّه: أحبه الله، أ م.

والاستعجال من جملة آفات القلب، فقد[406] يقصد العبد منزلة فيستعجل في نيلها وليس ذلك بوقتها فيفتر أو يترك الاجتهاد أو يقتل[407] في الجهد فينقطع، فكذلك[408] إذا أكثر الدعاء واستعجل الإجابة.

ومن الآفات الكبر والعظمة والكبرياء[409]، والكبرياء من الصفات المختصة بالرب فلا يجوز منازعته في خصوص صفته، وعلاجه أن يعرف الإنسان حقارته بالنسبة إلى سكان العالم الأعلى في العلم والعبادة والقدرة، وأن يتذكر أنه من نطفة قذرة وأن كل ما له من مال وجمال وأتباع فهي[410] أمور خارجة عن ذاته ولا تبقى له، وأنه جاهل بوقت زوالها وبأكثر أحواله وأنه عاجز عن أحقر شيء، ثم يتفكر في أن الأرض وجملة ما عليها، بل جملة العناصر وما فيها كل[411] شيء[412] بالنسبة إلى السماوات العلى، والإنسان[413] بالقياس إلى الملائكة كأحقر دودة أو بعوضة بالنسبة إلى الإنسان، بل التفاوت أكثر[414] من هذا بأضعاف. ومن أسباب الكبر العُجب وهو جهل محض، فيجب على الإنسان إذا استعظم[415] من نفسه شيئاً أن يتفكر أن[416] زواله مخوف على[417] القرب ويعجب من[418] إعطائه[419] ذلك بغير استحقاق لا من نفسه، وأن يقول لنفسه: كما أعطاني هذا بغير سبب فيوشك[420] أن يسلبه بغير سبب[421].

ومن الأدواء الصعبة الغضب، وهو جنون غير مستقر، فينبغي أن يكسر[422] باعتياد[423] الحلم مع التعرض[424] للمغضبات، وأن يخطر بالبال ما فيه من الرذيلة وما في كظم الغيظ من الثواب والفضيلة. ومن أسبابه القوية الكبر والعجب فبانكسارهما ينكسر.

---

406 فقد: وقد، ي.
407 أو يقتل: وفعله، أ؛ وفعلوا، م؛ أو يخلو، ج ي.
408 فكذلك: وكذلك، ف ي.
409 والكبرياء: -، ف ي.
410 فهي: فهو، أ م.
411 كل: كلا، ف ي.
412 ثم... شيء: -، م.
413 والإنسان: وإن الإنسان، ف ي.
414 أكثر: بأكثر، أ م.
415 استعظم: الستعظم، م.
416 أن: -، أ ج م.
417 مخوف على: على مخوف، م.
418 من: ممن، ي.
419 إعطائه: أعطائه، ج ف؛ إعطاء، م.
420 فيوشك: ويوشك، م.
421 بغير سبب: بغيره، أ م.
422 يكسر: يكس، أ ج ف م.
423 باعتياد: باعياد، أ؛ بأجياد، م.
424 التعرض: المعرض، أ م.

ومن أدواء القلب البخل، ودواؤه العلم بأن المال لله وإنما مكّنه منه ليصرفه في أهمّ أموره الدينية والدنيوية ويوصل ما فضل عن ذلك إلى الفقراء العاجزين من أبناء نوعه، وأوجب الله عليه ذلك عدلاً في عباده ورحمة لهم وعناية بهم، فلا ينبغي أن يكدّر عطاءه بالمنّ، بل يعتقد في من[425] أعطاه شيئاً منه أنه قد قبض حقاً من حقوقه، ويعتقد في نفسه أنه قد أوصل الحق إلى مستحقه والدين إلى مستوجبه، وإمساك المال إن كان للتنعم[426] وللشهوات[427] فالثواب[428] وحسن الأحدوثة[429] الذّ من[430] ذلك عند العقلاء، وإن كان ليخلفه لولده أو من يجري مجراه[431] فإنه يترك ولده أو غيره بخير ويُقدم[432] هو على الله بشرّ. وولده إن كان صالحاً فالله يكفيه، وإن كان فاسقاً فقد أعانه على المعصية. وكم من مات فقيراً وصار[433] ولده غنياً وبالعكس، ومن حمل نفسه على مداومة البذل تكلفاً صار ذلك له عادةً وطبعاً.

وحبّ الجاه أشد[434] من حبّ المال لأن التوصل بالجاه إلى المال أيسر من التوصل بالمال إليه، وهو من الصفات الإلهية فلذلك[435] كان محبوباً بالطبع، والسرّ في ذلك كون الروح مناسبة للأمر الإلهي، ولهذا لما تعذر على الإنسان الاستيلاء على السماوات[436] والملائكة والجبال والبحار اشتهى[437] الاستيلاء[438] على جميعها بالعلم، وكذلك[439] كان من يعجز عن وضع الأشياء العجيبة يشتهي أن يعرف[440] كيفية ذلك الوضع، ولما تصور[441] الإنسان أن يتسخر له أبناء نوعه أحبّ أن يستسخرهم فيملك[442] قلوبهم، ولهذا أحبّ أرباب الهمم[443] العالية أن يتسع[444]

---

425 في من: بمن، م.
426 للتنعم: بالتنعم، م.
427 وللشهوات: للشهوات، أ؛ بالشهوات، ي.
428 فالثواب: والثواب، م.
429 الأحدوثة: حدوثه، أ م.
430 الذ من: اللذة من، ج.
431 مجراه: -، أ م.
432 ويقدم: ويتقدم، أ م.
433 وصار: أو صار، أ م.
434 أشد: مكرر في ج.
435 فلذلك: ولذلك، م.
436 السماوات: السماويات، ف.
437 اشتهى: انتهى، م.
438 الاستيلاء: للاستيلاء، أ م ف.
439 وكذلك: ولذلك، أ م.
440 يعرف: يقف على، ي.
441 تصور: مقصود، م.
442 فيملك: ويملك، ج ي.
443 الهمم: الهم، م؛ إلهية، أ.
444 يتسع: يتبع، أ م؛ يتصبح، ج.

جاههم وينتشر[445] صيتهم حتى إلى بلاد لم تطأها[446] أقدامهم ولا رأوا[447] أهلها. والرفعة الحقيقية دالّة على شرف النفس وليست مذمومة، وإنما المذموم طلب الارتفاع الوهمي، فإنه أمر عارض لا بقاء له[448]. وأما الكمال الحقيقي وهو القرب إلى الله تعالى فهو ارتفاع محمود. وطلب الكمال الوهمي هو الوقوف من النفس عند الأمر[449] الأدون، والمستبصر يعلم[450] أنه لو سجد له أهل الأرض جميعاً لما بقي لا الساجد ولا المسجود[451] له، وكل أمر ينقرض حقير، وكل كبير لا يدوم صغير، وكل أمد[452] ينقضي قصير. ويكفي الإنسان من الجاه القدر اليسير الذي يتحرس[453] به من الظلم والعدوان[454]، وطلب ما[455] زاد على ذلك فضل لا يحتاج إليه. وقد يكون الكثير من ذلك بالقياس إلى شخص يسيراً بالقياس إلى[456] آخر وبالعكس.

ومن أعظم الآفات الرياء، وهو طلب المنزلة في قلوب الناس بالعبادات وأعمال الخير. ومن قصد بالعبادة العباد فقد اعتقد أنهم أقدر على نفعه وضره من الله، ويكفي المرائي أن الناس لو علموا ما في باطنه من قصد الرياء لمقتوه، وفي إخفاء[457] العبادة[458] السلامة من الرياء. وأنت إذا رأيت قدر طاعة الله تعالى ورأيت عجز الخلق وضعفهم وجهلهم فلا تلتفت[459] إليهم بقلبك، وازهد[460] في ثنائهم[461] ومدحهم وتعظيمهم الذي لا طائل[462] تحته، ولا تقصد بطاعتك شيئاً من ذلك، وتجنب القسم[463] بالله تعالى ولو كنت صادقاً، إلا أن[464] يلجئك إليه حكم شرعي،

---

445 وينتشر: -، أ م ج.
446 تطأها: تطاء، ج.
447 رأوا: راءا، ج.
448 لا بقاء له: لإبقاءه، ج.
449 الأمر: -، م.
450 يعلم: فعلم، أ م.
451 المسجود: السجود، م.
452 أمد: أيد، ج.
453 يتحرس: يحترس، م.
454 والعدوان: والعلم، م.
455 وطلب ما: وما، م.
456 شخص . . . إلى: -، أ م.
457 إخفاء: إغفاء، ج.
458 العبادة: العباد، أ م.
459 فلا تلتفت: فلما يلتفت، أ م.
460 وازهد: وزاهد، ف ي.
461 ثنائهم: ثوابهم، م.
462 طائل: يماثل، أ م.
463 القسم: بالقسم، ج ف؛ بالقسم قسم، ي.
464 إلا أن: الآن، م.

فتحرّ[465] حينئذ الصدق وتحذر[466] الكذب، فإن اليمين الكاذبة من أكبر الكبائر[467]، فلا تجتلب[468] بها نفعاً ولا تدفع بها أذىً دنيويين[469]. والخلق هيئة للنفس يصدر عنها الفعل بسهولة من غير روية ولا تكلف، وقد يكون التكلف طريقاً إلى تحصيل الخلق بالتعويد، وتفاوت الناس في الحسن باطناً كتفاوتهم في الحسن ظاهراً، بل أكثر، وقد يظن الإنسان بنفسه حسن الخلق وهو عاطل عنه.

## الفصل الرابع[470]
## في المقامات التي يفتقر إليها طالب[471] العبادة والعرفان

ومن المقامات التي يفتقر إليها طالب العبادة وعرفان الحق التوكل، وهو دوام حسن ملاحظة القضاء والقدر[472] في جميع الحوادث دون اقتصار النظر على الأسباب الطبيعية[473]. والعارف يتحقق أن الله هو الرزاق ولا فاعل[474] سواه، بل لا يرى في الوجود غيره، وإنما الكثرة منه في حق من يفرق نظره كالذي يرى من الإنسان عضواً فعضواً، فإن رآه جملة لم يخطر بباله الآحاد ورؤية الكثرة.

وارتباط الأسباب والمسببات لا يقدح في حقيقة توكل العبد على الله، بل إذا انضاف إليه الإيمان بالرحمة والجود والحكمة حصلت الثقة بالتوكل. ومن تحقق أن الله لم يبخل على الخلق أصلاً ولا أخر في صلاحهم أمراً وأنه أوصل الرزق إليهم وهم في بطون أمهاتهم، ثم أوصله[475] إليهم[476] وهم أطفال عاجزون، وأنه رزق الحيوانات على اختلافها وهداها إلى مصالحها وإلى ما فيه بقاء شخصها ونوعها، فجدير بأن يفوض أمره إليه ويثق به. واعلم أن الرزق المضمون هو ما تقوم[477]

---

465 فتحر: فتحرى، ي.
466 وتحذر: واحذر، م.
467 الكبائر: الكبار، م.
468 تجتلب: تجلتلب، ج ف ي.
469 دنيويين: بياض في ج؛ دنيوياً، ي.
470 الفصل الرابع: د، ج ف ي.
471 طالب: طلاب، ج ي.
472 والقدر: القدر، أ.
473 التوكل... الأسباب الطبيعية: راجع السهروردي: كلمة التصوف، ص 134:3–134:5؛ مقامات الصوفية، ص 71:1–71:2.
474 فاعل: فاضل، م.
475 أوصله: أوصل، أ م.
476 إليهم: -، م.
477 تقوم: يقوم، ي.

به البنية لا ما يفضل عن قدر الحاجة[478]. وإذا اشتبه عليك أمر لا تدري[479] صلاحه من[480] فساده فإذا فوضت أمرك إلى الله فإنك لا تقع إلا في صلاح وخير، فإن ظننت غير ذلك أو ظننت نقصاً في الوجود[481]، فاعلم من الأصول التي سبقت لك أن ذلك الشر أو النقص يرتبط به كمال أو[482] خير أعظم منه. والوالد الشفيق قد يضرب ولده ويسقيه الأدوية الكريهة ويحجمه ويفصده لمصلحته والولد كاره لذلك، وربما ظن أنه يطلب أذاه لا نفعه، والله أرحم بعباده من الوالد والوالدة[483]. وكيف لا يكون الأمر كذلك والشفقة والرحمة التي ركزها[484] في طبع الوالد والوالدة هي جزؤ يسير من جملة رحمته وشفقته. وليس من شرط التوكل ترك الكسب[485] أو ترك التداوي مثلاً، لأن ارتباط المسببات بالأسباب من سنة[486] الله وهي لا تتغير، بل يجب أن تعتقد أن الوسائط كلها كاليد[487] والطعام والبدن[488] وقدرة[489] التناول وإلهام السعي والتمكين منه جميع ذلك[490] وغيره من قدرة الله ومن آثار عنايته بخلقه وإرادته لمصالحهم، فلا تتوكل إلا على الله خالقها لا عليها، فإنه قادر على إزالتها وعدم[491] الانتفاع بها.

ومن فضائل التفويض طمأنينة القلب كطمأنينة قلبك بمن توكل في الخصومة عنك إذا علمت نصحه وخيرته. وعليك أن ترضى بكل مقضي من الله.

والرضا هو ملكة تلقى النفس لما يأتي به القدر من الحوادث الحرمانية[492] على وجه لا يتألم بوقوعه، بل مع ابتهاج لطيف نظراً إلى العلة السابقة العجيبة[493]. وقد يكون الشيء مرضياً به من وجه

---

478 واعلم أن الرزق...الحاجة: -، ف.
479 يدري: يدرك، م.
480 من: في، ج.
481 الوجود: الموجود، ج ف ي.
482 كمال أو: -، ج ف ي.
483 والوالدة: -، أ ف ي.
484 ركزها: ذكرناها، أ م.
485 الكسب: التكسب، ج.
486 سنة: سنن، م.
487 كاليد: كالعد، أ م.
488 والبدن: والبذر، أ م.
489 وقدرة: وقدر، أ م.
490 منه جميع ذلك: من جميعه، م.
491 عدم: إعدام، ف ي.
492 الحرمانية: الزمانية، ج ي.
493 والرضا...العجيبة: راجع السهروردي: كلمة التصوف، ص 134:6–134:7؛ مقامات الصوفية، ص 71.3–71:5.

مسخوطاً من آخر، كالمعصية عند من يرى أنها مقضية من عند[494] الله، فقد تكون مرضية[495] من[496] حيث أنها بقضائه[497] ومسخوطة من حيث أنها مبعدة عن جنابه.

وتمسك بالصبر في حالتي السراء والضراء. أما في حالة السراء فلئلا تطغى فتنسى مبدأك ومعادك، وأما في حالة الضراء فلئلا يمنعك الجزع والتلهف من التفرغ لعبادة[498] ربّك أو يدعوك[499] إلى عدم الإيمان بالرحمة والعناية.

ورأس الحكمة مخافة[500] الله. والخوف هو تألم النفس بمكروه أخطرت إمكان وقوعه في المستقبل.

والرجاء هو ابتهاجها[501] بملائم لها أخطرت إمكان حصوله في ما يستأنف[502]. وأنت إذا تذكرت كمال جلالة الله وعظمته وعظم سلطانه وهيبته، ثم شدّة غضب الذي لا يقوم له السماوات والأرض، ثم غاية غفلتك وكثرة ذنوبك وجفوتك مع دقّة أمره وحصر معاملته وإحاطة علمه وبصره بالعيوب والغيوب[503] وشدّة وعيده وأليم عقابه الذي لا يحتمل القلوب ذكره، فكيف مباشرته؟ أدى بك جميع ذلك إلى الخوف[504]، ولكن لا ينبغي أن يفرط بحيث يؤدي إلى القنوط، فإن ذلك مذموم، بل اذكر كثرة نعمة الله عليك في الحال وإمداده لك بأنواع الألطاف من غير استحقاق ولا سؤال.

واذكر سعة رحمته وسبقها غضبه وأنه أرحم الراحمين الرؤوف بعباده الغني الكريم، وما وعد من جزيل ثوابه وعظيم كرامته. فإذا واظبت على هذه الأذكار أفضت بك إلى الرجاء. فإذا جمعت بين هذين النوعين من الأذكار اعتدل خوفك ورجاؤك وكنت سالكاً لسبيل القصد[505] عادلاً عن الجانبين الممرضين، فلا تهلك بحرارة الخوف ولا ببرودة الرجاء. والرجاء يخالف التمني، فإن من

494 عند: -، أ ي.
495 مرضية: مقضية، أ.
496 عند الله . . . مرضية من: -، م.
497 أنها بقضائه: قضائه، أ م.
498 لعبادة: بعبادة، أ ف م.
499 أو يدعوك: ويدعوك، أ ف م.
500 مخافة: مخاف، أ.
501 ابتهاجها: ابتهاجا، أ م.
502 والخوف . . . يستأنف: راجع السهروردي: كلمة التصوف، ص 133:13، 133:11؛ مقامات الصوفية، ص 69:5–69:6، 69:3–69:4.
503 والغيوب: -، م.
504 الخوف: الحزن، م.
505 لسبيل القصد: للسبيل، ج.

لا يتعهد الأرض ولا يبث البذر[506]، ثم ينتظر الزرع، فهو متمنٍ[507] مغرور، إنما الراجي من حصّل كل غرض يتعلق باختياره ثم بقي، يرجو من الله الإتمام ودفعَ الموانع.

ومن أيقن أن الرحمة والخير مقصودان بالذات وأن الشرّ في العالم مقصود بالعرض غلب رجاؤه خوفه لا محالة.

ويجب أن يشكر الله على نعمه العظيمة ومننه[508] الكريمة. والشكر هو ملاحظة النفس لما نالت ممن[509] أنعم عليها من إعطاء ما ينبغي أو دفع ما لا ينبغي، كان من كمالات النفس أو البدن، وتحريك الآلة المعبرة[510] لإخبار[511] النوع بذلك[512].

ومن علم أن النعم كلها من الله وأن الوسائط مسخرون مقهورون لم يشكر إلا[513] الله وحده[514]، ومن اعتقد أن لغير الله مدخلاً في النعمة الواصلة إليه لم يصح حمده ولم تتم معرفته وشكره، وكان كمن أنعم عليه الملك فاعتقد أن لعناية[515] الوزير مشاركة في ذلك، ولو رأى توقيع الملك[516] بذلك الإنعام بقلمه لما التفت قلبه إلى شكر القلم ولا إلى الخازن والوكيل لأنهم مسخرون مقهورون، وإنما يجب شكر من أجرى الله الإحسان على يده كالوالدين وغيرهما لمجرد[517] تعبد شرعي لا لأن الإحسان حصل منهم[518] بالذات.

ولا شدة إلا وفي جنبها[519] نعم الله كثيرة، منها أن تلك الشدة زائلة غير دائمة، وأنها معرضة لمنافع في العاقبة تتلاشى مشقة هذه في جنبها كما به لا يبقي[520] ألم الفصد والحجامة ومرارة الدواء في جنب صحة النفس وسلامة البدن من الخطر، وذلك لأسرار لا يطلع على وجه الحكمة فيها. فلذلك لزم العبد أن يشكر على الشدة لما يتبعها من النعم كما يشكر على النعمة. وإذا كانت

---

506 البذر: البزر، ف ي.
507 متمن: متمني، ي.
508 ومننه: ومنه، ج ي.
509 ممن: من، ف ي.
510 المعبرة: المعتبرة، ف.
511 لأخبار: الأخبار، أ ج.
512 والشكر ... بذلك: راجع السهروردي: كلمة التصوف، ص 133:16–133:17؛ مقامات الصوفية، ص 70:1–70:3.
513 إلا: إلى، أ م.
514 وحده: مكرر في م.
515 لعناية: بعناية، أ م.
516 توقيع الملك: الملك يوقع، م؛ توقع الملك، أ.
517 لمجرد: بمجرد، ف ي.
518 منهم: حملهم، ج.
519 جنبها: جانبها، أ م.
520 به لا يبقي: يتلاشي، أ م ف ي.

القلوب مجبولة على حبّ من أحسن إليها، ولا محسن ولا منعم على الحقيقة إلا الله، فيجب أن يكون محبوباً لا محالة. وأيضاً لما كان الكمال محبوباً لذاته[521]، وكان كل كمال وجمال وبهاء في الوجود فهو من كمال الله وجماله وبهائه، بل لا نسبة لكمال غيره إلى كماله، وكيف ينسب المتناهي إلى ما لا يتناهى؟ فالعارف لا يحبّ إلا الله، فإن أحب غيره فيحبه[522] له، فإن كل شيء ينسب إلى المحبوب فهو محبوب، وشدة المحبة على حسب شدة المعرفة، وأنوار العظمة إنما خفيت لشدة جلائها وظهورها. ألا ترى[523] أن النور الذي به تظهر[524] الأشياء لو دام ولم يعرف الظلام لأنكر وجوده ولما[525] ميّز بينه وبين الألوان؟ ولو كان الأنوار[526] قدرة الله تعالى حجاب لأدرك في الحال من التفاوت[527] ما يضطرّ معه إلى الاعتراف بالقدرة والقادر، ولعمري إن[528] هذا فرض محال ومع فرض وقوع هذا المحال فلا يبقى من يدرك التفاوت[529] ولا من يدركه.

والمحبة هي الابتهاج بتصور حضرة[530] ذات ما، والشوق هو الحركة إلى تتميم هذه البهجة، وكل مشتاق فقد وجد شيئاً وعدم شيئاً، فإذا وصل بالكلية بطل الشوق والطلب[531]. وأنت يا طالب السلامة فاستكثر من الدعاء واجعله حصنك[532] الذي تتحصن به من المخاوف والجأ إليه في البأساء والضراء. وإذا دعوت في شيء ووصلت إليك مع الدعاء مضرّة فازدد حرصاً على الدعاء ورغبة في التضرع والبكاء، فربما كان ما[533] نالك[534] اكتسبت من سيىء أعمالك ومع ذلك[535] فالذي ألهمك الدعاء ووفقك[536] له لا بد أن يحسن العاقبة لك. والدعاء يعدّ للإجابة ولقبول الألطاف والأنوار كما يعدّ المزاج الصالح لقوة من القوى، أو لنفس[537] من النفوس لقبول

---

521 لذاته: بذاته، ف.
522 فيحبّه: لحبه، أ؛ فتحه، م.
523 ترى: + إلى، أ م.
524 تظهر: يظهر، ي.
525 ولما: لما، ج أ م.
526 الأنوار: الألوان، ج.
527 ولو كان... من التفاوت: كذا في كل المخطوطات.
528 ولعمري إن: -، أ.
529 ما يضطرّ... التفاوت: -، م.
530 بتصور حضرة: بحضرة، أ م.
531 والمحبّة... والطلب: راجع السهروردي: كلمة التصوف، ص 134.11–134:13؛ مقامات الصوفية، ص 72.4–72:6.
532 حصنك: حظك، أ م.
533 ما: بما، أ ج م.
534 نالك: -، م.
535 ومع ذلك: -، م.
536 ووفقك: وفعك، ج؛ وفعك ف.
537 أو لنفس: والنفس، م.

ذلك من واهبه، فلا تقنط إذا تأخرت الإجابة، فلكل شيء وقت مقدّر لا يتعداه[538]، فلعلّ[539] في التأخير أو في عدم[540] الإجابة خيرة، وأنت لا تعلم بها.

## الفصل الخامس[541]
## في تحسين الأخلاق وتعديلها

ويجب أن تحسن أخلاقك وتعدلها[542]. والأخلاق الذميمة كثيرة ولكن أصولها ترجع إلى ما ذكرته. ولا يكفيك تزكية النفس عن بعض الرذائل حتى تتزكى عن جميعها، لأن بعضها يرتبط بالبعض ويتقاضى البعض منها البعض. وكما أن الحسن الظاهر لا يكمل بحسن بعض الأعضاء ما لم يحسن جميعها، فكذلك الحسن الباطن. واستعمال التوسط بين الخلقين الضدين يحصل للنفس الهيئة[543] أو الجهة[544] الاستعلائية على البدن، فإن المتوسط غير مضاد لجوهر النفس ولا مائل به إلى جهة البدن، بل عن جهته، لأنه يسلب[545] عن الطرفين دائماً. والوساطة المذكورة هي العدالة، وهي عفّة تنسب إلى القوة الشهوانية، وشجاعة تنسب إلى الغضبية، وحكمة تنسب إلى العقلية، وهذه هي أصول الفضائل الخلقية. فالعفة ملكة التوسط بين الخمود والفجور، والشجاعة ملكة التوسط بين الجبن والتهور والحكمة[546] ملكة التوسط بين البلادة والجربزة[547].
ومن تفاريع الشهوانية القناعة والسخاء، وهما فضيلتان يكتنف[548] كل واحدة منهما رذيلتان[549].
ومن تفاريع الغضبية الصبر والحلم وسعة الصدر وكتمان السرّ والأمانة، ومقابلات هذه الخمس رذائل.[550]

---

538 يتعداه: يتعدر، أ ج م.
539 فلعلّ: وأجل، م.
540 عدم: العدم، أ ج ف.
541 الفصل الخامس: ٥، ج ف ي.
542 وتعدلها: وتعديلها، أ م.
543 الهيئة: الغيبة، م.
544 أو الجهة: -، أ ف ي.
545 يسلب: يسبب، أ م.
546 ملكة التوسط بين الجبن والتحور والحكمة: -، أ م.
547 البلادة والجربزة: الجربزة والغباوة، م؛ الغبادة والجربزة، ج ف؛ الجربزة والبلادة، أ. الوساطة...الجربزة: راجع السهروردي: كلمة التصوف، ص 129:9–129:13؛ مقامات الصوفية، ص 63:8–63:14.
548 يكتنف: يمشف، م.
549 رذيلتان: رذيلتين، م. ومن تفاريع...رذيلتان: راجع السهروردي: كلمة التصوف، ص 131:2–131:6، مقامات الصوفية، ص 65:1–65:6.
550 ومن تفاريع...رذائل: راجع السهروردي: كلمة التصوف، ص 131:7–131:19، مقامات الصوفية، ص 66:7–67:2.

ومن تفاريع العقلية الفطنة والبيان وإصابة الرأي والحزم والصدق والوفاء والرحمة والحياء وعظم الهمّة وحسن العهد والتواضع، فهذه إحدى عشرة فضيلة تختص بالحكمة بهذا المعنى، ومقابل[551] كل واحد منها رذيلة.[552] وليست النجاة موقوفة على الكمال البالغ. ولن يسلم الحسن المطلق لأحد من البشر، اللهم إلا على[553] الندور. فينبغي على كل واحد[554] إذا لم يمكنه أن يبلغ الغاية من كمال نوعه أن يجهد أن يكون إلى حسن[555] الصورة الباطنة[556] أميل، فهذا ما يتعلق بمصلحات ذات النفس وكيفية السياسة لها واستعمال العدل فيها. وهو وإن كان المقصود منه بالذات في هذا الموضع أن ينتفع به كل واحد من المستعملين له لإصلاح ذات نفسه[557]، فإنه أيضاً مستلزمٍ لنفع غيره من المصاحبين له والداخلين تحت رعايته. ومع هذا فإني بعد ذلك أستأنف[558] بحثاً مجملاً يختص بتدبير المصالح المنزلية[559] والمدنية. ومن الله التسديد والهداية.

## الباب الثاني من الجملة الثانية
## في إيراد أصول تتعلق بمصالح أصحاب الإنسان وأهله ورعيته بالسياسة المخصوصة بهم

### الفصل الأول[560]
### في فائدة الاجتماع ووجوب المعاضدة وقانونها

قد سبق فيما مضى جملة يستعان بها في أمور السياسة، والذي أزيده الآن على ذلك هو أن الناس لا ينتظم حال اجتماعهم ومعاشهم إلا إذا كانوا كالبنيان يشدّ[561] بعضه بعضاً. ولهذا قال الأوائل: الإنسان مدني بالطبع، فيجب على كل أحد أن يراعي أهله وإخوانه وجيرانه وأصحابه

---

551 ومقابل: ويقابل، أ م.

552 من تفاريع... رذيلة: راجع السهروردي: كلمة التصوف، ص 17:129–20:130؛ مقامات الصوفية، ص 17:63–21:64.

553 على: –، م.

554 واحد: أحد، أ ف ي.

555 إلى حسن: –، م.

556 الباطنة: الباطنية، أ م.

557 نفسه: + فقط، ف ي.

558 استأنف: مستأنف، أ م.

559 المنزلية: المنزلة، أ م.

560 الفصل الأول: أ، ج ف ي.

561 يشد: يشده، أ.

وأهل منزله ورعيته وأعوانه، وعليه في ذلك مواصلة[562] كبيرهم وصغيرهم، والمشاركة لهم بالمال والحال والجاه، والبرّ لهم والإشفاق عليهم والمواساة لفقيرهم والحرص على إنماء مال عينهم[563]، والحفظ لعينهم والقيام بحوائجهم[564] وتعليم جهّالهم[565] والتجافي عن[566] ذنوبهم وكتمان أسرارهم من غير انتظار مفاوضة[567] أو معاوضة[568] منهم. ومن كان له عائلة فينبغي[569] أن يكتسب لهم[570] بقدر حاجتهم ويحسن معاشرتهم بالإحسان إليهم وصرف الأذى عنهم، ولا يهمل مراعاة أحوالهم وتفقد شؤونهم وأن يحضنهم[571] من غير عنف ويلين لهم من غير ضعف. ولا يهازل أمته ولا عبده[572] فيسقط وقاره عندهم فيسوء انتفاعه[573] بهم.

وهذه السيرة موجبة للتوادّ[574] والتعاضد. وكم من جماعة لم يكن لهم أحوال ولا أقدار أقام أحوالهم ورفع أقدارهم اتفاقهم وتعاضدهم[575]. وكم من جماعة كانت أقدارهم[576] سامية وأحوالهم ظاهرة نامية، محق أحوالهم ووضع أقدارهم اختلافهم وتباعدهم، والتجربة[577] قد حققت ذلك في الممالك وغيرها.

## الفصل الثاني[578]

## في السيرة التي يجب أن يكون عليها الملك والوالي في ذات نفسه لينتفع بها في أحوال سياسته

ويجب على من استرعاه الله خلقه وجعله الوسيط بينه وبين عباده أن يبدأ أولاً بسياسة نفسه وسياسة منزله على القانون الحكمي الذي قد[579] ذكرت أصوله وجملة من فروعه، ثم يقصد

---

562 مواصلة: بمواصلة، ج ف ي.
563 عينهم: عنهم، أ م.
564 بحوائجهم: لحوائجهم، أ م.
565 جهّالهم: جاهلهم، ج ف ي.
566 عن: من، ج.
567 مفاوضة: مقايضة، م؛ مقارضة، ج ف.
568 معاوضة: معارضة، ج ف.
569 فينبغي: فيجب، ف ي.
570 لهم: بهم، م.
571 يحضنهم: يخيفهم، ج ف ي.
572 ولا عبده: وعبده، أ م.
573 انتفاعه: -، م.
574 للتوادّ: للترادد،ج؛ للتوادد ف ي.
575 لم يكن . . . تعاضدهم: -، ج ف ي.
576 أقدارهم: -، م.
577 والتجربة: وبتجربة، م.
578 الفصل الثاني: ب، ج ف ي.
579 قد: -، أ م.

بعد ذلك سياسة رعيته. فإن من طلب[580] طاعة غيره وطاعة نفسه عليه ممتنعة فقد طمع في غير مطمع، فما ينبغي على الملك اعتماده أن يفرق بين متملقيه احتيالاً لما لديه وبين من يخلص له النصيحة ولو بإبداء عيوبه، وأن يكون وقوراً في حركاته وسكناته طويل الصمت ومعتمداً للصدق في مقاله متجنباً للكذب في كل[581] كلامه غير مستشيط[582] ولا متوعد في معصيته بأكثر من عقوبتها، لئلا يظلم إن لم يخلف أو لا يبقى موثوقاً إلى كلامه إن أخلف[583]، وأن يتغضب لإظهار الهيبة من غير غضب، وأن يعاقب للأدب لا للغضب. وأن لا يستفزّه[584] السرور فينسب إلى ضعف العزيمة ولين الهمة.

وأن يكون صبوراً عند طروق الحوادث، كاتماً لسره لأن لا يهون[585] عند رعيته، مكثراً من مشاورة العقلاء النصحاء. ثم يستعمل كل واحدة من هذه الخلال الأربع في موضعه[586]، وهي الرحمة والقساوة والسماحة والإمساك.

ويقال: الناس على دين ملوكهم[587]. فإذا كان الوالي[588] عفيفاً في نفسه عفّت حاشيته أيضاً تشبهاً به وتقرباً[589] إليه، فامتنعت عن[590] كثير من الدنايا كقبح المكاسب وخبث المطالب وقبول الرشا وغير ذلك. وينبغي أن ينافس[591] الأكفاء في فضائلهم من غير أن يحسدهم عليها، ويتصفح في ليله ما فعله في نهاره، فإن الليل أجمع للفكر وأحضر للخاطر.

ويقتصد في كل أموره محترزاً[592] عن الإفراط والتقصير، ويقتدي بأحسن ما يتصفحه من أفعال غيره، ولا يغفل عن الحذر والاحتراس وفعل الأحوط في كل أحواله. ولا يأمر بمعروف إلا بدأ بفعله، ولا ينهى عن منكر إلا بدأ[593] بتركه. ولا يأنف من حق إن لزم أو حجة[594] إن قامت، وأن لا يتفرغ[595] للهوى[596] فتتفرغ الرعية لإفساد ملكه، ويراقب[597] وليه كمراقبة عدوه.

580 طلب: يطلب، ج ي.
581 كل: -، أ م.
582 المستشيط: المستشيظ،ف.
583 أخلف: أخلط، ج؛ خلف، أ م.
584 يستفزّه: يستقر، م.
585 لان لا يهون: لئلا يكون، أ؛ لئلا يكون يكون، م.
586 موضعه: موضعها، م.
587 ملوكهم: ملكهم، أ ف ي.
588 الوالي: الواسط، ج.
589 وتقرباً: يقررنا، م.
590 عن: من، ج ي.
591 ينافس: يناقش، أ م.
592 محترزا: متحرزا، ف ي.
593 بدأ: يبدؤ، ج.
594 لزم أو حجة: لم أوجهه، ج ف ي.
595 يتفرغ: يتضرع، ج.
596 للهوى: للهو، ج ي.
597 ويراقب: ومراقبة، أ م.

ولا ينبغي أن يهمل أمر الدين معولاً على قوته وكثرة أجناده، فإنهم إذا لم يعتقدوا وجوب طاعته كانوا أضرّ[598] عليه من العدو، ولو سمعوا[599] بنابح[600] نبح[601] عليه قوي طمعهم في احتياج أمواله من غير أن يأمن سطوتهم به. والدين أسّ والملك حارسه[602]، وأصل ما تبتني عليه السياسة العادلة بعد حراسة الدين وتخير[603] الأعوان أربعة[604] أشياء، أحدها الرغبة الداعية إلى التآلف وحسن الطاعة، وثانيها الرهبة الحاسمة بخلاف ذوي العناد[605] وسعي أهل الفساد، وثالثها الإنصاف والعدل الذي به يستقيم حال الرعية، وإذا رغب الملك عن العدل رغبت الرعية عن الطاعة، ورابعها الانتصاف الذي فيه إعزاز[606] الملك وتوفير الأموال.

## الفصل الثالث[607]
## في استصلاح الأعوان وسياسة الرعية

ولا يستغني الملك عن الكفاة ولا الكفاة عن الأنصار[608] ولا الأنصار[609] عن المادّة ولا المادة عن العدل. والأعوان بمنزلة الأعضاء التي لا قوام للجسد إلا بها ولا يقدر على التصرف[610] إلا بصحتها واستقامتها، فيجب على الملك تقويم عوجهم وإصلاح فاسدهم وإبادة من لا يرجى صلاحه منهم، وينبغي أن يصرف كل واحد منهم[611] فيما طبع له وجبل عليه، ولا يعطي أحدهم منزلة[612] لا يستحقها ولا ينقصه عن مرتبته التي يستوجبها، ولا يرفع السفلة[613] الأدناء، ومن لم ينهض بعبء[614] ما استكفي فليحذر توليته بشفاعة شفيع أو لرعاية حرمة. ولا يطمع في أخذ

---

598 أضرّ: احر، أ م.
599 سمعوا: سمع، ف.
600 بنابح: بنابغ، ج ي.
601 نبح: نبغ، ج ي.
602 حارسه: حارس، أ م.
603 وتخير: وبغير، أ م؛ بغير، ج.
604 أربعة: أربع، أ م.
605 العناد: الفساد، م.
606 إعزاز: عز، م؛ أعز، أ.
607 الفصل الثالث: ج، ح ف ي.
608 الأنصار: الاتصال، أ م.
609 الأنصار: الاتصال، أ م.
610 التصرف: -، م.
611 منهم: -، أ م.
612 منزلة: مكرر في ي.
613 السفلة: -، م.
614 بعبء: بعبد، ج؛ بعبي، م.

مال الرعية فإن أموال الرعية[615] كالبضاعة للملك إن قلّت وأخفيت فلم يتصرف فيها قل ربحه وإن كثرت وأظهرت[616] كثر[617].

ولا ينفق[618] من الوجوه الدينية[619] مزاحماً فيها للرعية والعوام، ولا من الوجوه المحظورة شرعاً ولا من غش الذهب والفضة، فضرر ذلك كله أكثر من نفعه. ويحتاج الملك إلى تفقد خمس طبقات الوزراء والقضاة وأمراء الأجناد وعمال الخراج، والأقربون إليه في خلوته[620] لمطعمه ومشربه وملبسه وغير ذلك، ولا يكثر من الصرف والاستبدال، فإنه إن[621] اعتقد كل وال أن[622] أيامه قصيرة عمل لمصلحة نفسه في يومه[623] ولم يلتفت إلى إصلاح غده[624]. ومتى أطمع الملك نفسه في أن يأخذ من عماله اليسير أطمعهم في أن يأخذوا منه الكثير. وقد قيل: من خاف شرك أفسد أمرك، بل يجب أن يكثر التطلع عليهم، فإن ظهر له أنهم خانوه طالبهم مطالبة منصف ثم كان من وراء تأديبهم بما تقتضيه المصلحة في ردع غيرهم. ولا يكلفهم فعل ما ليس في وسعهم[625]. ويجب أن لا يذهب عنه صغير ولا كبير من أخبار رعيته وأحوال أعوانه، من دنت داره[626] منهم ومن بعدت، بمداومة الاستبحاث[627] وبثّ أصحاب الأخبار الصادقين الخالين من أغراض أنفسهم فيما يخبرون به.

وكثيراً ما أدخلت السعاية الشبهة حتى تصور الأمين بصورة الخائن والمحسن بصورة المسيء، فيستضر من سعى إليه أكثر من[628] استضرار من سعى عليه[629]. وإذا علم الأعوان أنه لا يخفى على ملك شيء من أفعالهم أقلع الخائن عن خيانته وازداد الناصح نصحاً في ولايته. وينبغي أن يقيم[630] رعيته مقام[631] عياله[632] في ارتياد مرادهم وصلاح معاشهم وصلة فقيرهم[633] ومراعاة خلة

615 فإن أموال الرعية: -، أ م.
616 وأظهرت: أظهرت، ج؛ وظهرت، م.
617 كثر: -، ا م.
618 ينفق: يترفق، أ ج ف.
619 الدينية: المدنية، أ م.
620 خلوته: حلومه، م.
621 إن: إذا، ي.
622 أن: -، أ م.
623 في يومه: -، ج ف ي.
624 غده: غيره، ج.
625 وسعهم: مقدورهم، م.
626 داره: دان، ج.
627 الاستبحاث: الاستسهب، ج؛ الاستصحاب، ف ي.
628 أكثر من: -، أ م.
629 عليه: إليه، ج.
630 يقيم: تنعم، أ م.
631 مقام: -، أ؛ كلمة غير مقروءة في م.
632 عياله: -، ج؛ وعياله، م.
633 فقيرهم: فقرهم، أ م.

كريمهم، ويهتمّ كل الاهتمام بأمن سبلهم وتهذيب طرقهم من أهل[634] الدعارة[635] والمفسدين، فإن ذلك يكثّر جلبهم ويخصب بلادهم وعقبى ذلك فيه نفع[636] عظيم للمملكة. وأن يساوي بينهم وبين نفسه في الحق لهم وعليهم، ولا يقدم في أخذ الحق شريفاً على مشروف، ولا قوياً على ضعيف، ولا خاصياً على عامي، ولا مسلماً[637] على ذمي، بل يجب أن يجعل[638] للمستضعف فضل عناية منه. وإذا رأى العامة متعصبة[639] على طائفة قاصدة لأذاهم بسبب مذهب أو دين[640] أو انحياز إلى شخص أو فئة[641] فليمل في جانب الطائفة الضعيفة، ولا يمكّن القوية من الاستيلاء عليها ولا من أذاها بوجه، وقطع أطماع العامة[642] من التسلط واجب، ومتى لم يجسم[643] المادة فيه فربما أدى ذلك إلى ضرر في الملك عظيم، ولأن الضعيف إذا لم ير من العامة مشفقاً ولم يكن الوالي له ناصراً فيلتجيء بكليته إلى الدعاء إلى الله تعالى بكل نيته غير ملتفت إلى ما[644] سواه، فيجيبه الذي يجيب المضطر إذا دعاه[645] كما هو في القضاء السابق فينتقم له ممن آذاه وممن خذله[646]، لا سيما إذا كان المظلوم معاهداً لا محابئً[647]. ولهذا قال[648] أهل التجربة: إن أذى أهل الذمة يزيل الملك ويزيل النعمة، وليس هذا مخصوصاً بهم، بل بكل مظلوم لا ناصر له.

ولا يمكّن أحداً من الرعية[649] أن ينتصف لنفسه ولو بالحق، فإن ذلك يؤدي إلى انتشار الجور وأخذ ما لا يستحق[650]، بل يجب أن ينصفهم وينتصف لهم، ويجب أن يمنعهم من الربا والقمار ومن كل اكتساب لا يحصل فيه معاوضة الانتفاع من الجانبين، ففي إباحة ذلك لهم ضرر عظيم في فائدة الاجتماع والتعاضد المدني. وخير الملوك من أمنه البريء وعافه[651] السقيم، وشرّ الملوك

---

634 أهل: أجل، أ م.
635 الدعارة: الدعاة، أ م؛ الذعارة، ي.
636 نفع: وقع، أ م.
637 مسلماً: ملياً، أ م.
638 يجعل: -، م.
639 متعصبة: متغضبة، ج ي.
640 دين: نية، م.
641 فئة: دينه، م.
642 العامة: الطاقة، أ ج م.
643 عليها . . . يجسم: -، أ م.
644 ما: من، ج ي.
645 دعاه: دكاه، أ.
646 خذله: خذله، أ ج ف.
647 محابئَ: مخائنا، أ ج ف ي.
648 قال: -، م.
649 الرعية: + إلى، ي.
650 يستحق: يستحقه، م.
651 وعافه: وخافه، ج ي.

من أمنه السقيم وعافه[652] البريء، وأن[653] يصاحب أهل العلم والخير والعقل، فإن الإنسان[654] موسوم بسيما من قارن ومنسوب إليه أفاعيل من صاحب، وليحذر من مصاحبة المنافقين والمتسمين بالعلم أو بالتنسك الذين لا يقصد بهما وجه الله العظيم.

ويكافيء المحسن بإحسانه ليألف الناس الإحسان رغبة في إحسانه[655]، ومن لم يره من المخطئين للعفو أهلاً[656] ورآه للعقوبة مستوجباً فليعاقبه تأديباً له أو للمصلحة العامة، لا غضباً عليه وتشفياً به.

وليحافظ[657] على الوفاء بالعهد ولا يخون من خانه فقد جرب أن عقوبة الغادر معجلة، وليحذر من دعاء المظلوم ولو كان كافراً، ولا يتهاون بصغار[658] الأمور فربما أدت إلى كبارها[659]. وليعلم أن العقل وحسن التدبير أنفع من كثرة الجند[660] والمال، ولهذا كان الإنسان متسلطاً على كثير من الحيوان الذي هو أشدّ منه قوة وبطشاً وجرأة.

## الفصل الرابع[661]
## في فائدة العدل والشفقة ووجه وجوبهما[662] وبيان الحكمة فيهما[663]

والعدل من سنة الله في خلقه وبه قامت السموات والأرض، يعرف ذلك من اطلع على أسرار الخليقة ومن تأمل خلق الإنسان في أعضائه وأرواحه وأخلاطه وقواه[664] الحساسة والطبيعية والحيوانية وكيفياته المتضادة[665] وما فيها من المعاضدة[666] وحسن النظام وانتفاع[667] الرئيس منها

652 وعافه: وخافه، ج ي.
653 وأن: ومن، م.
654 فإن الإنسان: والإنسان، أ م؛ فالإنسان، ي.
655 رغبة في إحسانه: رغبته، أ م.
656 للعفو أهلاً: أهلاً للعفو، ج ف ي.
657 وليحافظ: ليحافظ، ف ي.
658 بصغار: بصغائر، ف ي.
659 كبارها: كبائرها، ج ف م ي.
660 الجند: الجد، أ م.
661 الفصل الرابع: د، ج ف ي.
662 وجوبهما: وجوبها، أ م.
663 فيهما: فيها، أ؛ منها، م.
664 وقواه: وقوة، أ م.
665 وكيفياتها المتضادة: وكيفيته المضادة، ج.
666 المعاضدة: المعاحذة، م.
667 وانتفاع: وارفاع، أ؛ وارتفاع، ج م.

بالمرؤوس[668] والمرؤوس[669] بالرئيس، يرى من العدل ما يبهره[670] عجائبه. ويتحقق ذلك من علوم الأطباء وإن كان لا سبيل لهم إلى الإحاطة بجميعه، وكذلك خلق غير الإنسان كأنواع الحيوان والنبات والمعادن وأصنافها وأشخاصها، وكذلك العناصر وسائر ما[671] في السماوات والأرض. وقد شبّه العالم بإنسان كبير والإنسان بعالم صغير، وقد بيّن كيفية ما في ذلك من العدل في الكتب المبسوطة. والجور إذا وقع في العالم فهو كالمرض للشخص الواحد أو كأيام الوباء العام، وكما اقتضت الحكمة الربانية[672] والعناية الإلهية والرحمة الشاملة أن تكون الصحة أكثر من المرض، فلذلك[673] أوجبت على رعاة[674] الأمة اعتماد العدل وترك الجور ليكون العدل هو الغالب على الجور، فلذلك كان الجائر مضاداً لله في حكمه كاذباً عليه في خبره[675] مغيراً[676] لشريعته مخالفاً له في خليقته[677]، فجدير لمن[678] اقتضت رحمته تغليب العدل على الجور أن لا يطيل أمد[679] رعايته[680]، ولهذا حكم أرباب التجارب بأن الملك يبقى على الكفر ولا يبقى على الظلم، وما ذلك إلا بالعناية[681] من الحكيم[682] العادل على الوجه الذي بينته[683] لا غير.

وقد روي أن الخلق كلهم عيال الله، وأن أحبّ الخلق إلى الله أحوطهم لعياله. وفي التورية ما معناه: واختر[684] لصاحبك ما تختاره لنفسك، وتحتوي هذه الكلمة على أكثر مكارم الأخلاق ومصالح السياسات. ويجب أن لا تقتصر شفقتك على نوع البشر خاصة، بل وعلى كل ذي روح حساس، فإن عناية البارئ كما اتصلت بك فهي متصلة به، وما أباحته[685] الشريعة من استخدام بعض الحيوانات وقتل بعضها إما للتغذية[686] أو لدفع الأذى فيجب أن يستعمل الشفقة

---

668 منها بالمرؤوس: فيها المرؤوس، أ م.
669 والمرؤوس: -، أ م.
670 يبهره: سره، أ م.
671 ما: من، م.
672 الربانية: -، م.
673 فلذلك: لذلك، أ م.
674 رعاة: دعاة، أ ف م.
675 خبره: -، م.
676 مغيرا: مغير، أ ج؛ مصير، م.
677 خليقته: خلقه، أ م.
678 لمن: بمن، م ي.
679 أمد: أيد، ج ي.
680 رعايته: رعاته، ج.
681 بالعناية: للعناية، أ ج ف ي.
682 الحكيم: الحليم، أ م.
683 بينته: يبنيه، م.
684 واختر: اختر، ج ي.
685 وما أباحته: وإباحته، أ ف م.
686 للتغذية: لتغذية، ف ي.

عليها في جميعه، فيستخدم أرفق استخدامٍ ويقتل أسهل قتل، ولعل لها فيما أبيح منها من الأعواض[687] والمصالح ما لا[688] تحيط به علماً. ولا تستبعدن ذلك بعد مشاهدتك ما في خلق[689] أقلّ بعوضة من الحيوان من الحكم العجيبة والإلهامات الغريبة، ولا تلتفت[690] إلى حكم من جزم بأن لا نفس[691] مجردة لها، فليس في العلوم الحكمية ما يدل على ذلك، بل فيها وفي الأقاويل المأخوذة من مشكاة النبوة ما يوجب خلاف ذلك أو يرجحه. وإذا[692] وجدت في كلام بعضهم أن لا نفس مجردة لها فافهم منه أنه لم تثبت لها عنده تلك النفس، لا أنه ثبت[693] أن ليست لها، وبين المفهومين فرق عظيم.

وقد جرت عادة المتقدمين بذلك كما في قولهم أن آخر[694] الأفلاك هو التاسع مع تجويزهم الوفاء من الأفلاك، وإنما أرادوا أنه لم يثبت عندهم فلك عاشر، لا أنه يثبت عندهم عدمه[695] في نفس الأمر. وكثير من أقاويلهم تجري على هذا الأسلوب، فلا تستنكر أن نفوس الحيوانات العجم لا تعدم وأن لها أعواضاً بعد الموت.

## الفصل الخامس[696]
## في آداب الولاة الذين من قبل وال آخر وفي ما به يختم الكتاب

وإذا كان متولي الرعية[697] معيَّناً[698] من قبل سلطان مستقل، أو من قبل متول آخر أعلى مرتبة منه، فيجب عليه النصيحة له وكتمان سرّه، وأن لا يطلب ما عنده بمسألة، ولكن بالإحسان في الخدمة والمعاشرة، وأن لا يعتدّ ويتبجح بكفايته البالغة، وأن لا يظهر عيوب مخدومه ولا يقارب من أبعده ولا يباعد من قرّبه، وأن لا يعادي الأمر الأغلب لا ظاهراً ولا باطناً، وأن يتجنب

---

687 الأعواض: الأعراض، أ ف م.
688 لا: -، أ م.
689 في خلق: خلق في، ج ي.
690 تلتفت: تليف، أ.
691 بأن لا نفس: بأنفس، م؛ الأنفس، أ.
692 وإذا: فإذا، أ م.
692 ثبت: يثبت، م.
694 آخر: أواخر، أ.
695 عدمه: كلمة غير مقروءة في أ م.
696 الفصل الخامس: ٥، ج ف ي.
697 الرعية: + أو غيرهم، أ ي.
698 معيَّنا: معين، ج م؛ -، ي.

مواقع الظنة[699] وحدوث الغيبة[700]، وأن لا يمدح أو يذم من لم ينل أخباره على حقائقها. وينبغي أن يودع[701] قلوب الرعية شدة المحبة له[702] وإذا أتى[703] بصواب رأي أو بأمر مستحسن فلا يجتهد في تقبيحه وتغيير رأي الملك المستقل أو غيره فيما رضيه[704] منه حذراً من تقدمه عند الملك ومن يجري مجراه بما سبقه إليه أو فاق[705] فيه عليه.

وهذه جمليات لها تفاصيل طويلة والاعتماد على جودة الحدس[706] وحسن القريحة[707]، فمع فقدهما لا ينفع شيء مما قيل، ومع وجودهما يسهل[708] تفريع ما قلت وتفصيله إن أعانت العناية الربانية عليه. وجملة الأمر أن غاية الكمال في ما يكون عليه الإنسان في ذات نفسه وفي مشاركته[709] مع من سواه هو التشبه بالله تعالى بحسب الممكن، وتحت[710] هذا أسرار[711] عظيمة لا يحتمل هذا المختصر شرحها، وقد أشير[712] في التوراة أيضاً إلى هذا المعنى، وأشار إليه أكابر الحكماء المتألهين[713].

فهذا ما رأيت أن أذكره من الفوائد على حكم ما اقتضته[714] الحال الحاضرة وأكثره يلتقط من كلام من تقدم، إنما الأقلّ النزر[715] هو الذي أدركته بفكري أو سنح لخاطري[716]، وفيه مباحث غريبة كثيرة النفع، ولولا ضعف بخاري في البصر يمنع من كثرة الكتابة وطول المطالعة وموانع أخرى لبسطت الكلام بزيادة مسائل وبراهين دقيقة ذكرتها في غير هذه الرسالة، إنما الغرض من هذه ذكر مهمات العقائد والأعمال لا غير، وفي كتب[717] من تقدم كفاية وأنا معترف بالتقصير

---

699 الظنة: المظنة، ج ي.
700 الغيبة: الريبة، ج ف ي.
701 يودع: يتودع، ف ي.
702 المحبة له: للحنة، م؛ المحبة، أ.
703 أتى: + أحد، ي.
704 فيما رضيه: رصيفة، م؛ رصيفته، أ.
705 فاق: فارق، ج ف ي.
706 الحدس: الحسن، أ م.
707 القريحة: القريح، م.
708 يسهل: -، ج ف ي.
709 وفي مشاركته: ومشاركته، أ م.
710 وتحت: وبحث، أ ف م.
711 أسرار:+ وأغوار، ف ي.
712 أشير: اشر، ج م.
713 الحكماء المتألهين: العلماء والمتألهين، ج ي؛ العلماء المتأهلين، ف.
714 اقتضته: اصفته (؟)، ج.
715 النزر: + منه، ي.
716 لخاطري: خاطري، أ م.
717 كتب: -، م.

وقلّة البضاعة من العلم، وأسأل الله تعالى العفو والعافية، وأن يوفقنا جميعاً لإصلاح[718] العلم والعمل، ويعصمنا من الزيغ والزلل، وأن يجعلنا لسبله من المنتهجين، وبرضائه في دار الأبد من المبتهجين، إنه لطيف كريم رؤوف رحيم[719].
والحمد لله رب العالمين وصلواته على ملائكته المقربين وأنبيائه المرسلين وأوليائه المخلصين وخصوصاً على محمد وآله الطاهرين. وأختم بالخير[720].

في آخر م:

قد أمر باستنساخ هذه النسخة الشريفة مقرب ظلّ السلطان المستشار والمؤتمن حاجي محمد شريف خان مستشار لشكر، وتمت في يوم الأحد في إحدى وعشرين شهر محرم الحرام سنة أربع وثمانين ومأتين بعد الألف بيد الأقل تراب أقدام الفقراء محمد بن حسين (؟) الحاج حاجي ناد علي في أرض الأقدس.

في آخر ج:

وقد تم تسويده في شهر صفر سنة ستين وألف من الهجرة المقدسة والحمد لله وحده.

---

718 لإصلاح: لصاح، ج ي.
719 رؤوف رحيم: -، أ م.
720 وصلاته ... بالخير: -، م؛ واختم بالخير: -، ج ي؛ هو الذي أدركته بفكري ... وأختم بالخير: -، ف.

## *Text 2*

### *Ithbāt al-mabdaʾ wa-ṣifātihī wa-l-ʿamal al-muqarrib ilā llāh*

The edition is based on the single known manuscript of the text found in thc collcction of Muḥammad Alī Rawḍātī in Isfahan. It was available to us on the basis of the microfilm copy extant in the Central Library of Tehran University (microfilm no. 2712). The manuscript is partly corrupt, but it was possible to make a number of ementions on the basis of a comparison with Ibn Kammūna's *Kalimāt wajīza mushtamila ʿalā nukat laṭīfa fī l-ʿilm wa-l-ʿamal* (= Text 1), of which the text presented here is a more concise version.

رسالة شريفة في إثبات المبدأ[1] وصفاته والعمل المقرب إليه

بسم الله الرحمن الرحيم وبه ثقتي

قال العبد الفقير إلى رحمة الله تعالى سعد بن منصور بن سعد بن الحسن [بن] هبة الله بن كمونة عفا[2] الله عنه في دنياه وأخراه:

كمال نفس الإنسان هو إما من جهة قوتها النظرية بالعلم أو من جهة قوتها العملية بالعمل[3]، وكل العلوم والأعمال على اختلافها وكثرة تشعبها ترجع إلى كمال هاتين القوتين. والقوة البشرية عاجزة عن الإحاطة بجميع ذلك، بل عن استقصاء فن واحد منه، فكيف جميعه؟ إنما يجب على الراغب في الكمال أن يحصل منه أبهى ما يمكنه تحصيله مما هو الأهم منه والأولى بتجريد العناية له.

ولا شك أن أصل العلوم هو معرفة الحق الأول ووحدانيته وما له من صفة الجلال والإكرام. وأفضل الأعمال ما كان مقرباً إليه ومزلفاً لديه، وهذا أيضاً بحر واسع طالما انكسرت سفينة الأفكار في تلاطم أمواجه، وجنابه قدسي وقل من اهتدى إلى طرقه وأنهاجه. فمن لم يتمكن من التغلغل في بحار تلك الحقائق والتوغل في تلك السبل والمضائق فلا أقل من أن يعتمد على أن ينظر بعين الاعتبار إلى هذا العالم[4] وما فيه من إتقان الصنعة وحسن النظام، وما يشتمل

---

[1] المبدأ: المبدىء، الأصل.
[2] عفا: عفى، الأصل.
[3] بالعمل: بالعلم، الأصل.
[4] العالم: العلم، الأصل.

عليه من التأليف اللطيف والترتيب العجيب والإحكام البالغ، وما في بدن الإنسان وغيره من الحيوان من الحكمة العظيمة في أعضائه ومنافعها وارتباط النفس بها وآثار قواها[5] فيها. فإنه إذا اعتبر بذلك فلا بد وأن يؤمن إيماناً تاماً بأن لهذا العالم مدبراً موجوداً حكيماً قادراً مريداً للأصلح والأفضل. وهذا المدبر إن استغنى عن الموجد فهو الإله جلّ جلاله، وإن احتاج إلى الموجِد فلا بد من الانتهاء إلى الموجد المحيط علمه الكاملة قدرته البالغة حكمته، فإن موجد الأثر المحكم لا بد وأن يكون أبلغ إحكاماً من الأثر الصادر عنه، فله الكمال الأعلى. وهو الجواد بالحقيقة، واسم الجواد على غيره مجاز، لأنه أفاد الخير والأنعام من غير غرض وفائدة ترجع إليه. وهو الغني التام لكونه غير متعلق بشيء خارج عنه، لا في ذاته ولا في صفاته، وكل ما عداه فقير إليه. وهو الملك الحق، لأن ذات كل شيء هي له لأنها[6] منه أو مما منه وجوده وكماله، ولا يستغني عنه شيء في شيء، وكل شيء بالقياس إلى ذاته باطل وبه حق.

ولا يعزب عن علمه شيء من مخلوقاته، إذ الفاعل بالقصد والإرادة لا بد وأن يعلم ما خلق. ونحن فلا نفهم شيئاً من نعوته إلا بالمقايسة إلى ما نفهمه من أنفسنا، لكن ما نفهمه في حقه هو من الكمال والزيادة إلى حد لا سبيل لنا إلى تصوره لأن مثل تلك الزيادة لا توجد[7] في حقنا، بل هي زيادة لا تتناهي. فهكذا ينبغي أن يعلم حال علمه وقدرته وغيرهما، وكل نعت فيه لا نظير له فينا فلا [سبيل] لنا إلى فهمه. وما ذكر من نعوت جلاله إنما هو بقدر ما في وسعنا أن نعلمه لا بقدر ما يستحقه لذاته. ولا تساوي حقيقته المقدسة حقيقة شيء من مخلوقاته، وإلا لكان[8] المخلوق بكونه مخلوقاً أولى من كونه خالقاً، ولا[9] ندّ له. وإذ هو عالم فليس بمركب لأن العلم بالجزء متقدم على العلم بالكل تقدماً بالذات. فمن يعلم ذاته يعلم جزءه أولاً، وكل من علم شيئاً فهو عالم بأنه عالم به، وذلك يتضمن علمه بذاته، فيكون قد علم ذاته قبل علمه بجزئه الذي هو معلوم قبل ذاته، وذلك محال.

وإذ ليس بمركب فليس بجسم. وإذ ليس بجسم فليس بمتحيز، لأن كل متحيز ليس جسماً فهو لا يتجزأ، وخالق الكل يتعالى عن ذلك. وإذ لا ندّ له ولا هو مركب فلا صاحبة له ولا ولد، وإلا لانفصل منه ما يصير مساوياً له في النوع أو الجنس. ولا يحلّ في غيره إذ الحلول لا يتصور إلا إذا كان تعين الحال بتوسط المحل. وإذ لا يحلّ في شيء فلا ضد له[10] يجامعه في محله، وأيضاً

5 قواها: قويها، الأصل.
6 لأنها: لأن، الأصل.
7 توجد: يوجد، الأصل.
8 لكان: لم يكن، الأصل.
9 ولا: فلا، الأصل.
10 له: + لا، الأصل.

لا ضد له بمعنى أنه يساويه في القوة ما يمانعه. ولا يتعلق ببدن كالنفس فإن قدرته أوسع وأعم من أن تتخصص[11] ببدن يصدر عنه. ولا يجوز أن ينعدم، وإلا لم يكن وجوده من ذاته فافتقر إلى موجد، وحينئذ يكون ذلك الموجد هو الإله لا هو.

وأيضاً، هو وحداني لا شرط له في ذاته ولا مضاد، وما سواه فتابع له فلا يتصور له مبطل، بل هو الأول والآخر. ويعلم من ارتباط أجزاء العالم بعضها ببعض أن صانعه واحد لا غير، إذ لو كان له صانعان لتميز صنع كل واحد عن صنع غيره، فكان ينقطع الارتباط والتعاون فيختل النظام، وكان الذهن السليم يستغني بها عن غيره من الأدلة. ويدلنا على هذا التعاون افتقار أعراضه إلى جواهره باعتبار وافتقار جواهره إلى أعراضه باعتبار آخر، وكذا متحيزاته ومجرداته وعنصرياته وفلكياته وحيوانه ونباته ومركباته محتاجة إلى بسائطه، وبسائطه محتاجة بعضها إلى بعض في تكوّن المركبات منها، وأنواع الحيوانات وأشخاصها محتاج بعضها إلى بعض، وكذا أعضاء الشخص الواحد منها. ولا سبيل لنا إلى استقصاء جميع وجوه الارتباط، فالعالم كشخص واحد مركب من نفس ذات قوى كثيرة ومن بدن مؤلف من أعضاء متشابهة وغير متشابهة ذوات قوى وأفعال مختلفة وغير مختلفة، يستغني[12] بعضها ببعض وينتفع بعضها ببعض انتفاعاً، بعضه محسوس وبعضه معقول. فلو استبدّ أحد الصانعين بتدبيره فهو الإله لا غيره، وإن اجتمعا وكان كل واحد منهما كافياً في التأثير والتدبير في العالم لاستغنى العالم بكل واحد[13] منهما عن كل واحد منهما، فتعطلا عن التأثير معاً على تقدير تأثيرهما معاً، واختصاص الواحد بالتأثير في البعض والآخر بالتأثير في البعض الآخر متعذر فيما نحن فيه لأن الذي يفعل شيئاً فيتبع وجوده وجود آخر أو ينتفع الآخر فلا محالة له تأثير في شيئين، فلا يكون الثاني مستبدّاً بالبعض الذي فرض استبداده به، وكون أحدهما يفعل في زمان والآخر في زمان آخر ظاهر الاستحالة عند تأمل ذوي العقول السليمة، وكيف فرض الحال فهو عند التأمل يؤدي إلى محال. وإن أمكن عالم آخر فوجوده[14] يقتضي إيجاده، فإله العالم لا شريك له ولا ظهير، ولا معلّ ولا مشير، سبحانه وتعالى عن توهمات المشبهين والمشركين.

فهذه نكت نافعة في معرفة الله جلّ جلاله وما يجب له وينتفي عنه، ما أسعد من يتحققها وأهدى من يسترشد بها. ومن تحقق ما بين وأوضح من عناية البارئ بمخلوقاته ورحمته لهم علم أن العناية في[15] خلق الإنسان ليس أمراً زائلاً لأنه كلما كان أجلى كان خوف زواله أقر، فكيف

[11] تتخصص: يتخصص، الأصل.
[12] يستغني: يستبقي، الأصل.
[13] واحد: واحدة، الأصل.
[14] فوجوده: فجوده، الأصل.
[15] في: من، الأصل.

يصلح عناية[16] مثل هذا الحكيم في أقل بعوضة من الحيوان، فكيف في خلق الإنسان الذي يظهر في خلقه من آثار العناية به ما يقتصر منه آخر العجب؟ فلا بد له من نفس لا تعدم هي غير هذه[17] البنية الفانية، بل البنية إنما أحكمت لتكون آلة لها في الاستكمال، وغايتها هو النعيم الدائم. وقد علمت أن كل ما يدرك ذاته فهو غير مركب، فالنفس غير مركبة.

وإذا تأملت الملكات التي لا تتجزأ[18] بالتجزئة الاتصالية كملكة الفطنة[19] والعلم والشجاعة والجبن والتهور وغيرها مما يقبل الشدة والضعف ولا يقبل الانقسام الاتصالي تحقق لك أن محل هذه هو النفس لا يكون جسماً ولا شيئاً حالاً في الجسم حلول السريان فيه، إذ كل ما محله كذا فهو قابل للانقسام[20] ويجب قبول محله [له]. ولا يجوز أن تكون النفس حالة في غيرها، وإلا لكان وجودها هو وجودها لغيرها ولا تستقلّ بفعل من الأفعال، بل ينسب الفعل إلى محلها بها كما ينسب الفعل إلى محل القدرة والإرادة بهما لا إليهما، فيقال: فعل بالإرادة، ولا يقال: الإرادة فعلت بمحلها، وهذا أمر واضح عند عقول أهل الحقائق. فالنفس جوهر بسيط، وإذا استقرت بسائط الجواهر في العالم تجدها قابلة للعدم، بل المتقدم هو التأليف والتركيب. وربما اكتفى ذو الفطانة بهذا في أن النفس لا تقبل العدم باستعداد يحصل لها[21] للحكم الكلي من اعتبار حال الجزئيات كما في التجربة. وتعلق النفس بالبدن إنما هو تعلق تدبر وتصرف واستكمال، لا تعلق حلول. وكما أن فساد عضو واحد لا يقتضي عدم النفس كذلك فساد سائر الأعضاء، وليس الموت لازماً به جملة الأعضاء، وفقد الآلة لا يقتضي فقد مستعملها. واعتبر كيف بدنك دائماً في التحلل والتبدل وكيف تعلم أنك أنت الذي كنت منذ ثلاثين سنة مثلاً، وهذه تنبيهات عساها تكون مقنعة لكل ذي لب، فالقدر المورد إلى ههنا قد حصل منه الإيمان بحال المبدأ[22] والمعاد على سبيل الإجمال من غير تعمق في البراهين الدقيقة والتقريرات العميقة والمقدمات المتشعبة والمسالك المستصعبة.

فمن أراد ذلك فلا بد له من تحصيل هذه المطالب وما ينضاف إليها من التبيين والاشتغال بما في الكتب المطولة إن ساعدته الفطانة والتوفيق، فهذا ما يتعلق بالعلم الأعلى.

---

16 عناية: غاية، الأصل.
17 هذه: هذا، الأصل.
18 تتجزأ: تتجزى، الأصل.
19 الفطنة: العطية، الأصل، والتصحيح عن كلمات وجيزة.
20 للانقسام: الانقسام، الأصل.
21 لها: له، الأصل.
22 المبدأ: المبدء، الأصل.

وأما العلم المقرب إلى الحضرة القدسية المقتضي للسعادة الأبدية فمن لم يقدر منه على ما يرتقي إلى المقامات الموصلة إلى علم المكاشفات والأسرار العزيزة، ولم يتمكن أيضاً من استقصاء ما دون ذلك فينبغي أن لا يخل بشيء مما أذكره في هذه اللمعة، وهو أن يشتغل بوظائف العبادات الواجبة، فإنها جارية مجرى رأس المال الذي به أصل التجارة، ويستكثر من النوافل ما أمكنه، فهي الربح الذي به الفوز بالدرجات. وما فضل من أوقاته فالأولى أن يصرفه في العلم الذي يزيد في بصيرته بعيوب نفسه وفي معرفته بعبادة ربه، ويقلل من الرغبة في الدنيا ويزيد في الرغبة في الآخرة، لا ما ألّب الناس عليه وسموه علماً. فإن لم يقدر على العلم فليشتغل بنفع الناس أو بالاكتساب على عياله.

واعلم أن روح العبادة الإخلاص وحضور القلب. وقد يرتفع من أحوال الجوارح أنوار إلى القلب، كما ينحدر من معارف القلب آثار إلى الجوارح، وترك المناهي أشد على الإنسان من فعل الطاعات. وإنما تعصي بجوارحك التي هي نعمة من الله تعالى عليك، فاحذر أن تستعين بنعمته على عصيانه، فلا تنظر بعينك إلى محرّم ولا إلى شيء من مخلوقات الله بعين الاحتقار. ولا تصغ بأذنك [إلى] الغيبة والفحش أو الخوض في الباطل وذكر مساوئ الناس، واحفظ لسانك من الكذب ولو هزلاً. وإذا أردت أن تعرف قبحه فانظر إلى استحقارك لصاحبه، واستقبح من نفسك ما تستقبحه من غيرك، ولا تخلف الوعد وحافظ على الوفاء بالعهد. ولا تذكر إنساناً بما يكرهه لو سمعه لا تصريحاً ولا تعريضاً، واطرح[23] المراء[24] والجدال والمناقشة في الكلام، ولا تزك نفسك ولا تمدحها فذلك نقص[25] من قدرك[26] [عند] الناس. ولا تلعن شيئاً مما خلق الله تعالى إنساناً أو غيره، ولا تدخل بين العباد وبين [الله]، فلا إثم في ترك لعن إبليس فكيف غيره؟ ولا تدع على أحد وإن ظلمك ووكّل[27] أمره على الله. واترك[28] السخرية والاستهزاء وكثرة المزاح والبطن فاحفظها من تناول الحرام والشبهة. واحرص على طلب الحلال واقتصر على دون الشبع فإن الشبع يثقل الأعضاء وينصر جنود الشيطان. واحفظ فرجك من كل ما حرّم الله واستعن على ذلك بترك الشبع[29] والنظر والفكر في المحرم. واحفظ يديك من أن تؤذي بهما أحداً ورجليك من أن تسعى بهما إلى الحرام أو تمشي بهما إلى ظالم، فهذا ما ينبغي أن يحفظ به الجوارح الظاهرة.

---

23 واطرح: واصطرح، الأصل، والتصحيح عن كلمات وجيزة.
24 المراء: المرء، الأصل، والتصحيح عن كلمات وجيزة.
25 نقص: يفيض، الأصل، والتصحيح عن كلمات وجيزة.
26 قدرك: مددك، الأصل.
27 ووكل: وكل، الأصل.
28 واترك: وترك، الأصل.
29 الشبع: السمع، الأصل، والتصحيح عن كلمات وجيزة.

وأما رذائل القلب الباطنة فكثيرة وعلاجها غامض[30]، فمنها قوة الغضب وهو جنون غير مستقر، فينبغي أن ينكسر[31] باعتياد الحلم مع التعرض للمغضبات وما فيه من الرذيلة وما في كظم الغيظ من الثواب والفضيلة، وسببه الأعظم الكبر[32] فبانكساره ينكسر. وحقيقة الكبر أن يرى الإنسان نفسه فوق غيره وهو منازعة لله تعالى في خصوص صفته، وعلاجه أن يعرف الإنسان حقارته بالقياس إلى سكان العالم الأعلى في العلم والعبادة أو القدرة وأنه من نطفة قذرة، وأن كل ما له من مال وجمال وأتباع فهي أمور خارجة عن ذاته ولا تبقى له، وأنه جاهل بوقت زوالها وبأكثر أحواله، وأنه عاجز عن أحقر شيء. والعُجب جهل محض وهو سبب الكبر. فيجب على[33] الإنسان إذا استعظم شيئاً أن يتفكر أن زواله مخوف على القرب ويعجب من إعطاءه ذلك من غير استحقاق لا من نفسه[34]. وأن يقول لنفسه: كما أعطاني هذا بغير سبب فيوشك أن يسلبه بغير سبب.

ومن الرذائل الحسد، وهو يضرّ الحاسد ولا يضرّ المحسود، وهو حبّ زوال نعمة الله تعالى عن عباده. والعلم بذلك كاف للعاقل في إزالته. والخلق كلهم عباد له، فيجب محبتهم لأنهم عباد المحبوب وعياله. والبخل داء، دواءه العلم بأن المال لله، مكنه منه ليصرفه في[35] أهمّ أموره، فإمساكه إن كان ليتنعم بالشهوات فالثواب وحسن الأحدوثة ألذّ. وإن كان ليخلفه لولده فإنه يترك ولده بخير ويقدم على الله بشر، وولده إن كان صالحاً فالله تعالى يكفيه، وإن كان فاسقاً فقد أعانه على المعصية. ومن حمل نفسه على مداومة البذل[36] تكلفاً[37] صار ذلك له عادةً. وحبّ الجاه أشدّ من حبّ المال لأن التوصل به إلى المال أيسر من التوصل بالمال إليه، وهو من الصفات الإلهية فلذلك كان محبوباً بالطبع لسرّ خفي في مناسبة الروح[38] للأمور الإلهية. ولذلك لما تعذر على الإنسان الاستيلاء على السماوات[39] والملائكة والجبال والبحار اشتهى الاستيلاء على جميعها بالعلم، فإن من يعجز عن وضع الأشياء العجيبة يشتهي أن يعرف كيفية الوضع. ولما تصور أن يتسخر له الإنسان أحبّ [إليه من] أن يستسخره ويملك قلبه، ولهذا يحب الإنسان

30 غامض: رامض، الأصل.
31 ينكسر: يكثر، الأصل.
32 الكبر: الأكبر، الأصل.
33 على: أن، الأصل.
34 لا من نفسه: لا بنفسه، الأصل، والتصحيح عن كلمات وجيزة.
35 في: إلى، الأصل.
36 البذل: البدن، الأصل.
37 تكلفا: مكلفا، الأصل.
38 الروح: الرفع، الأصل، والتصحيح عن كلمات وجيزة.
39 السماوات: السموات، الأصل.

أن يتسع جاهه وينتشر صيته إلى بلاد لا تطأها [أقدامهم][40] ولا يرى أهلها. والرفعة الحقيقية هي القرب إلى الله تعالى وهو محمود، وإنما المذموم طلب الكمال الوهمي فإنه أمر عارض لا بقاء له، ولو سجد لك أهل الأرض جميعاً لما بقي الساجد والمسجود له. ويكفيك من الجاه القدر اليسير لتتحرس به من الظلم والعدوان، وطلب ما زاد على ذلك جهل ورذيلة.

وحبّ الدنيا رأس كل خطيئة. وكل ما لا يبقى معك بعد الموت فهو من الدنيا، وإنما خلقت للتزوّد ههنا إلى الآخرة. فمن اقتصر منها على قدر الضرورة فقد أنجح، ومن اشتغل بلذاتها ونسي المقصد منها هلك، ومن ظن أنه يلابسها ببدنه ويخلو عنها بقلبه[41] فهو مغرور. ومن أعظم الآفات الرياء، وهو طلب المنزلة في قلوب الناس بالعبادات وأعمال الخير. ومن قصد بالعبادة العباد فقد اعتقد أنهم أقدر على نفعه وضره من الله، ويقرب علاجه من علاج حبّ الجاه. ويكفي المرائي من الناس لو علموا ما في بطنه من الرياء لمقتوه[42] وفي إخفاء العبادة السلامة من الرياء. والخلق هيئة للنفس يصدر عنها الفعل بسهولة من[43] غير رؤية وتكلف، لكن التكلف قد يكون طريقاً إلى تحصيل الخلق بالتعويد، وتفاوت الناس في الحسن باطناً كتفاوتهم في الحسن ظاهراً. وقد يظن الإنسان بنفسه حسن الخلق وهو عاطل عنه. ومن مواقع الغرور أن يغضب مثلا لغير الله ويظن أنه يغضب لله.

ومبدأ[44] طريق السالكين ومفتاح سعادة المؤيدين [هو][45] التوبة، وهي الرجوع من طريق البعد إلى طريق القرب، ومعنى قبولها [هو][46] أن يحصل في قلبك استعداد القبول لتجلي أنوار المعرفة. ورأس الحكمة مخافة الله تعالى، ولا ينبغي أن يفرط بحيث يؤدي إلى[47] القنوط فذلك مذموم. والرجاء يخالف التمني، فإن من يتعهد الأرض ولا يبث البذر ثم ينتظر الزرع فهو متمن[48] مغرور، إنما الراجي من حصل كل غرض يتعلق باختياره، ثم بقي يرجو من الله تعالى دفع الموانع. ومن نظر إلى رحمة الله ورأفته بخلقه وعنايته بهم غلب رجاؤه خوفه وأيقن أن الرحمة والخير[49] مقصودان[50] بالذات، فإن الشرّ في العالم مقصود بالعرض.

---

40 [أقدامهم]: الزيادة عن كلمات وجيزة.
41 بقلبه: نعليه، الأصل، والتصحيح عن كلمات وجيزة.
42 لمقوه: لمسوه، الأصل.
43 عن: من، الأصل، والتصحيح عن كلمات وجيزة.
44 ومبدأ: ومبدء، الأصل.
45 [هو]: الزيادة عن كلمات وجيزة.
46 [هو]: الزيادة عن كلمات وجيزة.
47 يؤدي إلى: يورث، الأصل.
48 متمن: متمني، الأصل.
49 الخيرَ: + هي، الأصل.
50 مقصودان: مقصودة، الأصل.

وكمال الزهد في الدنيا هو الزهد[51]، فمن[52] اعتدّ بالزهد فيها وتركها[53] فقد عظمها. والدنيا عند ذوي البصائر كلاشيء، إذ الدنيا شيء متناه والآخرة لا نهاية لها. وزهد المحققين استحقار ما سوى الله، وزهد غيرهم معاملة[54]. وجميع ما يلقى العبد في هذه الحياة إما أن يوافق هواه أو يخالفه، وفي الحالتين يحتاج إلى الصبر. أما في الموافق فلئلا يطغى فينسى[55] مبدأه[56] ومعاده، وأما في المخالف فظاهر. ومن علم أن النعم كلها من الله تعالى وأن الوسائط مسخرون مقهورون لم يشكر إلا الله تعالى وحده. ومن اعتقد أن لغير الله دخلاً في النعمة الواصلة إليه لم يصح حمده ولم تتم معرفته وشكره، وكان كمن خلع عليه السلطان، أي الملك، [ولو رأى توقيع الملك بذلك الإنعام بقلمه لما التفت قلبه][57] إلى شكر القلم ولا إلى الخازن والوكيل لأنهم مسخرون مضطرون. والنية والعمل بهما تمام العبادة، والنية خير الجزئين لأنها تجري مجرى الروح والعمل مجرى الجسد. والنية ميل القلب إلى الخير فاستكثر منها، وثواب العمل على قدرها. فالنوم بنية عود النشاط إلى فعل الخير خير من العبادة على الملال.

وعليك بالإخلاص والصدق. ومن جملة الصدق تحقق القلب بأن الله تعالى هو الرزاق وأن لا فاعل سواه، بل لا يرى في الوجود غيره، وإنما الكثرة منه في حق من يفرق نظره كالذي يرى من الإنسان عضواً فعضواً، فإن رآه جملة لم يخطر بباله الآحاد ورؤية الكثرة، وارتباط الأسباب والمسببات لا يقدح في حقيقة توكل العبد على الله، بل إذا انضاف إليه الإيمان بالرحمة والجود والحكمة حصلت الثقة بالتوكل. وكل ما يظن في الوجود من نقص فيرتبط به كمال وخير أعظم، ولم يبخل الله على الخلق أصلاً ولا أخّر في صلاحهم أمراً، فهو حقيق بأن تفوض أمرك إليه وتثق به. وليس من شرط التوكل ترك الكسب أو ترك التداوي مثلاً لأن ارتباط المسببات بالأسباب من سنة الله تعالى وهي لا تتغير، بل يجب أن تعتقد أن اليد والطعام والبذر وقوة التناول جميع ذلك من قدرة الله تعالى، فلا تتوكل إلا على خالقها لا عليها. والكمال محبوب لذاته وكل كمال وجمال وبهاء في الوجود فهو من كمال الله تعالى وجماله وبهائه، بل لا نسبة لكمال غيره إلى كماله وكيف ينسب المتناهي إلى ما لا يتناهى.

---

51 الزهد: كذا في الأصل.
52 فمن: فإن من، الأصل.
53 إعتدّ بالزهد فيها وتركها: اعتدّ تركها، الأصل.
54 معاملة: مقابلة، الأصل.
55 فينسى: فيسؤ، الأصل.
56 مبدأه: مبدءه، الأصل.
57 [ولو توقع... قلبه]: الزيادة عن كلمات وجيزة.

والعارف لا يحبّ إلا الله تعالى، فإن أحب غيره فيحبه له، إذ قد يحب المحب عبد المحبوب وكل شيء ينسب إليه. وأيضاً، فالقلوب مجبولة على حب من أحسن إليها ولا محسن ولا منعم إلا الله، وشدة المحبة على حسب شدة المعرفة، وإنما خفيت معرفة الله مع جلائها لشدة ظهورها كالنور الذي به تظهر الأشياء، فإنه لو دام ولم يعرف الظلام لأنكر وجوده ولما ميّز بينه وبين الألوان ولو كان الأنوار قدرة الله حجاب لأدرك في الحال من التفاوت ما يضطر معه إلى الاعتراف بالقدرة والقادر.

ومن أحبه فلا بد أن يرضى بقضائه وقد يرضى بالمعصية من يرى أنها مقضية من الله تعالى من حيث [أنها] بقضائه، وتشخصها من حيث أنها مبعدة عن جنابه وليس الرضا[58] بمناف للدعاء فإن الإجابة مقتضية لمستحقها فاستكثر من الدعا لتسعد به لقبول الألطاف والأنوار، ولا تقنط إذا تأخر الإجابة، فلكل شيء وقت مقدر[59] لا يتعداه، ولعل في التأخير خيرة[60]، وينبغي أن لا تغفل عن ذكر الموت وتتفكر في من مضى من الأقران ومن قبلهم وما كانت عليه حال واحد واحد منهم من جاه وعزّ ومال وأعوان، ففي ذلك منافع كثيرة وأصل الغفلة عنه طول الأمل، وهو عين الجهل، بل العاقل إذا أمسى[61] لا يحدث نفسه بأنه لا بد وأن يصبح فإن لم يمت فجأة فقد يعرض[62] الموت فجأة[63] فاستعد لحياتك لموتك وحاسب نفسك في كل يوم، ولا تؤذِ[64] ولو نملة، فعناية البارئ متصلة بها كاتصالها بك. واختر لصاحبك ما تختاره لنفسك، واعدل في نفسك وأهلك ورعيتك، فبالعدل قامت السماوات[65] والأرض يعرف ذلك من اطّلع على أسرار الخليقة

وبالجملة فغاية[66] الكمال التشبه بالله بحسب الممكن وتحت هذه الأسرار والفاز[67].

فهذا ما ألّفه تمثيلاً للتقدم المولوي مؤلفه عفا الله عنه وأعانه على مرضيه، ونسأل من الله تعالى بأن ينفعنا بما علمنا ويعلمنا[68] ما ينفعنا وأن يؤمننا في الدنيا والآخرة، وأن يوصلنا إلى غاية السعادة

58 الرضا: الرضى، الأصل.
59 مقدر: مقتدر، الأصل.
60 خيرة: الخيرة، الأصل. والتصحيح عن كلمات وجيزة.
61 أمسى: مسى، الأصل.
62 يعرض: عرض، الأصل.
63 فجأة: + فقد عرض الموت (ولعله مشطوب)، الأصل.
64 تؤذِ: تؤذين، الأصل.
65 السماوات: السموات، الأصل.
66 فغاية: فعناية، الأصل.
67 وتحت هذه الأسرار والفاز: كذا في الأصل.
68 ويعلمنا: يعلمنا، الأصل.

والكمال، وأن يجعل حالنا[69] خير حال، إنه ولي الخيرات ومنتهى المدعيات. وصلى الله على خير خلقه محمد وآله وأصحابه خير صحب وآله وسلم تسليماً كثيراً، وحسبي الله ونعم الوكيل.

في آخر اللأصل:
كتبه ابن جليل الله الموسوي ابو الحسن اللّهم طوّل عمر أيامه وعفا عنه....

[69] حالنا: مالنا، الأصل.

## *Text 3*

### *Taqrīb al-maḥajja wa-tahdhīb al-ḥujja*

The edition is based on the following manuscripts:
ص = Ayasofya 2447/2, ff. 77b–87a
أ = Ayasofya 2446/4, ff. 294b–308a
س = Esad Efendi 1933/2, ff. 49b–55a

بسم الله الرحمن الرحيم[1]

قال مولانا وسيدنا وشيخنا عز الدولة سعد بن منصور بن سعد بن الحسن[2] بن هبة الله بن كمونة سقى الله صوبه الغمام:

هذا تلخيص في الغاية القصوى من الإيجاز اقتصرتُ فيه من العلوم الحكمية على ما هو من المهمات عند طالبي الفوز بالسعادة الأبدية والنجاة من الشقاوة الأخروية المنوطتين[3] على ما أجمع عليه الأنبياء صلوات الله عليهم أجمعين والحكماء المحققون بالعلم بالله تعالى وصفاته وأفعاله الذي يجب على كل عاقل بمجرد هذا الإجماع الاجتهاد في اكتسابه عملاً بالأحوط، ولو كان متردداً في تصديقهم فضلاً عن المؤمن، وعنونتُه بتقريب المحجة وتهذيب الحجة. فأقول:

العلم منه تصور وهو الإحاطة بكنه حقيقة الشيء إن كان تاماً وتمييزها عن كل ما عداها في الذهن من غير تلك الإحاطة إن كان ناقصاً، ويكتسبان بالأعرف من المكتسب إذا صدق كل من المعرّف والمعرّف على كل ما صدق عليه الآخر مع تقييده في التام بكونه مركباً من جميع أجزاء المكتسب المتصورة تصوراً تاماً. ومنه تصديق وهو الحكم بمتصور على متصور وتسمى قضية، فمنها الحملية وهي التي حكم فيها بارتباط مخبَر به ثبوتياً أو سلبياً، وهو المحمول، بمخبَر عنه كذلك، وهو الموضوع، وهي الموجبة، أو بسلب ذلك الارتباط، وهي السالبة. والموجبة يتوقف صدقها على وجود موضوعها في الأعيان إن حكم بثبوت محمولها له كذلك[4]، بخلاف السالبة. ومتى كان الموضوع جزئياً، وهو ما يمنع معناه صدقه على كثيرين، فهي المخصوصة، وإلا فهي الكلية إن كان

---

[1] الرحيم: + وبه نستعين، ص.
[2] الحسن: الحسين، أ.
[3] المنوطتين: المنوطين، أ س ص.
[4] كذلك: لذلك، س.

الحكم على كل أفراده، والجزئية إن كان على بعضها. والمهملة المساوية للجزئية في الصدق إن لم يتعرف لكمية الموضوع. والحكم بكيفية الارتباط أو سلبه يسمى[5] جهة. فإن كانت الدوام بدوام الموضوع فهي الدائمة إن لم تقيد بالوجوب والضرورية إن قيدت به. وإن كانت الدوام بدوام وصفه[6] المعبر به عنه فهي العرفية، وإن كانت الارتباط أو سلبه بالفعل من غير تعرض[7] لقيد آخر أو مع قيد في بعض أوقات وصف الموضوع فهي المطلقة. وإن كانت سلب الضرورة عن طرفي الوجود والعدم أو عن الجانب المخالف فهي الممكنة. ومنها المتصلة وهي التي حكم فيها بصدق قضية تسمى التالي على تقدير صدق[8] أخرى تسمى[9] المقدم لعلاقة بينهما تقتضي ذلك وهي الموجبة أو بلا صدقها[10] وهي السالبة، إما حكماً في كل أزمنة صدقِ المقدمِ وهي الكلية أو في بعضها غير معين وهي الجزئية، أو معيناً وهي المخصوصة أو مهملاً مساوياً للجزئي في الصدق. ومنها المنفصلة وهي التي حكم فيها بالتعاند بين قضيتين أو أكثر وهي الموجبة، أو باللاتعاند وهي السالبة، إما في الصدق فقط وهي مانعة الجمع أو في الكذب فقط وهي مانعة الخلو أو فيهما وهي الحقيقة، إما عناداً دائماً وهي الكلية، أو في وقت معين وهي المخصوصة، أو غير معين وهي الجزئية أو مهملاً في قوة الجزئي.

وإيجاب القضية وسلبها تسمى كيفية، وكليتها وجزئيتها كمّية. وكل قضية تناقضها الموافقة في كل الأمور مسلوباً في إحداهما[11] ما أوجب في الأخرى على ما أوجب تحقيقاً أو تقديراً. وحكمهما عدم الاجتماع على الصدق والكذب. ويلزم من صدق الحملية الموجبة غير الممكنة صدق حملية موجبة جزئية مطلقة مركبة من طرفيها مع تبديل كل واحد منهما بالآخر بدليل إنتاج المحال من نقيض الدعوى مع الأصل.

ومن صدق الحملية السالبة الكلية الدائمة أو العرفية صدق حملية سالبة كلية موافقة لها في الجهة مركبة من طرفيها مع التبديلِ المذكور، وإلا صدق نقيض الدعوى ولزمه مناقض الأصل، ويسمى هذا الكلام وما قبله عكساً.

ومن صدق الحملية الموجبة الكلية الدائمة أو العرفية سالبة كلية موافقة لها في الجهة موضوعها مقابل محمول الملزومة ومحمولها موضوع الملزومة بدليل لازم الموجبة غير الممكنة. ومن صدق

---

5 يسمى: تسمى، أ.
6 وصفه: وصفة، س.
7 تعرض: يعرض، أ.
8 صدق: -، أ.
9 تسمى: يسمى، س.
10 بلا صدقها: تلاصقها، س ص.
11 إحداهما: احديهما، أ س.

المتصلة الموجبة الصادقة المقدم صدق متصلة موافقة في الكمية والمقدم مخالفه في الكيفية مناقضه في التالي ومنه سواء كان مقدمها صادقاً أو لم يكن صدق منفصلتين موافقتين في الكمية والكيفية إحداهما[12] مانعة الجمع مركبة من المقدم ونقيض التالي، والأخرى مانعة الخلو من نقيض المقدم والتالي ومن صدق المنفصلة الحقيقية الصادقة الجزئين باعتبار منعها للجمع متصلة مقدمها أي واحد من جزئيها وتاليها نقيض الآخر باعتبار منعها[13] للخلو متصلة مقدمها نقيض أي واحد من جزئيها وتاليها الآخر.

والقضايا المنضم بعضها إلى بعض إذا استلزمت أمراً سمي المستلزم قياساً، وكل واحد منها مقدمة، وهيئة تركيبها شكلاً، واللازم قبل تركيب القياس مطلوباً، وبعده نتيجة. ويجب أن تكون يقينية الصدق إن كان كل واحد من المقدمات كذلك، ويسمى ذلك القياس برهاناً، وما عداه لا يفيد اليقين. وقد يتركب القياس من حملتين منتجتين[14] لحملية، ويسمى موضوع المطلوب الحد الأصغر ومحموله الأكبر. وما يتكرر في المقدمات ويسقط في النتيجة الأوسط، فإن كان المطلوب يشارك بعض المقدمات في أحد جزئيه والبعض في الآخر، فالمقدمة التي فيها الأصغر تسمى صغرى والتي فيها الأكبر كبرى.

فإن كان الأوسط محمولاً في الصغرى موضوعاً في الكبرى فهو الشكل الأول، وإن كان بالعكس فهو الرابع، وإن كان محمولاً فيهما فهو الثاني، وإن كان موضوعاً فيهما فهو الثالث. وكلما كان الأوسط بالتصريح أو اللزوم إما موضوعاً للطرفين للأصغر بالإيجاب ولأيهما[15] كان بالعموم، وإما مسلوباً عن الأصغر ثابتاً لكل الأكبر، أنتج نتيجة تابعة لأخسّ المقدمتين في الكمية والكيفية، بشرط أن يكون جهة صغرى الأول والثالث غير ممكنة، ويكون الدوام صادقاً على صغرى الثاني، أو العرفية على كبراه مع اقتران الضرورة بممكنه. ويكون سالبة الرابع منعكسة وليس فيه ممكنة. والجهة تابعة في الأول والثالث لكبراها إلا العرفية هي كموضوعية الأصغر، وفي الثاني إن لم يكن فيه ضرورية أو دائمة للصغرى، وإن كان فيه أحدهما كانت دائمة. وفي الرابع يظهر بعكس النتيجة اللازمة عند عكس الترتيب، أو بالنتيجة اللازمة عند عكس المقدمتين. والبيان فيما عدا الرابع إما بعكس المقدمة المخالفة لنظم الأول أو بالخلف، وهو ضم نقيض النتيجة إلى المقدمة المخالفة لينتج نقيض الأخرى.

---

12 إحداهما: احديهما، أ س.

13 منعها: منها، أ.

14 منتجتين: منتجا، أ، منتجه، س ص.

15 ولأيهما: ولأهما، أ.

وما لم يشتمل على هذه الشروط تبين عقمه بصدقه مع صدق إيجاب ما تفرض نتيجته له في مادة وصدق سلبها في أخرى. وقد يُركب القياس من متصلتين كقولنا: كلما كان كل «أ» «ب» فـ«ج» «د» وكلما كان «ج» «د» فـ«ه» «ر» والمنتج[16] دائماً إما ليس «أ» «ب» أو «ه» «ر» مانعة الخلو لأن مانعة الخلو اللازمة من المقدمة الأولى أي جزئها صدق مع الثانية حصل المطلوب. ومن منفصلتين مثل إما «أ» «ب» أو كل «ج» «د» وإما كل «د» «ط» أو «ه» «ر» ومعتبر فيها منع الخلو المنتج إما «أ» «ب» أو كل «ج» «ط» أو «ه» «ر» مانعة الخلو كلية إن كانت المقدمتان كليتين وجزئية إن كان أحدهما جزئياً بدليل أن الصادق من الأولى مع الثانية إن كان الجزء الغير المشترك حصل المطلوب، وإن كان المشترك فأي جزء صدق معه من الثانية حصل أيضاً.

ومن حملية ومتصلة مثل كلما كان «ه» «ر» فكل «ج» «ب» وكل «ب» «أ» المنتج دائماً إما ليس «ه» «ر» أو كل «ج» «أ» مانعة الخلو بما به بين المركب من متصلتين. ومن حملية ومنفصلة مثل كل «ج» «ب» ودائماً إما كل «ب» «أ» أو «ه» «ر» معتبر فيهما منع الخلو المنتج دائماً إما كل «ج» «أ» أو «ه» «ر» مانعة الخلو. ومن[17] متصلة ومنفصلة مثل كلما كان «أ» «ب» فـ«ج» «د» ودائماً إما «ج» «د» أو «ه» «ر» مانعة الجمع فدائماً إما «أ» «ب» أو «ه» «ر» كذلك لأن معاند لازم الشيء معاند لملزومه في الجمع.

ومتى استثنى في المتصلة عين مقدمها أنتج عين تاليها، ولما كان انتفاء الملزوم لازماً لانتفاء اللازم انتج استثناء نقيض تاليها نقيض مقدمها. والمنفصلة الحقيقية يلزم من وضع كل واحد من جزئها رفع الآخر، ومن رفعه وضع الآخر، ومانعة الجمع يلزمها الأول دون الثاني، ومانعة الخلو بالعكس. ويجب أن يقاس ما لم يذكر على ما ذكر، وأن تجرد المعاني عما تحيله الألفاظ مع التدرب باستعمال هذه القوانين ليؤمن الغلط من الألفاظ[18] إن شاء الله تعالى.

وبعد هذا أقول: المعلوم ينقسم إلى معدوم وموجود وإلى ممكن وواجب وإلى عرض وجوهر. والوجود هو نسبة الماهية إلى خارج الذهن بلفظة «في»، وهذه النسبة اعتبار عقلي مناف للعدم[19] الذي هو سلبها. ولا وجود لها في الأعيان من حيث هي نسبة وإلا لكان حلولها في المحل من حيث هو نسبة كذلك، فيلزم عدم النهاية في أمور موجودة معاً ولها ترتيب ويسمى

---

16 والمنتج: المنتج، أ ص.
17 ومن: وهو، أ.
18 من الألفاظ: - ، س ص.
19 للعدم: العدم، أ.

بالتسلسلِ وهو محال. وإلا لحصل جملتان غير متناهيتين إحداهما[20] أنقص من الأخرى بمرتبة واحدة مثلاً. فإن صدق على الناقصة أنها لو أطبقت على الزائدة بمعنى مقابلة كل مرتبة من هذه بمرتبة من تلك استغرقها كان الزائد مساوياً للناقص وإلا وجب انقطاعها من الجانب الغير متناهي فانقطعا معاً. ويجوز في الذهن أن يكون للشيء صفات غير متناهية بمعنى أن الذهن لا يقف عند حد ككون الواحد نصف الاثنين وثلث الثلاثة وهلمّ جراً.

ومفهوم الماهية في العقل مغاير لمفهوم الوجود، وإلا ما حصل العلم بها حال الشك في وجودها، والجزم بصدق الوجود على الشيء عند الجزم بكذب العدم عليه يدل على أنه مفهوم مشترك بين الموجودات. والماهية الواحدة إن التأمت من عدة أمور يفتقر بعضها إلى بعض فهي المركبة وإلا فهي البسيطة. ولا يجوز اشتراك موجودين في الأعيان في تمام الماهية، وإلا لكان اختصاص أحدهما بالفارق إن كان لماهيته أو لوازمها وقع الاشتراك فيه فلا يكون فارقاً. وإن كان لأمر[21] آخر عاد الكلام في اختصاصه به إن كان صفة له أو تخصيصه بالفارق إن لم يكن، ولزم التسلسل. وهذا المطلوب مع البرهان الدال عليه مما حصلتُه بنظري.

والممكن هو الذي يفتقر في وجوده إلى غيره، وذلك الغير هو العلة، والواجب ما ليس كذلك. والعلة إن كانت جملة ما يتوقف عليه الشيء فهي التامة وإلا فهي الناقصة، ويلزم من وجود التامة وجود معلولها ومن عدمها عدمه. إذ لو تخلف عنها لافتقر[22] اختصاص صدوره بوقت دون غيره إلى مخصص لاستحالة ترجيح أحد المتساويين من غير مرجح، فكان معتبراً في العلية، فلا تكون تامة. ويجب أن تكون هي مع معلولها في زمان واحد، إذ لا معنى لتأثيرها فيه إلا وجوده بها. فإن اتصفت بالمؤثرية حال وجودها كان المعلول موجوداً في تلك الحال، أو حال عدمها كان المعدوم علة للموجود، وبطلانه بديهي، وعليه يقاس حال المؤثر في عدم الشيء. ولا تجتمع علتان تامتان على معلول واحد بالشخص، وإلا لزم إما تحصيل الحاصل إن ترتب الحكم على العلة الأخرى أو تخلف المعلول عن علته التامة إن لم[23] يترتب. وفي[24] الناقصة يلزم من عدمها عدم[25] المعلول ولا يلزم من وجودها وجوده. فإن كانت جزءاً من المعلول فهي الصورية إن كان بوجوده يكون المعلول بالفعل، والمادية إن لم يكن كذلك. وإلا فهي إما الفاعلية إن كانت مؤثرة في وجوده أو الغائية إن كانت التي لأجلها الشيء المتمثلة في ذهن الفاعل لا الموجودة بعد وجود المعلول.

20 إحداهما: احديهما، أ س.
21 لأمر: لآمر، أ ص.
22 لافتقر: لا افتقر، أ.
23 إن لم: ألم، أ س ص.
24 وفي: في، س.
25 عدم: -، أ.

ولا ينفكّ عدم المعلول عن عدم علته الناقصة لاستحالة تخلفه عن علته التامة، وافتقار كل من الشيئين إلى الآخر من جهة واحدة بلا واسطة أو بواسطة أو وسائط، وهو المسمى بالدور محال، لأن المفتقر إلى المفتقر إلى الشيء مفتقر إلى ذلك الشيء، فيفتقر الشيء إلى نفسه.

ولا يجوز صدور البسيط عن المركب، لأنه إن استقل واحد من أجزائه بالعلية لا يمكن إسناد المعلول إلى الباقي، وإلا إن كان له تأثير في شيء من المعلول وللباقي تأثير في باقيه كان المعلول مركباً، وإن لم يكن لشيء منها تأثير فيه. فإن حصل لها عند اجتماع أمر زائد هو العلة، فإن كان عدمياً لم يكن مستقلاً بالتأثير في الوجود وإلا لزم التسلسل في صدوره عن المركب إن كان بسيطاً، وفي صدور البسيط عنه إن كان مركباً. وإن لم يحصل بقيت مثل ما كانت قبل الاجتماع، فلا يكون الكل مؤثراً.

ويجب أن تكون علة الحادث مركبة لوجوب حدوثها أيضاً، وإلا لكان صدور الحادث في وقت دون ما قبله ترجيحاً من غير مرجح. فلو كانت بسيطة لوجب لأجل حدوثها حدوث علتها ولأجل بساطتها ولزم التسلسل. وكل حادث مركب، وإلا لكانت علته بسيطة. وهذه الدعاوي الثلاث، أعني أن المركب لا يكون علة البسيط، وأن علة الحادث يجب أن تكون مركبة، وأن كل حادث مركب، مع البراهين الدالة عليها مما لم أجده في كلام من تقدم.

ولا بد من وجود واجب، وإلا لكان كل موجود ممكناً فيفتقر إلى علة إما واجبة وهو المطلوب، أو[26] ممكنة، فيعود الكلام فيها ويلزم إما الدور أو التسلسل. وتجب[27] أزليته، وإلا لكان حادثاً فافتقر إلى علة فكان ممكناً، وأبديته، وإلا توقف وجوده على عدم سبب عدمه فلم يكن واجباً. وهو بسيط لافتقار المركب إلى أجزائه التي هي ممكنة لاحتياج بعضها إلى البعض، وإلا لم يحصل منها ماهية واحدة فيكون ممكناً. وتأثيره في العالم يجب أن يكون أزلياً لأنه إن كان علة تامة فالأمر ظاهر، وإلا فكل ما لا بد منه في التأثير إن كان حاصلاً في الأزل وجب أزليته، وإلا فشيء منه حادث فيفتقر إلى علة حادثة، وكذلك إلى[28] ما[29] لا نهاية، إما دفعة وهو التسلسل أو على سبيل التعاقب لا إلى أول فيتحقق أزلية التأثير. ويلزم من انتهاء جميع الممكنات في سلسلة الحاجة إلى الواجب أن يكون كل حيوان مجبوراً في كل ما يصدر عنه من الأفعال الإرادية لوجوب كل ممكن بغيره.

---

26 أو: و، س.
27 وتجب: ويجب، ص.
28 إلى: ما، أ.
29 ما: -، أ س.

والمفتقر إلى ذات أخرى يكون شائعاً[30] فيها بالكلية حتى يقوم بالفعل هو العرض، والذي يقابله هو الجوهر. والعرض إما أن يكون غير قار الذات كالحركة والزمان أو لا. فإن كانت ماهيته معقولة بالقياس إلى الغير فهو النسبة، وإلا فإما أن يكون قابلاً للتجزئ لذاته أو غير قابل، والأول هو الكمّ. فإن أمكن أن يفرض بين أجزائه حدّ مشترك هو نهاية لأحدهما وبداية الآخر فهو المقدار، وإلا فهو العدد. والثاني إن كان مفهومه أنه شيء ليس بمنقسم فهو النقطة إن كان ذا وضع والوحدة إن لم يكن. وإن كان له مفهوم آخر فهو الكيفية، وهي إما محسوسة بأحد الحواس الخمس كالألوان، أو مختصة بذوات الأنفس كالعلم، أو عارضة للكمّ كالزوجية، أو استعداداً نحو الانفعال كالمراضية أو نحو اللا [31]انفعال كالصحاحية. وأكثر هذه الأعراض اعتبارات ذهنية لا وجود لها في الأعيان، كالوحدة التي لو كانت موجودة لكانت شيئاً واحداً فلها وحدة أخرى وهلمّ جراً.

وكذلك كل صفة يجب تكرر نوعها عليها لا إلى[32] نهاية وكالنقطة التي هي نهاية الخط، والخط الذي هو نهاية السطح، ونهاية الشيء انقطاعه وعدمه كالوجوب والإمكان، وإلا كان[33] وجوب الوجود[34] للماهية وإمكانه متأخراً عن ذلك الوجود لوجوب تأخر الصفة عن وجود الموصوف. وذو الفطانة يتنبه بالقدر المذكور على غيره.

وكل عرض موجود فهو منقسم بانقسام محله، وإن لم[35] يوجد في شيء من الأجزاء المفروضة في المحل شيء منه لم يكن حالاً فيه. وإن وجد[36] فإما الحال بتمامه في كل واحد منها فيكون العرض الواحد في الحالة الواحدة في أكثر من محل واحد أو في بعضها، فلا مدخل لما عداه في المحلية، وإما بعضه في كل واحد منها أو في بعضها، فينقسم بانقسامه.

ولا تتصور الحركة فيما هو بالفعل من جميع الوجوه، وهي قد تكون في الكمّ كالنمو والذبول، وفي الكيف كالاستحالة من كيفية إلى أخرى، وفي الأين كالانتقال من مكان إلى مكان، وفي الوضع كما تدور الكرة على نفسها ولا تخرج عن مكانها. ولا بد من حصول حركة واحدة في الأزل متصلة الاستمرار، حدوث أجزائها بحسب اعتبار الذهن فقط، لما أنه ليس لها جزء في نفس الأمر بحيث يكون مجموع الحوادث العنصرية تابعاً لها في هذه الأمور حتى أن ما يقال

---

30 شايعاً: مشايعاً، أ س.
31 اللا : إلا، س ص.
32 لا إلى: إلى لا، س ص.
33 وإلا كان: والإمكان، أ.
34 والإمكان وإلا كان وجوب الوجود: - ، س.
35 وإن لم: وإلا ألم، ا س ص.
36 وجد: وجود، س ص.

له من الأمور الموجودة منها أنه عدم أو وجد بعد العدم، فليس ذلك العدم هو الحقيقي بمعنى ارتفاع الماهية عن الخارج بل الإضافي، كما تنتقل الصفة عن الموصوف فيقال: عُدمت عنه. أو كما يقال للغائب عن الحسّ أنه عدم عنه، وهذا على الحقيقة تغير وانتقال. ولولا أن الأمر كذلك في هذه الحركة وما يتبعها للزم من وجوب أزلية تأثير الواجب في معلوله الأول دوام معلول معلوله، وكذا إلى أن لا ينتهي الأمر إلى حصول حادث البتة. ومن وجوب حدوث علة كل حادث التسلسل، ومن وجوب عدم العلة عند عدم المعلول عدم الواجب عند عدم أي شيء كان من الحوادث، والتوالي الثلاثة باطلة. وهذه الدعوى مع هذا الإيضاح في تقريرها مما لم أعرف [من] سبقني[37] إليه.

والعلم يستحيل عليه الانقسام بذاته أو بغيره لأنه متعلق بالبسائط لا محالة لتوقف العلم بالماهية المركبة على العلم بأجزائها، فلو انقسم فجزؤه إن تعلق به كله كان كل معلوم مركباً أو بكله فكان الجزء هو الكل أو لا بشيء منه[38]. فإن لم يحصل العلم عند اجتماع الأجزاء لم يكن هناك علم. وإلا أن انقسم الحاصل لزم التسلسل وإن لم ينقسم حصل المطلوب. والحياة هي كون الذات بحيث لا يمتنع عليها أن تعلم وتفعل. والإرادة هي كون الفاعل عالماً بفعله إن كان ذلك العلم سبباً لصدوره عنه مع كونه غير مغلوب ولا مستكره[39]. والقدرة هي كون الحي بحيث يصح منه الفعل والترك بحسب الدواعي المختلفة. والحكمة هي كونه عالماً بالأشياء على ما هي عليه، أو كونه بحيث يصدر عنه الأفعال المرتَّبة المحكمة الجامعة لكل ما يحتاج إليه. والجود هو إفادة الغير ما هو مضطر إليه مع القصد إلى ذلك.

والجوهر إن كان متحيزاً وهو كونه بحيث يصح أن يشار إليه بأنه هاهنا أو هناك فهو الجسم وإلا فهو المجرد. والجسم لا يجوز تركبه من أجزاء كل واحد منها لا يقبل التجزئة بالفعل ولا بالفرض، وإلا كان المفروض بين اثنين منها إن منع الطرفين عن التلاقي انقسم إلى ما منه إلى أحدهما وإلى ما منه إلى الآخر، وإلا تداخلت الأجزاء بالكلية فلا يحصل من اجتماعها مقدار. فإذن لا تنتهي في القسمة إلى حد إلا وهو بعدد ذلك قابل للانقسام الفرضي.

ولا توجد الانقسامات الغير المتناهية بالقوة دفعة واحدة بالفعل، وإلا لكان مجموع عدد متناه منها زائداً على مقدار الواحد، فكان التفاوت في زيادة الحجم كالتفاوت في زيادة العدد. فكما أن نسبة الأجسام بعضها إلى بعض نسبة متناهٍ إلى متناهٍ[40] لوجوب تناهي كل جسم ببرهان

37 سبقني: سبقي، أ س ص.
38 إن تعلق به ... لا بشيء منه: كذا في كل النسخ.
39 مستكره: مكره، أ.
40 متناهٍ إلى متناهٍ: متناهي إلى متناهي، س ص.

التطبيق، وكذا نسبة أجزائها الموجودة بالفعل والجسم إن لم تكن[41] حقيقته مركبة من أجسام مختلفة الطبائع في الحس فهو البسيط وإلا فهو المركب. وكل منهما إذا قدر خلوه عن كل ما يصح خلوه عنه فلا بد له من مكان ومقدار وشكل، ووضع كل منهما معيَّن لاستحالة حصوله في كل الأماكن وعلى جميع المقادير والأشكال والأوضاع، أو حصوله خالياً عن شيء منها، وذلك المعين هو الطبيعي له. ولا يجوز أن يكون للجسم مكانان طبيعيان، لأن عند حصوله في أحدهما إن طلب الآخر فالثاني هو الطبيعي ودون الأول، وإن لم يطلبه فالأمر بالعكس. وعند مفارقته لهما إن طلب أحدهما فقط فهو الطبيعي لا غيره، وإن طلبهما معاً استحال توجهه إليهما دفعة أو إلى[42] أحدهما دون الآخر لأنه ترجيح بلا مرجح. وإن لم يطلب واحداً منهما فليس شيء منهما طبيعياً له.

والأجسام البسيطة منها فلكية ومنها عنصرية. والعناصر التي منها تتركب[43] المتولدات في عالمنا أربعة: الأرض والماء والهواء والنار، بدليل أنه إذا وضع المركب منها في القرع والإنبيق حصل منه جوهر أرضي ومائي وهوائي، ولا بد في اختلاطها من حرارة طابخة، والجسم الطابخ بالطبع هو النار. وهذه الأربعة منفعلة من الأجرام[44] الفلكية، ويتولد من تأثير تلك وانفعال هذه موجودات شتى مثل حوادث الجو وغيرها. ونفس الإنسان وهي الموجود الذي يشار إليه بـ«أنا» يجب أن تكون جوهراً مجرداً بسيطاً لأنها عالمة، والعلم غير مركب فيكون الموصوف به كذلك. وكل متحيز مركب، فليست بمتحيزة، فلا تحلّ في الجسم وإلا[45] انقسمت بانقسامه، ولا في المجرد، وإلا[46] كان له تصرف في البدن بذاته فهو النفس، أو بعرض فيه كان التصرف منسوباً إليه بذلك العرض لا إلى العرض وحده فكان هو النفس أيضاً. ولكونها بسيطة يجب أن لا تكون[47] لها علة مركبة لاستحالة صدور البسيط عن المركب، فلا يكون الجسم ولا ما يحلّ فيه علة لها لوجوب انقسامهما. ويجب أن تكون أزلية، وإلا كانت مركبة، وأبدية، وإلا لعدم القديم. فإن كان[48] سبب عدمه وجود أمر فلا بد وأن يتبعه عدم علة ذلك القديم التي هي قديمة كذلك إلى أن يلزم عدم الواجب. وإن كان عدم أمر عاد المحال إن كان المعدوم قديماً وأوجب عدمه السابق

[41] تكن: يكن، أ س ص.
[42] أو إلى: وإلى، س.
[43] تتركب: يتركب، أ.
[44] من الأجرام: للأجرام، س ص.
[45] وإلا: ولا، أ.
[46] وإلا: + إن، أ س ص.
[47] تكون: يكون، أ ص.
[48] كان: كانت، أ س ص.

عليه حدوث القديم إن كان حادثاً. وبراهين هذه الدعاوي الثلاث، وهي أنه ليس للنفس علة مركبة، وأنها أزلية، وأنها أبدية مما لم اكتسبه من غيري أيضاً. وقد بسطتُ الكلام فيها وفي غيرها مما استنبطتُه بفكري في تعاليق غير هذا الموجز.

وهي المدركة بجميع أصناف الإدراكات لجميع أصناف المدركات بدليل حكمها بالبعض على البعض ووجوب إدراك الحاكم على شيء بشيء لكل منهما. وهي الموصوفة بالشهوة والنفرة واللذة والألم والإرادة[49] والقدرة والفعل، وإلا لما لزم من إدراكها حصول هذه الأشياء، ولاختلاف أبدان الناس في أمزجتها الأصلية والظاهرية. وما يحصل لهم بالكسب والاجتهاد تأثير قوى في اختلافهم في أخلاقهم وذكائهم وبلادتهم وجميع ما يصدر عنهم. والسبب الأصلي في هذا الاختلاف هو اختلاف النفوس في ماهياتها بالبرهان العام على امتناع اشتراك كل موجودين. ومن النفوس من هي مائلة بأصل فطرتها إلى الجناب الأعلى يذهلها عن الجسمانيات أدنى تذكر فليستولي[50] عليها الوجد والجنين فيغشاها من النفحات الإلهية أمور لا يتصور كنهها إلا من وجدها من نفسه. ومنها من ليس لها ذلك بالفطرة إما مع قبولها له بالاكتساب أو مع عدم هذا القبول. وترك الفضول مع إصلاح الضروريات والعزلة عن الناس مقرونة بالفكرة في دقائق حكمة الله تعالى والإخلاص في التوجه إليه مع الاستعانة في بعض الأوقات على المطلوب بالألحان المتناسبة من المؤثرات[51] القوية في نيل اللذات الإلهية لمن كان مستعداً لها[52]، وذلك كله معروف بالتجربة لمن وجده، وبالتواتر والحدس لمن هو مستعد له.

وقد يظهر عن العارفين أحوال خارقة للعادة تسمى باعتبار اقترانها بالتحدي مع عدم المعارضة معجزات، وباعتبار خلوها عن ذلك كرامات كالخوارق المتواترة عن الأنبياء والأولياء عليهم أفضل الصلاة والسلام.

ومن نظر إلى تناسب العالم وترتيبه وانتظامه ودقائق الحكمة الموجودة فيه علم بالبديهة أن له مدبراً عالماً حياً مريداً قادراً حكيماً جواداً. وبواسطة[53] علمه تتبين أزليته وأبديته وتنزهه عن العرضية والتحيز والتركيب [...] كتب بالبراهين التي اخترعتُها على إثبات هذه المطالب في النفس وأفعاله لغاية لا محالة لما فيها من الحكمة الظاهرة لكل معتبر وإلا لم يترجح الجود على مقابله[54]

49 والإرادة: -، أ؛ مشطوباً في ص.
50 فليستولي: فيستولى، أ.
51 المؤثرات: المتواترات، أ س ص.
52 لها: -، أ س.
53 وبواسطة: وبواسطته، أ.
54 مقابله: مقابلة، أ.

وهو الغاية لا غير[55] لوجوب أن يكون إفادة الغير أولى به من عدم إفادته وإلا لم يترجح على غيرها فكانت عائدة إليه على تقدير عودها إلى غيره[56].

ومدبر العالم جلّت عظمته واحد لأن أجزاء العالم مرتبط بعضها بالبعض، بمعنى انتفاعه به كما يظهر مما مضى. فهو شخص واحد مركب منها، فلو وجد إلهان فإن كان أحدهما مستبدّ بتدبيره اجتمع علتان تامتان على معلول واحد بالشخص أو ببعضه والآخر بالأخر وجب استبداده بتدبيرهما لخلقه أحدهما لأجل[57] الآخر. فلو كان الثاني مؤثراً في شيء منهما عاد المحال، وإن لم يكن واحد منهما مستبداً به ولا شيء منه لم يكونا إلهين. وهذا برهان غريب في التوحيد اخترعتُ تقريره على هذا الوجه وختمتُ به هذا المختصر الذي من اطلع على فوائده وأسراره وتأمل ما ذكروه في المطولات ثبتت له منزلته. ومن الله تعالى أسأل وإليه أضرع في أن يوفقني للهدى[58] والرشاد ويؤمنني من عقابه في يوم المعاد ويرزقني سعادة الأبد التي وعد بها من آمن وعمل صالحاً من العباد بمنّه ورحمته إن شاء الله تعالى.
تمّ.

في آخر ص:
تمت كتابة المختصر بتوفيق الله عزّ وجلّ في العشر الآخر من جمادى الأولى[59] سنة سبعة وعشرين وسبع مائة والحمد لله ربّ العالمين
قوبل حسب الطاقة وبحسب الكتاب الذي نقل منه.

---

55 غير: غيره، ص.
56 كتب بالبراهين ... إلى غيره: كذا في كل النسخ.
57 والآخر بالأخر وجب استبداده بتدبيرهما لخلقه أحدهما لأجل: - ،أ.
58 للهدى: الهدى، أ س.
59 الأولى: الأول، ص.

## *Text 4*

*Min taʿālīq al-Ḥakīm ʿIzz al-Dawla Abi l-Riḍā Ibn Kammūna*

The edition is based on the only known manuscript, ms. Esad Efendi 3642/7, ff. 166a–167b

### [تعريفات ابن كمونة]

### من تعاليق الحكيم عز الدولة أبي الرضا ابن الكمونة

وجه تقسيم الألفاظ:

اللفظ إما أن يكون متكثراً أو متّحداً، والأول إن تكثر مفهومه فهو المتباين كالسماء والأرض وكالسيف والصارم وكالصارم والمهنَّد وكالناطق والفصيح وكالأسامي المشتقة وكالأسامي المنسوبة. وقد يظنّ في بعض هذه أنها مترادفة لاتفاق موضوعاتها وليس كذلك. وإن اتّحد مفهومه فهو المترادف كالليث والأسد. والثاني إن كان جزئياً فهو العَلَم كزيد، وإن كان كلّياً فلا يخلو إما أن يكون مفهومه كثيراً أو واحداً. فإن كان مفهومه كثيراً فإن كان اللفظ لهما لا على السوية بل لأحدهما أولاً وللآخر ثانياً، كالفرس للطبيعي والمصوَّر، فإذا قيس اللفظ إليهما جميعاً فهو المتشابه، وإذا قيس إلى المعنى الأول منهما فهو الحقيقة، وإذا قيس إلى الثاني منهما فهذا على قسمين، لأنه لا يخلو إما أن تكون دلالته على المعنى الثاني أقوى أو لا، فالأول هو المنقول، كالصوم والصلاة، والثاني [166ب] هو المجاز والمستعار. فالمجاز هو يُطلق في الظاهر على شيء والمطلق عليه في الحقيقة غيره، مثل ﴿وَسْأَلِ الْقَرْيَةَ﴾[1]، والمستعار هو الذي استعير للشيء من غيره من غير نقل إليه بالكلّية، كما يقال للبليد حمار. وإن كان اللفظ موضوعاً لهما على السوية فهو المشترك، كالعين للدينار وللجاسوس، وكالكلب للحيوان المعروف وللشعرى. وإن كان ثَمّ تشابه في معنى بعيد عن الفهم لا يخرج مثله عن الاشتراك المحض ولا يدخل في المتشابه، إذ المعتبر فيه هو التشابه في أمر قريب إلى الفهم لا بعيد عنه. وإن كان مفهوم اللفظ الكلي واحداً فإن كان حصوله في جزئياته الخارجية والعقلية لا على السواء بمعنى أن يكون لبعضها أولاً أو في بعضها أولى به أو أشدّ في بعضها فهو المشكّك، كالوجود على الواجب والممكن

[1] سورة يوسف (12):82.

وعلى الجوهر والعرض، وكالبياض على الثلج والعاج. وإن كان حصوله في جزئياته المذكورة على السواء بالمعنى المقابل للمعنى المذكور في المشكك فهو المتواطئ كالحيوان على الإنسان والفرس والثور. [167 أ] والفرق بين الغضب والغمّ أن مع الغضب الطمع في الوصول إلي البغية بالانتقام من المسيء ومع الغمّ اليأس من ذلك. الفرق بين الغمّ والهمّ أن الغمّ عارض يعرض لفقد محبوب النفس أو فوت مطلوبها، والهمّ غموم مترادفة متأكدة. وقال بعضهم: الغمّ من حادث كان والهمّ من حادث يكون. الفرق بين الفزع والخوف أن الفزع هو من الشيء الذي يلحق الإنسان بغتةً، والخوف هو من الأمر الذي يتوقع وروده. الفرق بين الخوف والغمّ أن الخوف هو مجاهدة الأمر المخوف قبل وقوعه، والغمّ ما يلحق الإنسان منه بعد وقوعه. البخل من خواص القوة الفكرية لأن اتحاد الغريزي بنقيض عند التوبيخ ويعاود لمقاومة الحجة[2]. الشره والنهم من خواص القوة الشهوانية، والسخاء والبخل والرضا من خواص الناطقة، لأن السخاء وضع الشيء بحيث يحسن وضعه. الجبروت جبر الناقص بالكامل وقيل: جبر المادة بالصورة. الوهم لفظ مشترك بين ثلاثة معانٍ: (أ) مرجوح [167ب] الظن، (ب) إدراك المعاني الجزئية المتعلق بالأشخاص المحسوسة كصداقة زيد، (ج) الاعتقاد الباطل أو الرأي الباطل. القطب عند المنجّمين كوكب يدور عليه الفلك، وعند الحكماء نقطة موهومة تدور عليها الفلك. الجهة خطّ مستقيم يميّز بمحاذاة خطّ الاستواء من المشرق إلى المغرب بحيث يجوز عليه الكعبة باستقامته. التنبيه، قالوا[3]، جذب[4] النظر إليه لعلم مما سبق. المقدمة جملة من الكلام يتوقف عليه الغرض المقصود من الكتاب. البيان هو المنطق الفصيح المعرّب عما في الضمير. الصدر لغةً مقدم المجلس واصطلاحاً المرجوع إليه في الأمور المستأذن فيها. النظام لغةً ما ينظم فيه أو ما يُنظَمُ، واصطلاحاً الإنسان الذي ينتظم به أمور الملك.

[2] البخل من... لمقاومة الحجة: كذا في الأصل.
[3] قالوا: قالو، الأصل.
[4] جذب: جرد، الأصل.

## *Text 5*

### *Correspondence with Naṣīr al-Dīn al-Ṭūsī*

The edition of Ṭūsī's reply to Ibn Kammūna which is the only extant part of this correspondence is based on the following three manuscripts:
آ = Ākhūnd 1187/61, ff. 267a–b [pp. 495–496]
م = Majlis 2938, ff. 165–166
ر = Ragıp Paşa 1461, ff. 258b–259a [253b–254a]

In addition, the edition of ʿAbd Allāh Nūrānī (in *Collected Texts and Papers on Logic and Language*. Eds. M. Mohaghghegh and T. Izutsu. Tehran 1974, pp. 285–86) has been consulted. This edition which is supposedly based on a single manuscript (Majlis 2938) contains only the first part of the text. Possibly due to a technical error, the second part is replaced by a completely different text.

الذي خطر ببال الداعي المخلص على الحجة المذكورة وعلى[1] إثبات الجواز في صورة النزاع عرضه على الآراء المولوية حتى يتبين ما هو الصواب فيها، وهو هذا:

أقول: قوله في الحجة: الجواز في صورة النزاع لا يستلزم ارتفاع الواقع، وكل ما لا يستلزم ارتفاع الواقع فهو واقع، مؤلف من حمليتين[2]. وأثبت الصغرى منهما بانعكاس قولنا: لو كان الجواز مستلزماً لكان منتفياً لانتفاء ارتفاعِ الواقع، عكس النقيض. وتلك متصلة مؤلفة من متصلتين هكذا: لو كانِ الجواز كلما كان ثابتاً لزمه ارتفاع الواقع لكان كلما كان ارتفاع الواقع منتفياً كان الجواز منتفياً. وأورد في عكس نقيضه متصلة بسيطة، وذلك غيرِ صحيح. ثم إن صح ذلك فالصغرى هاهنا شرطية هكذا: لو كان الجوازِ في صورة النزاع ثابتاً فهو لا يستلزم ارتفاع الواقع[3].

ثم بدلها حملية[4] بقوله: وإذا لم يكن مستلزماً لارتفاع الواقع على تقدير ثبوته لا يكون مستلزماً لارتفاع الواقع بالضرورة. وهذا غير صحيح، لأن موضوع الحملية يجب أن يكون ما يقال عليه

---

1 وعلى: على، النسخة المطبوعة.
2 الحملتين: الجملتين، أ م ر والنسخة المطبوعة.
3 الواقع: بالواقع، النسخة المطبوعة.
4 حملية: الحملية، أ م ر والنسخة المطبوعة.

الموضوع مما لا يمتنع أن يقال عليه، وإن كان الجواز لا يصح حصوله في صورة النزاع لا يمكن أن يجعل الجواز في صورة النزاع موضوعاً للحملية مع جواز كونه مستلزماً لشيء على تقدير ثبوته. وأما الكبرى فقوله في بيانها[5]: إن كل ما هو غير واقع فهو مستلزم لارتفاع الواقع، إنما يصح إذا قيد بقولنا: ما دام غير واقع، وحينئذ تكون مشروطة،[6] والاستلزام إن جعل جزاء المحمول صارت القضية مطلقة، ولا ينعكس عكس النقيض، كما إذا قلنا: كل كاتب يلزمه أن يحرك يده، فإذا قلنا في عكسه: كل ما لا يلزمه أن يحرك يده فهو ليس بكاتب، لم يكن صادقاً، بل الصواب أن يجعل الاستلزام فيه جهة، وحينئذ يكون العكس قولنا: كل ما ينافي ارتفاع الواقع فهو واقع. ثم إنا إذا سلمنا انعكاس القضية على ما ذكره صار القياس شرطياً هكذا: لو كان الجواز في صورة النزاع ثابتاً فهو لا يستلزم ارتفاع الواقع، وكل ما لا يستلزم ارتفاع الواقع [فهو واقع]، وأنتج: لو كان الجواز في صورة النزاع ثابتاً فهو واقع، وهذا لا يفيد المطلوب. هذا خطر بباله وعرض على أرائه متوقعاً بيان ما فيه من الصواب والخطأ ليجعله مضافاً إلى سوابق فوائده، لا زالت أيامه منتصبة لنجاح مطالبه في الدارين بمحمد وآله.

---

5 بيانها:، إثباتها، النسخة المطبوعة.

6 ومن هنا تتفاوت النسخة المطبوعة.

## *Text 6*

### *Correspondence with Najm al-Dīn al-Kātibī*

The edition is based on the following five manuscripts:
ص = Ayasofya 4862, ff. 269b–271a
ه = Ākhūnd 1187/67, ff. 290b–292b
م= Majlis 630/2, pp. 9–11
س = Majlis-i Sinā 286/5, ff. 39b–41b
ر = Ragıp Paşa 1461, ff. 276b–278b
ب = Pertev Paşa 617, ff. 173a–176b

As the variations in the manuscripts show, there are two recensions of the text. One is obviously closer to the original version written by Ibn Kammūna, as is evident from the way the author is introduced there, which corresponds to the way he usually introduces himself at the beginning of his writings. Manuscripts ه and ر belong to this recension. The other recension, in which the name of Ibn Kammūna is mentioned more respectfully, was presumably prepared by a later generation of scholars. Manuscripts ب, ص, م and س belong to this recension. Manuscripts ص and م were apparently copied by the same hand. In manuscript ص the scribe left out many words, but in the margin numerous explanatory notes have been added; most of these notes are part of the main text as it appears in manuscript ب ; one of these notes at the end of text is also to be found in manuscript م.

استفاد[1] العبد المخلص سعد بن كمونة[2] من[3] مجلس مولانا علامة العالم ملك العلماء[4] قدوة الفضلاء مفتي الفرق جامع علوم المتقدمين والمتأخرين[5] نجم[6] الملة والدين أدام الله أيامه وحرس

---

[1] استفاد: قال الحكيم الفيلسوف عز الدولة ابن كمونة استفاد، م؛ قال الفيلسوف عز الدولة ابن كمونة استفاد، ب؛ قال الفيلسوف عز الدولة ابن كمونة استفاد، ص؛ وفي حاشية ر: فائدة صور أسئلة وأجوبة بين المولى العلامة نجم الدين وبين الفاضل المحقق عز الدولة.
[2] سعد بن كمونة: ابن كمونة، س.
[3] من: -، ه ر.
[4] العلماء: الحكماء، ه.
[5] مفتي الفرق جامع علوم المتقدمين والمتأخرين: مفتي الفرق جامع العلوم، م؛ قبلة الحكماء، ب ص.
[6] نجم: + الدولة، ص م.

مدّته وأمتع أهل العلم بطول حياته[7]: إن ثبوت الإمكان لا يلزم منه إمكان الثبوت فلا يلزم من[8] صدق بعض[9] «ج»[10] «ب» بالإمكان العام إمكان[11] صدق بعض «ج» «ب» بالفعل، لأن الأول حكم بثبوت الإمكان والثاني حكم بإمكان الثبوت، ومستند المنع من اللزوم أن الحادث ثبت[12] إمكان وجوده في الأزل ولا يمكن ثبوت وجوده[13] في الأزل، ففي هذه المادة قد ثبت الإمكان ولم يمكن الثبوت. وهذا المنع في غاية الدقة والمستند في غاية الحسن. وقد قلت مستفسراً ومستوضحاً: إن الإمكان لا يمكن تعقله إلا مضافاً إلى شيء يكون إمكاناً له، فالإمكان الثابت في القضية الموجهة[14] ليس إلا إمكان ثبوت المحمول للموضوع. فإذا حكمنا بثبوت ذلك الإمكان فقد حكمنا بإمكان ذلك الثبوت لا محالة[15]، فكيف يصدق أحدهما دون[16] صدق الآخر؟ ومستند المنع إنما كان يصح جعله مستنداً لو صدق حكمنا بثبوت إمكان وجود الحادث في الأزل ولم[17] يصدق بدون[18] إمكان ثبوت ذلك[19] الوجود في الأزل[20]، وليس كذا، فإنا إن جعلنا قيد «في الأزل» متعلقاً بوجود الحادث كانا كاذبين[21]، وإن جعلناه[22] متعلقاً بالإمكان كانا صادقين، وإنما يصدق الأول ولا يصدق الثاني إذا جعل قيد «في الأزل» متعلقاً تارةً بالإمكان وتارةً بوجود الحادث. وإذا عني به ذلك لم يكن مطابقاً لما ادعيناه.

لا يقال: إذا ثبت في الأزل إمكان وجود الحادث ولم يمكن ثبوت[23] وجود الحادث في الأزل ففي الحالة المعبر عنها بالأزل قد ثبت الإمكان ولم يمكن الثبوت، فجاز صدق الأول دون[24] الثاني في

7 وحرس مدّته وأمتع أهل العلم بطول حياته: -، ب ص.
8 من: منها، ب.
9 بعض: -، م.
10 ج: + هو، ص.
11 إمكان: -، ه ر.
12 ثبت: يثبت، ه.
13 ثبوت وجوده: وجود ثبوته، م؛ وجوده ثبوته، ب ص.
14 موجهة: الموجبة، ه س ر.
15 محالة: + لأنه إذا صدق بعض ج ب بالإمكان فقد تحقق الإمكان حينئذ بثبوت المحمول للموضوع بالفعل وكل شيء تحقق فيه الإمكان يكون ممكناً فثبوت المحمول للموضوع بالفعل، ب وإضافة في هامش ص.
16 دون: بدون، م ب ص ه.
17 ولم: فلم، ه.
18 بدون: -، ه ر ب ص.
19 ثبوت ذلك: -، م.
20 ولم يصدق ... في الأزل: -، س.
21 كاذبين: + لأن وجود مع الأزلية ‡ ... ‡ فيستحيل وجود الحادث في الأزل، ب وإضافة في هامش ص.
22 جعلناه: مكرر في ص.
23 ثبوت: -، س.
24 دون: بدون، م ص.

ذلك الحال[25]، فظهر صحة المستند، لأنا نقول[26]: المدعى أنه إذا صدق ثبوت الإمكان لشيء[27] صدق إمكان ثبوت ذلك الشيء في الجملة، وبرهنّا عليه، وهذا أعم من أنه إذا صدق ثبوت الإمكان لشيء في شيء آخر، سواء كان[28] ذلك الآخر هو الأزل[29] أو غيره، صدق إمكان ثبوت ذلك الشيء في ذلك الآخر، ولا يلزم من دعوانا صدق الأعم أن يكون الأخص منه صادقاً، فظهر الفرق. فالمسئول[30] من إنعام مولانا أسبغ الله ظلّه سطر[31] ما يتضح به للعبد حل هذه الشبهة والسلام[32].

فأجاب[33] بما صورته هذه:

أفاد المولى العلامة أفضل العصر[34] قدوة الحكماء المحققين[35] عز الدولة فخر الملة أدام الله[36] معاليه بأنكم قلتم: إن ثبوت الإمكان لا يلزم منه إمكان الثبوت، فلا يلزم من صدق بعض[37] «ج» «ب» بالإمكان العام إمكان صدق بعض «ج»[38] «ب» بالفعل، لأن الأول حكم بثبوت الإمكان والثاني حكم بإمكان الثبوت، ومستند المنع من اللزوم أن الحادث ثبت[39] إمكان وجوده في الأزل ولا يمكن ثبوت وجوده[40] في الأزل[41]، ففي هذه المادة قد ثبت الإمكان ولم يمكن الثبوت.

أقول[42]: نحن نقرر ما قلناه، ثم ننظر فيما[43] قاله المولى أدام الله[44] أيامه عليه[45]، فنقول: الكلام

---

25 في ذلك الحال: -، ه ر.

26 لأنا نقول: لأني أقول، س ه ر.

27 لشيء: -، م.

28 كان: -، س.

29 الأزل: الأول، م س.

30 فالمسئول: والمسئول، س ه ر.

31 سطر: بنظر، م س ب.

32 والسلام: -، م س ص ه.

33 فأجاب: + رحمه الله، ه ر س.

34 المولى العلامة أفضل العصر: -، ب ص.

35 المحققين: -، ب ص.

36 الله: -، ب ص.

37 بعض: -، م.

38 ج: + هو، ب ص.

39 ثبت: ثبتت، ه ص.

40 وجوده: وجود، ه.

41 ولا يمكن ثبوت وجوده في الأزل: -، م.

42 أقول: نقول، س.

43 فيما: ما، س.

44 الله:-، ب ص.

45 عليه: -، ب ص.

على البرهان الذي ذكرناه لبيان انعكاس السالبة الكلية الضرورية[46] كنفسها في الجهة والكمّ حيث قلنا: إذا صدق بالضرورة لا شيء من «ج» «ب» وجب أن يصدق بالضرورة لا شيء من «ب» «ج»، وإلا لصدق نقيضه وهو قولنا: بعض «ب» «ج» بالإمكان العام مع الأصل. وإذا صدق هذا مع الأصل صدق لازمه، وهو قولنا: أمكن أن يصدق بعض «ب» «ج» بالفعل معه، لأن صدق الملزوم مع الشيء يقتضي صدق[47] لازمه معه، لكن ذلك محال لأن صدقه بالفعل مع الأصل ينتج المحال، فإمكان صدقه[48] معه[49] يكون ملزوماً لإمكان المحال لأن إمكان الملزوم ملزوم لإمكان اللازم وإمكان المحال محال.

فنقول: لم قلتم بأنه إذا صدق مع الأصل قولنا: بعض «ب» «ج»، بالإمكان العام أمكن صدق الفعلية معه، أي إذا صدق نسبة الباء إلى بعض أفراد الجيم موجهة بالإمكان مع الأصل أمكن صدق هذه النسبة بالفعل[50] معه[51]؟ وما البرهان على ذلك؟ نعم، يلزم من صدق هذه النسبة موجهة بالإمكان مع الأصل صدق[52] إمكان[53] هذه النسبة بالفعل معه، ولا يلزم[54] من صدق إمكان هذه النسبة معه صدق إمكان هذه النسبة بالفعل معه[55]، لأنه لا يلزم من صدق إمكان[56] الشيء مع[57] شيء آخر أن يكون ذلك الشيء ممكن الصدق معه، بل جاز أن يكون ممتنع[58] الصدق معه. ألا ترى أن[59] إمكان وجود كل حادث صادق مع وجود[60] جميع الموجودات[61] الأزلية على معنى أن صدقه لا ينفكّ عن وجود شيء منها وصدق وجود شيء من الحوادث مع وجود شيء من الموجودات الأزلية بهذا المعنى محال؟

---

46 السالبة الكلية الضرورية: السالبة الضرورية الكلية، س؛ الكلية السالبة الضرورية، ب ص؛ الكلية الضرورية، م.
47 صدق: -، ٥.
48 فإمكان صدقه: فإمكانه، س.
49 معه: فإنه، م.
50 بالفعل: فعلية، ٥ ر م.
51 معه: -، م س؛ أي إذا صدق نسبة الباء... بالفعل معه: -، ص.
52 صدق: -، ٥ ر.
53 أي إذا صدق نسبة الباء... من صدق الإمكان: -، ب.
54 ولا يلزم: لأنه لا يلزم، ص.
55 ولا يلزم من صدق... بالفعل معه، ب.
56 هذه النسبة معه... من صدق الإمكان: -، ص.
57 هذه النسبة معه... إمكان الشيء: -، م.
58 ممتنع: ممكن، ٥ ر.
59 ألا ترى أن: إذ، م.
60 وجود: -، ب ص م.
61 الموجودات: الممكنات، س.

أما قوله أدام الله[62] أيامه: إن[63] الإمكان لا يمكن تعقله إلا مضافاً إلى شيء يكون إمكاناً له، فهو قول حق، لكن ليس له[64] مدخل في جواب هذا المنع.[65]

وأما قوله أدام الله أيامه[66]: فالإمكان الثابت في القضية الموجهة[67] ليس إلا إمكان[68] ثبوت المحمول للموضوع، فإذا حكمنا بثبوت ذلك الإمكان فقد حكمنا بإمكان ذلك الثبوت، قلنا: إن عنيتم بهذا الكلام أنا إذا حكمنا باجتماع ثبوت النسبة الموجهة بالإمكان مع الأصل فقد حكمنا بإمكان تلك النسبة بالفعل معه، فلا نسلّم ذلك[69] فإنه عين ما وقع فيه النزاع[70]. وإن عنيتم به أنه يلزم من صدق اجتماع الأول مع الأصل[71] صدق إمكان[72] اجتماع[73] النسبة الفعلية معه[74]، فهو مسلّم لكن لا ينفعكم لما بينّاه.

وأما قوله أدام الله[75] أيامه: ومستند المنع إنما يصح جعله مستنداً لو صدق حكمنا بثبوت إمكان وجود الحادث في الأزل[76] ولم يصدق إمكان ثبوت ذلك الوجود في الأزل، وليس كذا، فإنا إن جعلنا قيد «في الأزل»، متعلقاً بوجود الحادث[77] كانا كاذبين، وإن جعلناه متعلقاً بالإمكان كانا صادقين[78]، وإنما[79] يصدق الأول ولا يصدق الثاني إذا جعل قيد[80] «في الأزل» متعلقاً[81] تارةً بالإمكان وتارةً بوجود الحادث. وإذا عني به ذلك لم يكن مطابقاً لما ادعيناه، قلنا: نحن قد[82]

---

62 الله: -، ب ص.
63 إن: -، ب ص.
64 له: -، ه.
65 أما قوله . . . المنع:-، م.
66 أدام الله أيامه: -، ب ص.
67 الموجهة: الموجبة، ه م س ر.
68 إلا إمكان: الإمكان، ب ص.
69 فلا نسلم ذلك: فلا نسلم، س.
70 فيه النزاع: النزاع فيه، م .
71 الأصل، الأول، ه ر.
72 إمكان: -، م.
73 إمكان اجتماع: اجتماع إمكان، ه ر ص.
74 معه: مع، ب ص.
75 الله: -، ب ص.
76 الأزل: الأزلية، ه.
77 ولم يصدق إمكان ثبوت ذلك الوجود في الأزل . . . متعلقاً بوجود الحادث: -، ب ص؛ بوجود الحادث: تارة بالإمكان وتارة بوجود الحادث، م.
78 كانا صادقين:-، س.
79 وإنما: فإنما، س.
80 قيد: قيدا، م.
81 متعلقاً:-، س.
82 قد:-، س.

ذكرنا المستند على الوجه الذي هو واجب ولا نحتاج إلى هذا التقسيم، فعليكم البرهان على[83] دعواكم، وهي أنه لم قلتم بأنه[84] يلزم من صدق النسبة الموجهة بالإمكان مع الأصل إمكان أن تصدق تلك النسبة بالفعل معه؟ وما ذكرناه في المستند تبرُّع منا لا أنه[85] واجب علينا حتى تكلموا عليه على أنا نقول: الذي يدل على أنه لا يلزم من صدق النسبة الموجهة بالإمكان مع الأصل إمكان صدق هذه بالفعل، هو أنه لو كان صدقها بالفعل ممكناً معه لما يلزم من[86] فرض صدقها بالفعل معه[87] محال، وقد لزم لأنا لو فرضنا صدقها بالفعل[88] مع الأصل ينتظم[89] قياس منتج لسلب الشيء عن نفسه بالضرورة، وإنه محال.

وأما قوله: لا يقال: إذا ثبت في الأزل إمكان وجود الحادث ولم يمكن ثبوت وجود الحادث في الأزل[90] ففي الحالة المعبر عنها بالأزل قد ثبت الإمكان ولم يمكن الثبوت، فجاز صدق الأول دون[91] الثاني في ذلك الحال، فظهر صحة المستند[92]، قلنا: هذا ما ذكرناه في المستند، والمستند لا يطلب عليه البرهان ولا يصحح[93] بحجة، بل نذكر المانع ذلك ليعرف أن[94] منعه ليس منع[95] مكابرة ولا منع عن جزاف من غير تحقيق، وإذا كان كذلك فلا يجوز للمستدل أن يجيب عن المستند، بل عن المنع الوارد على مدّعاه[96]، ومع ذلك فالذي قلتموه في جوابه، وهو قولكم: إذا[97] صدق ثبوت الإمكان لشيء صدق إمكان ثبوت ذلك الشيء، لا أحقق معناه كما ينبغي. فإن عنيتم بثبوت الإمكان لشيء ثبوت المحمول للموضوع موجهاً بالإمكان، ولصدق[98] إمكان ثبوت[99] ذلك الشيء في الجملة صدق إمكان هذه النسبة بالفعل[100]، فمسلّم. وإن عنيتم بالثاني إمكان

---

83 هنا انقطعت نسخة س.
84 بأنه: إنه، ر ه.
85 لا أنه: لأنه، ص س ر ب.
86 من صدق النسبة الموجهة ... معه لما يلزم من: -، ب ص؛ إضافة في هامش م
87 بالفعل معه: معه بالفعل، ه ر.
88 بالفعل: -، ه ر.
89 ينتظم: + منها: ب ر ص ه.
90 في الأزل إمكان وجود الحادث ولم يمكن ثبوت وجود الحادث في الأزل: وجود الحادث في الأزل، ه م س ر؛ في الأزل إمكان وجود الحادث ولم يمكن ثبوت وجود وإمكانه في الأزل، ب.
91 دون: بدون، ه م س ر.
92 ففي الحالة المعبر ... المستند: إلى آخره، م.
93 يصحح: يصح، ب ص.
94 أن: أنه، ه ر.
95 منع: يمنع، ب ص.
96 مدعاه: ما ادعاه، ب ص.
97 إذا: + ثبت، ب ص.
98 ولصدق: وصدق، ه ر.
99 ثبوت: -، ه ر.
100 بالفعل: -، ه ر.

صدق هذه النسبة بالفعل فهو[101] ممنوع. والحاصل[102] أنه[103] ليس عليكم في هذا المقام إلا البرهان على مقدمة واحدة فقط، وهو[104] أنه متى صدقت النسبة الموجهة بالإمكان مع الأصل يلزم أن يكون صدق هذه النسبة بالفعل ممكناً معه لا غير.

فكتب إليه[105] ثانياً:

وقفت على الفوائد الصادرة عن جناب مولانا علامة العالم قدوة العلماء مفتي الفرق مفخر العراقين[106] تاج الأمة فخر الملة[107] نجم الدولة والحق والدين[108] ضاعف الله[109] جلاله وأدام مجده وحرس أيامه[110]. والذي آل إليه الأمر بعد تلك الفوائد الجمّة والمباحث الشريفة التي لا حاجة إلى تطويل العبد باستفسار[111] ما اشتبه عليه منها هو أنه ليس إلا بيان[112] أنه يلزم من صدق بعض «ب» «ج» بالإمكان العام مع الأصل أنه يمكن بالإمكان العام أن يصدق بعض «ب» «ج» بالفعل معه. وقد خطر لي في تقريره أنه إذا صدق مع الأصل بعض «ب» «ج» بالإمكان العام لزمه صدق أنه يمكن بالإمكان العام[113] صدق بعض «ب» «ج» بالفعل معه، وإلا لكان الصادق معه: ليس يمكن بالإمكان العام أن يصدق بعض «ب» «ج» بالفعل[114]، ويلزم[115] من ذلك صدق أنه يمتنع أن يصدق بعض «ب» «ج» بالفعل معه. وإذا صدق ذلك معه لزم أن يصدق معه أيضاً[116] بالضرورة: لا شيء من «ب» «ج». وذلك ظاهر، فيلزم أن يصدق معه بالضرورة: لا شيء من «ب» «ج» مع صدق بعض «ب» «ج» بالإمكان العام، فيلزم اجتماع النقيضين على الصدق مع صدق[117] الأصل، هذا خلف. وهذه اللوازم كلها لا أجد شيئا منها إلا

[101] بالفعل فهو: -، ب ص.
[102] والحاصل: فالحاصل، ٥ ر.
[103] أنه: -، ب ر ص ٥.
[104] وهو: وهي، ٥.
[105] فكتب إليه: فأجاب الشيخ عز الدولة إذا، م؛ فأجاب عز الدولة، ب ص.
[106] العراقين: العراق، م.
[107] فخر الملة: -، م.
[108] علامة العالم... نجم الملة والحق والدين: نجم الملة والدين، ب ص.
[109] الله: + تعالى، ر م ٥.
[110] وأدام مجده وحرس أيامه: -، ب ص.
[111] باستفسار: استفسار، ر؛ -، س.
[112] إلا بيان: على الإتيان، ٥ ص.
[113] العام: -، ٥.
[114] و إلا لكان الصادق... بالفعل: -، م؛ + معه: ص.
[115] ويلزم: + أيضاً، ر.
[116] أيضا: -، س ر م.
[117] صدق: -، ب م؛ الصدق، ص.

بديهياً، وهي المؤدية[118] إلى المطلوب. والمستمد من إنعام[119] المولوي[120] بيان موضع الخلل في[121] هذه والسلام.[122]

فأجاب المولى العلامة نجم الملة والدين رحمه الله[123]: وقف الداعي لدولة المولى[124] ملك ملوك الحكماء[125] والعلماء المحققين عز الدولة فخر الملة قدوة الأفاضل أدام الله معاليه[126] على الفائدة[127] التي أفادها[128]، فوجدها في غاية الحسن والدقة[129] فيعيدها ثم ينظر فيها.

قوله أدام الله أيامه[130]: إذا صدق بعض «ب» «ج» بالإمكان العام مع الأصل وجب أن يصدق مع الأصل أنه يمكن بالإمكان العام أن يصدق بعض «ب» «ج» بالفعل، وإلا لكان الصادق معه ليس يمكن بالإمكان العام أن يصدق بعض «ب» «ج» بالفعل[131]. ويلزم من ذلك صدق أنه[132] يمتنع[133] أن يصدق بعض «ب» «ج» بالفعل معه. وإذا صدق ذلك معه لزم أن يصدق معه أيضاً بالضرورة: لا شيء من[134] «ب» «ج»[135]. وذلك ظاهر، فيلزم أن يصدق معه النقيضان، وإنه محال[136]. قلنا: هذه المقدمات كلها[137] مسلّمة إلا قولكم: إذا صدق مع الأصل أنه يمتنع أن يصدق بعض «ب» «ج» بالفعل لزم أن يصدق معه بالضرورة لا شيء من «ب» «ج». وما البرهان على ذلك؟ بل يلزم وجوب صدق لا شيء من «ب» «ج»[138] دائماً مع الأصل، لأن

---

118 المؤدية: مؤدية، ٥.
119 إنعام: الإنعام، ر ٥.
120 المولوي: المولى، م ص.
121 في: من، ب ٥.
122 والسلام: -، ب م ص م.
123 المولى العلامة نجم الملة والدين رحمه الله: طاب مثواه، ب م ص.
120 المولى: المولوي، م.
125 الحكماء: -، م.
126 الدولة المولى ملك الملوك . . . أدام الله معاليه: -، ب ص.
127 الفائدة: الفوائد، م.
128 أفادها: + المولى، ب ص.
129 الحسن والدقة: الدقة والحسن، ب ص م.
130 أدام الله أيامه: -، ب ص.
131 وإلا لكان الصادق معه . . . بعض ب ج بالفعل: -، ب ص.
132 انه: أن، ص م س ر ب.
133 يمتنع: يمنع، ٥ م س ر.
134 بالفعل معه وإذا صدق ذلك معه . . . لا شيء من ج ب: -، ب ص.
135 بالفعل معه . . . من ب ج: -، ص.
136 مع الأصل أنه يمكن بالإمكان العام . . . وأنه محال: إلى آخره، م.
137 كلها: -، ص.
138 وما البرهان على ذلك... من ب ج: -، ب ص؛ إضافة في هامش م

امتناع صدق أحد[139] النقيضين مع الشيء يستلزم وجوب صدق النقيض الآخر معه[140] لامتناع الخروج عن[141] النقيضين، لا أن يصدق بالضرورة: لا شيء من «ب» «ج» معه. فإن امتناع صدق الموجبة الفعلية[142] الجزئية مع الشيء لا يستلزم صدق السالبة الضرورية الكلية معه، بل وجوب صدق السالبة الكلية الدائمة معه التي هي نقيضها. هذا ما وقع للداعي لدولته عليه[143]، فلينعم المولى بالنظر[144] فيه أدام الله فضائله وعلوّه[145].

فكتب[146] إليه[147] ثالثاً:

ما أفاده مولانا علامة العالم قدوة العلماء والفضلاء ملك ملوك الحكماء مفخر العراقين مفتي الفرق جامع علوم[148] الأوائل والأواخر[149] أسبغ الله ظلّه وضاعف مجده وأمتع بدوام أيامه[150] من أن اللازم من قولنا: يمتنع أن يكون بعض «ب» «ج» بالفعل، هو وجوب أن يصدق لا شيء من «ب» «ج»[151] دائماً وأن قولنا: بالضرورة لا شيء من «ب» «ج» غير لازم عنه، عندي فيه نظر[152]، هو ذا أنا عارض[153] له على الجناب المولوي أسبغ الله ظلّه[154]، وذلك[155] أن الدوام في نفس الأمر لا ينفكّ عن الوجوب البتة لأن كل شيء ما لم يجب وجوده عن[156] علته لم يوجد ولم يستمر وجوده، وما لم يجب عدمه لم يعدم ولم يستمر عدمه، وقد برهن على ذلك في الحكمة والعقل كما أمكنه أن يحكم بالدوام مع قطع النظر عن الوجوب، لا جرم كانت الدائمة في المفهوم[157]

---

139 أحد: -، ب ص م.
140 معه: منه، ه ر.
141 عن: من، ب ص.
142 الفعلية: العامية، ب ص.
143 لدولته عليه: -، ب ص؛ عليه: -، م
144 بالنظر: النظر، ه.
145 أدام الله فضائله وعلوه: -، ب.
146 فكتب: + الشيخ عز الدولة، م؛ عز الدولة، ب ص.
147 إليه: -، ص.
148 علوم: -، م.
149 قدوة العلماء والفضلاء ملك ملوك الحكماء مفخر العراقين مفتي الفرق جامع الأوائل والأواخر: -، م.
150 وأمتع بدوام أيامه: -، م؛ علامة العالم . . . بدوام أيامه: العلامة، ب ص.
151 ب ج: ج ب، م.
152 فيه نظر: -، ه.
153 عارض: كذا في حاشية م، وفي كل النسخ: معرض.
154 أسبغ الله ظله: -، ب ص.
155 وذلك: + هو، ر ص.
156 عن: عند، ر م.
157 في المفهوم: -، م.

أعم في[158] الموضوع[159] من الضرورية، وإذا كان الأمر كذا فمتى لاحظ العقل في الدوام وجوب[160] الذي لا ينفكّ عنه في نفس الأمر فقد لاحظه[161] من حيث هو ضروري، وصارت جهة الدوام من حيث أن وجوب ذلك الدوام ملاحظ في العقل[162] جهة الضرورة بعينها. وحيث الأمر هكذا فقولنا: لا شيء من «ب» «ج» دائماً الملاحظة لوجوب صدقه هو بعينه بالضرورة لا شيء من «ب» «ج» أو هو مساوٍ له، وكيف ما كان يحصل المطلوب على ما هو غير خاف اللهمّ إلا أن نعني بالضرورة[163] مصطلحَ[164] المحققين من أرباب فن[165] المنطق، وحينئذ يكون النزاع لفظياً لا فائدة فيه، و العبد يسأل أن ينظر في هذا وأن يقام عذره في كثرة[166] الإضجار والتثقيل، فإن الذي يحمله على هذا كون فوائد مولانا أعز الله أنصاره[167] مغتنمة ودقائق أنظاره ولطائف أفكاره يُتنافس[168] على استفادتها، ورأيه[169] الكريم المولوي[170] أعلى، والسلام على أهل الحق ومن اتّبع الهدى[171].

[158] في: -، ب.
[159] في الموضوع: -، ص ه.
[160] وجوب: وجوبه، ب ه ر ص.
[161] لاحظه: لاحظ، ب ص.
[162] العقل: + هي، ه ر.
[163] بالضرورة: + عنى ما هو، ص؛ غير ما هو، ب ه.
[164] مصطلح: + بين، ه ر؛ غير ما هو مصطلح، م.
[165] فن: -، ب ص؛ إضافة في هامش م.
[166] كثرة: أثر، ه ر.
[167] أعز الله أنصاره: -، ب ه ر ص.
[168] يتنافس: تنافس، ب ه.
[169] ورأيه: ورأيها، ص.
[170] المولوي: -، ب م ص م.
[171] والسلام على أهل الحق و من اتّبع الهدى: -، ه ر؛ وفي نسخ ب ص وإضافة في هامش م: + وأما ما ذكره في المستند أن أريد به الإمكان بحسب الذات فلا نسلم. إنه لا يمكن صدق [صدق: ثبوت، ب ص] وجودها في الأزل وإنما يلزم ذلك أن لو لم يكن ممكناً بحسب الذات فإن أراد بالإمكان [بالإمكان، ص: به الإمكان، م] بحسب الذات [+ والغير، ب ص] والذي يلاقِه [والذي يلاقه، م: فلا نسلم، ب ص] إمكان وجودها في الأزل إنما يلزم ذلك لو لم يكن ممتنعاً بحسب الغير وإن كان ممكناً في ذاته وهو محال [محال، م: ممنوع، ب ص] لم قلتم بأنه ليس كذلك لا بد له من بيان.

## *Text 7*

### *Correspondence with Kamāl al-Dīn al-Baḥrānī*

The edition is based on the following manuscripts:
ر = Ragıp Paşa 1461, ff. 346b:7–347a:2
ص = Ṣadr al-Dīn al-Maḥallāṭī 17/6, f. 45a
ه = Ākhūnd 1187/76, ff. 348b:8–349a:1

فائدة لعز الدولة ابن كمونة[1]: العلة التامة للمجموع المركب من أفرادٍ كل واحد منها ممكن لا بد وأن تكون[2] علة لكل واحد من تلك الأفراد، وإلا لكانت علة بعضٍ تلك الأفراد غير تلك العلة. وذلك الغير، أي علة ذلك البعض، إما أن يكون واجباً لذاته أو ممكناً لذاته[3]، والأول ملزوم للمدعى، والثاني محال لأنه يلزم اجتماع وجود العلة التامة مع وجود ذلك الغير الذي هو ممكن بالإمكان الخاص وعلة ذلك البعض من الأفراد، وهو محال لأن وجود ذلك الغير يستلزم جواز وجوده بحسب ذاته جواز عدمه، فيلزم من هذا أن يكون وجوده يستلزم جواز عدمه. ولو جاز اجتماع وجود العلة التامة للمجموع المذكور مع وجود ذلك الغير الذي هو علة لذلك البعض من الأفراد للزم[4] جواز اجتماعه مع لازمه، لأن اجتماع الشيء مع الملزوم يستلزم اجتماعه مع اللازم، لكن اجتماع العلة للمجموع مع عدم ذلك الغير محال، لأن جواز اجتماع العلة التامة مع عدمه يستلزم جواز وجود العلة التامة للمجموع بدون ذلك المجموع لضرورة[5] عدم الجزء الذي هو معلول ذلك الغير.

أجاب عنه العلامة المحقق كمال الدين البحراني[6] رحمه الله بأن الذي يلزم من جواز اجتماع وجود العلة التامة مع وجود ذلك الغير المستلزم لإمكان وجوده المستلزم لإمكان عدمه هو جواز اجتماع تلك العلة مع إمكان عدم ذلك الغير لا مع عدمه، وإمكان عدمه لذاته لا ينافي وجوب وجوده لعلته، فلا يلزم تخلف ذلك المجموع عن تمام علته. فلا معنى إذن لقوله: لكن اجتماع العلة للمجموع مع عدم ذلك الغير محال، لأن اجتماعها إنما هو مع إمكان عدمه اللازم لذاته، ولا يستلزم ذلك اجتماعها مع عدمه.

---

[1] فائدة لعز الدولة ابن كمونة: فائدة لابن كمونة، ص.
[2] تكون: يكون، ص ه.
[3] لذاته: كذا، ر ه.
[4] للزم: يلزم، ر ص ه.
[5] لضرورة: ضرورة، ر ص ه.
[6] البحراني: النجراي، ه.

*Text 8*

*Correspondence with Fakhr al-Dīn al-Kāshī*

The edition is based on the single known manuscript of the correspondence:
Esad Efendi 3642/8, ff. 168a–169a

[فيه ما سأل عنه الكاشي وأجاب عنه ابن كمونة]
سؤال سأله فخر الدين الكاشي من العلامة ابن الكمونة

استدلت الحكماء على إثبات العقل الفعال بالذهول والنسيان العارضين للنفس حالتي إمكان ملاحظتها للمعقولات التي غابت عنها من غير تجشم كسب وافتقار ملاحظتها باكتساب جديد كما في الأول لاحتياجها في الملاحظة شيئاً غيرها حافظاً لتلك الصور التي تكون موجودة فيه بالفعل حالة الذهول، غير موجودة حالة النسيان لامتناع اتحادهما. وإلا لزم عدم الذهول أو الاحتياج بتجدد الكسب واستحالة تجزئتها بجزئين أحدهما حافظ والآخر مدرك كما قالوه، فحصل منهما موجود مفارق يرتسم بجميع الصور الموجودة بالفعل غير جسم وجسماني ونفس وهو المسمى بالعقل الفعال. وإذا ثبت هذا فنقول: لا شك أن النفس كما تحكم بأحكام كلية صحيحة وافقة فكذلك تعتقد اعتقادات كلية فاسدة تارةً تذهل عنها وتارةً تنساها وتعود إليها مع اكتساب جديد ومع عدمه. فعلى هذا يلزم أحد أمرين[1]، إما ارتسام صور تلك المعقولات المعتقدة الفاسدة في ذلك الموجود بالفعل، وإما وجود موجود آخر حافظ لها [لا] جسماً [168ب] ولا جسمانياً ولا بنفس مثله، وهما باطلان. أما الأول فلفقد شيء من الأحكام غير مطابق لما في نفس الأمر ولعدم الفرق بين الحق والباطل لامتيازهما عن الآخر بالمطابقة وعدمها. وأما الثاني لخلاف معتقدهم وجهلهم[2] إياها والله أعلم.

المأمول من جنابه الأعلى أن ينعم العبد ببيان وافٍ وأن يشير إلى مفهومي الخارج والذهن وإلى أن الخارج، هل هو مرادف لنفس الأمر أم لا؟ ليضاف إلى سائر الألطاف ويسحب ذيل العفو على هفواته.

[1] أمرين: الأمرين، الأصل.
[2] وجهلهم: وجهله، الأصل.

الجواب، والله الموفق، نقلاً من خط ابن كمونة

إن المبادئ العالية التي يعبّر عنها بالملائكة والعقول المجردة وغير ذلك معلوم عندها أن زيداً مثلاً يعتقد في وقت كذا اعتقاداً كلياً فاسداً هو كذا، وهذا اعتقاد صحيح مطابق للواقع متضمن أو مستلزم لذلك الاعتقاد الفاسد. فإذا عقل زيد ذلك، ثم ذهل عنه، ثم استرجعه فإنما أمكنه استرجاعه لانحفاظ ذلك الاعتقاد الفاسد في تلك المجردات على الوجه المذكور، ولا يلزم من هذا أن يُنسب إلى تلك المبادئ المجردة تصديق غير مطابق للواقع وهو الذي يسمى بالجهل [169أ] المركب، بل هي متصورة لذلك التصديق الفاسد على وجه التبعية للعلم بأن زيداً يوقع ذلك التصديق، وهذا من الاحتمالات الدافعة للتشكك المذكور. وأما استفساره، حرس مجده، عن مفهومي الخارج والذهن، وهل الخارج مرادف لنفس الأمر؟ فالمرتسم في المدرك أو المتمثل عنده هو الذي يعبر عنه بأنه في الذهن، وهذا فقد يكون مطابقاً وقد لا يكون. والمطابق قد تكون مطابقته لما هو خارج عن المدرك، وقد يطابق ما هو مدرك أيضاً إما عين ذلك المدرك أو غيره. فالخارج عن المدرك هو المراد بقولهم: في الخارج، والمطابق لما في الذهن هو المسمى بما في نفس الأمر، سواء كان في الخارج أو لم يكن هو على هذا. فكل ما في الخارج فهو في نفس الأمر، وليس كل ما في نفس الأمر فهو في الخارج، فبينهما عموم وخصوص مطلقان. والأعم منهما هو ما في نفس الأمر، والأخص هو الذي في الخارج على هذا الاصطلاح. وهاهنا احتمالات أخرى يندفع بها التشكك أيضاً لا حاجة إلى ذكرها. والله الهادي بمنّه وجوده ورحمته.

## *Text 9*

### *Ibn Kammūna's Poetry*

The edition is based on the following two manuscripts:
م = Majlis 630/3, p. 20.
ج = *Jung-i Mahdawī. Tārīkh-i kitābat az 753 qamarī be ba'd*. Tehran: Markaz-i Nashr Dānishgāhī, 1380/2002, p. 193.

للحكيم عز الدولة ابن كمونة[1]

تَحَصّنْتُ مِنْ وَقْعِ الخُطوب بِخَالَقِي ... وَذَلِكَ حصن للمُحقّ حَصينُ
وَقَطّعتُ أَطْمَاعِي عَنِ النَّاسِ كُلِّهِمْ ... فوجهِيَ عَنْهُم مُلْفَتٌ وَمَصونُ
وَمَا أنا بِالدُّنْيَا ضَنينٌ وإنَّمَا ... بِديني وَعِرْضي وَالحَيَاءِ ضَنينُ
رَكَنْتُ[2] إِلى دَارِ الْخُلُودِ وَلَيْسَ لي ... إِلى زُخْرُفِ الدُّنْيَا المُضِلِّ رُكُونُ
إِذَا حَصَلَ القُوتُ المُقِيم لِبَيْتي ... فَتَرْكُ فُضُولِ العَيْش بَعْدُ يَهُونُ[3]
وَلَمْ أَرجُ إلاّ مَنْ وَثِقْتُ بِأَنَّهُ ... لِقُوتي وَقُوتِ العالَمِين ضَمينُ
تَكَفَّلَ لِلطِّفْلِ الرَّضِيعِ رِزْقَهُ ... وَغَذَّاهُ في الاَحْشَاءِ وَهْوَ جَنِينُ[4]
فَإِنْ[5] كُنْتُ أَنْحو مَقْصِداً عَن بابِهِ ... فَذاكَ جُنُونٌ وَالجُنُونُ فُنُونُ

---

[1] للحِكيِم عز الدولة ابن كمونة: للحكيم الكبير العالم الحبر ابن كمونة البغدادي، م.
[2] رَكَنْتُ: وكنت، ج م.
[3] إِذِا حِصل ... يهون: - ، م.
[4] تَكَفَّلَ ... جَنِينُ: -، ج.
[5] فإن: لئن، ج.

# APPENDICES

## Appendix A *The Firkovitch/Shapira Codex (Sharḥ al-Talwīḥāt)*

The State Library Berlin and the Russian National Library (RNL) St. Petersburg own three fragments that originally belonged to the same copy of Ibn Kammūna's *Sharḥ al-Talwīḥāt*. With the exception of at least four folios that are missing, the fragments constitute the complete text of parts I and II of the commentary. In the following table, the original order of the text is restored. In the right column reference is given to the available editions of the commentary by Sayyid Ḥusayn Sayyid Mūsawī containing all three parts and by Hossein Ziai & Ahmed Alwishah containing part II only. In the left column the respective portions of the codex as preserved in mss. Berlin Ms. Or. Fol. 1321, RNL II Firk. Yevr.-Arab. I 3137, and RNL II Firk. Yevr.-Arab. I 3138 are indicated.

| Part I: Manṭiq | |
|---|---|
| YA I 3137, f. 124 | ed. Mūsawī, pp. 2:1–3:1 |
| YA I 3137, ff. 1–7 | ed. Mūsawī, pp. 3:1–21:9 |
| Lacuna | ed. Mūsawī, pp. 21:9–23:11 |
| YA I 3137, ff. 8–9 | ed. Mūsawī, pp. 23:11–28:2 |
| YA I 3137, ff. 42–49 | ed. Mūsawī, pp. 28:2–45:18 |
| YA I 3137, f. 19 | ed. Mūsawī, pp. 45:18–48:3 |
| YA I 3137, ff. 10–18 | ed. Mūsawī, pp. 48:3–71:16 |
| Lacuna | ed. Mūsawī, pp. 71:16–73:16 |
| YA I 3137, f. 28 | ed. Mūsawī, pp. 73:16–75:23 |
| YA I 3137, ff. 50–88 | ed. Mūsawī, pp. 75:23–163:7 |
| YA I 3137, f. 114 | ed. Mūsawī, pp. 163:7–165:7 |
| YA I 3137, ff. 20–27 | ed. Mūsawī, pp. 165:7–181:2 |
| YA I 3137, f. 113 | ed. Mūsawī, pp. 181:2–182:23 |
| YA I 3137, ff. 32–41 | ed. Mūsawī, pp. 182:23–204:20 |
| YA I 3137, ff. 99–107 | ed. Mūsawī, pp. 204:20–226:21 |
| YA I 3137, f. 31 | ed. Mūsawī, pp. 226:21–228:19 |
| Ms. Or. Fol. 1321, ff. 1a:1–29b:1 | ed. Mūsawī, pp. 228:19–291:15 |

| Part II: Ṭabīʿiyyāt | |
|---|---|
| Ms. Or. Fol. 1321, ff. 29b:2–30b | ed. Mūsawī, pp. 293:1–297:4<br>eds. Ziai & Alwishah, pp. 3:1–9:2 |
| Ms. Or. Fol. 1321, ff. 31–49 | ed. Mūsawī, pp. 298:26–340:25<br>eds. Ziai & Alwishah, pp. 13:2–88:4 |
| YA I 3137, ff. 89–98 | ed. Mūsawī, pp. 341:1–364:1<br>eds. Ziai & Alwishah, pp. 88:4–125:3 |
| Ms. Or. Fol. 1321, ff. 50–99 | ed. Mūsawī, pp. 364:1–496:3<br>eds. Ziai & Alwishah, pp. 125:3–339:6 |
| YA I 3138, f. 1 | ed. Mūsawī, pp. 496:3–499:1<br>eds. Ziai & Alwishah, pp. 339:6–344:2 |
| YA I 3138, ff. 6–7 | ed. Mūsawī, pp. 499:1–504:21<br>eds. Ziai & Alwishah, pp. 344:2–353:8 |
| YA I 3138, ff. 2–5 | ed. Mūsawī, pp. 504:21–516:5<br>eds. Ziai & Alwishah, pp. 353:8–373:4 |
| YA I 3138, ff. 8–9 | ed. Mūsawī, pp. 516:5–521:16<br>eds. Ziai & Alwishah, pp. 373:4–382:12 |
| YA I 3137, f. 116 | ed. Mūsawī, pp. 521:16–524:12<br>eds. Ziai & Alwishah, pp. 382:12–387:9 |
| YA I 3137, f. 115 | ed. Mūsawī, pp. 524:12–526:21<br>eds. Ziai & Alwishah, pp. 387:9–391:11 |
| YA I 3137, f. 30 | ed. Mūsawī, pp. 526:21–529:12<br>eds. Ziai & Alwishah, pp. 391:11–396:3 |
| YA I 3137, ff. 117–123a | ed. Mūsawī, pp. 529:12–543:2<br>eds. Ziai & Alwishah, pp. 396:3–418:13 |
| Lacuna | ed. Mūsawī, pp. 543:2–544:8<br>eds. Ziai & Alwishah, pp. 418:3–420:13 |
| YA I 3137, f. 29 | ed. Mūsawī, pp. 544:8–547:6<br>eds. Ziai & Alwishah, pp. 420:13–424:12 |
| Lacuna | ed. Mūsawī, pp. 547:6–556:9<br>eds. Ziai & Alwishah, pp. 424:12–440:9 |
| YA I 3137, f. 111 | ed. Mūsawī, pp. 556:9–559:5<br>eds. Ziai & Alwishah, pp. 440:9–444:14 |
| YA I 3137, ff. 108–109 | ed. Mūsawī, pp. 559:5–563:15<br>eds. Ziai & Alwishah, pp. 444:14–451:16 |
| YA I 3137, f. 112 | ed. Mūsawī, pp. 563:15–565:14<br>eds. Ziai & Alwishah, pp. 451:16–455:5 |

| | |
|---|---|
| YA I 3137, f. 110 [lower margin heavily damaged] | ed. Mūsawī, pp. 565:14–567<br>eds. Ziai & Alwishah, pp. 455:5–458 |
| Lacuna | ed. Mūsawī, p. 568<br>eds. Ziai & Alwishah, pp. 459–461 |

## Appendix B *Kitāb fī ithbāt al-nubuwwa li-Muḥammad ʿalayhi al-salām*

The following edition is based on the single known manuscript of the text (ms. Hamidiye 1447/3, ff. 81b–84b). Readings from the manuscript that have been emended are indicated as الأصل. Readings by Moshe Perlmann[1] that disagree with the text of the manuscript are indicated as ب. Scriptio defectiva has been modernized without notification.

### كتاب في[2] إثبات النبوة لمحمد عليه السلام

بسم الله الرحمن[3] الرحيم

أحمد الله تعالى حمد مسترشد بنور هدايته، مفتقر إلى إرشاده وعنايته، وأصلي على القديسين[4] من ملائكته وأنبيائه، خصوصاً على أفضلهم وأكملهم محمد المصطفى وآله وأصحابه. [و]بعد[5]، لما تأملت في الرسالة المنسوبة إلى الحكيم الفاضل عز الدولة[6] سعد بن منصور بن[7] كمونة في تنقيح [الأبحاث] للملل[8] الثلاث وجدت فيها كثيراً من التعسّفات الفاسدة والتعصّبات الباردة في أفضلية شريعة موسى عليه السلام[9] على جميع الشرائع وتفضيل طائفته[10] على جميع الطوائف، وليس عنده حجة وبينة بل حكايات مستندة إلى تواتر منقطع[11]. ومع شهرته بالحكمة وتصانيفه في

---

[1] Moshe Perlmann: "'Proving Muhammad's Prophethood', A Muslim critique of Ibn Kammuna." [Hebrew] In *Hagut Ivrit Ba'Amerika. Studies on Jewish Themes by Contemporary American Scholars*. Eds. Menahem Zohori, Arie Tartakover and Haim Ormian. Vol. III. Tel Aviv 1974, pp. 75–97.

[2] في: -، ب.

[3] الرحمن: الحمن، ب.

[4] القديسين: القدسيين، الأصل ب.

[5] [و]بعد: فبعد، ب.

[6] الدولة: الدين، ب.

[7] بن: -، الأصل؛ (ابن)، ب.

[8] للملل: الملل، الأصل، ب.

[9] عليه السلام: -، ب.

[10] طائفته: طائفة، الأصل؛ طائفته، ب.

[11] منقطع: منقطعة، الأصل ب.

العلوم[12] الحكمية، كيف يكون مظنّةً[13] للتقليد[14]، والعناد من غير العقل والانتقاد؟ فإن الحكيم يجب أن يعتصم بحبل الفطانة والرشاد، ولا يكون قانعاً[15] بما أخذه من الآباء والأجداد. وكان خاطري في أثناء مطالعتها ملتفتاً إلى أجوبة ما رمزه في خلال كلامه، وإلى دفع أجوبة اعتراضات الحكيم العلامة مؤيد الدين سمؤال[16]، حتى أن الزمان ساعدني فحررتها مرتبةً على بابين[17]: الباب الأول في أحكام الإسلام، والباب الثاني في مذمّة اليهود.

## الباب الأول [في أحكام الإسلام] وفيه عشرة[18] فصول

### الفصل الأول في بيان معتقد المسلمين

وصورة اعتقادهم أن الله تعالى[19] حي واحد عالم مريد قادر سميع بصير متكلم بلا تجسيم ولا تشبيه ولا شريك له ولا نظير ولا صاحبة ولا ولد، وأنه أرسل رسله بالهدى وأرسل محمداً إلى[20] كافة الخلق، وهو خاتم النبيين لا نبي بعده، وأنه أخبر عن الله تعالى أنه أمر[21] بالوفاء بالعهد[22] وبرّ الوالدين وأمر بالعفّة والتقوى والشجاعة والجهاد[23] في سبيل الله والحكمة والعدالة، وبالجملة أمر بارتكاب أمهات الفضائل والاجتناب عن الرذائل والأخلاق المذمومة. وأخبر بأن الله يبعث مَن في القبور[24] ويحاسب الناس يوم القيامة على عقائدهم وأعمالهم ويجازي الناس على قدر استحقاقهم، ﴿فَمَنْ يَعْمَلْ مِثْقَالَ ذَرَّةٍ خَيْراً يَرَهُ وَمَنْ يَعْمَلْ مِثْقَالَ ذَرَّةٍ شَرّاً يَرَهُ﴾.[25] والناس كلهم فريقان، فريق في الجنة وفريق في النار. وأهل الجنة متنعمون[26] فيها نعيماً مخلداً غير منقطع،

---

[12] في العلوم: بالعلوم، ب.
[13] مظنّة: مطية، الأصل؛ مطيته، ب.
[14] للتقليد: التقليد، ب.
[15] قانعاً: قانعة، الأصل؛ قانتاً: ب.
[16] سمؤال: سمؤل، الأصل ب.
[17] بابين: ما بين، ب.
[18] عشرة: كذا في الأصل.
[19] تعالى: -، ب.
[20] إلى: على، الأصل ب.
[21] أنه أمر: اقامة، ب.
[22] بالعهد: والعهد، الأصل ب.
[23] والشجاعة والجهاد: الشجاعة بالجهاد، ب.
[24] مَن في القبور: رسل النبوة، ب.
[25] سورة الزلزلة (99):7–8.
[26] متنعمون: يتنعمون، ب.

وهناك من النعيم «ما لا عين رأت ولا أذن سمعت ولا خطر على قلب بشر»[27]. وأهل النار قسمان، قسم يدخلون فيها مخلداً مع الزبانية والأنكال و﴿كُلَّمَا نَضِجَتْ جُلُودُهُم بَدَّلْنَاهُمْ جُلُوداً غَيْرَهَا﴾[28]، وقسم يدخلون النار لكن[29] لا يخلدون فيها، بل يخرجون منها إلى الجنة بالشفاعة والعفو والرحمة الواسعة.

## الفصل الثاني في إثبات النبوة لمحمد عليه السلام[30] على نهج الحق والعقل

وتقريره أن محمداً ادّعى النبوة وظهرت منه على وفق دعواه معجزة، ومن ادّعى النبوة وظهرت[31] منه على وفق دعواه معجزة فهو نبي، فنتج[32] من الشكل الأول أن محمداً نبي وهو المطلوب. أما بيان الصغرى، أن التواتر الصحيح يشهد على وجوده ودعوته وأن القرآن ظهر عليه وهو معجز، لأن العرب الذين هم الغاية في الفصاحة والبلاغة عجزوا عن المعارضة مع توفر دواعيهم على الإتيان بالمعارضة، لأنه كلّفهم بترك[33] أديانهم وملة أجدادهم، وأوجب عليهم ما أتعب أبدانهم كالصلاة والصوم وأنقص أموالهم كالحج والزكاة، ويطلب منهم المعارضة [82أ] عزّ من قائل ﴿قُلْ لَّئِن[34] اجْتَمَعَتِ الإِنسُ وَالْجِنُّ[35] عَلَى أَن يَأْتُواْ بِمِثْلِ هَذَا الْقُرْءَانِ لاَ يَأْتُونَ[36] بِمِثْلِهِ وَلَوْ كَانَ بَعْضُهُمْ لِبَعْضٍ ظَهِيراً﴾[37]، وقوله تعالى[38] ﴿أَمْ[39] يَقُولُونَ افْتَرَاه قُلْ[40] فَأْتُواْ بِعَشْرِ سُوَرٍ مِثْلِه[41]﴾[42]، وقوله تعالى ﴿وَإِن كُنتُمْ فِي رَيْبٍ مِمَّا نَزَّلْنَا عَلَى عَبْدِنَا فَأْتُواْ بِسُورَةٍ مِّن مِّثْلِهِ﴾[43]. ثم إذ[44] لم يتمكنوا من المعارضة بالكل[45] والبعض والواحد التي هي أسهل، اتفقوا على المقاتلة[46] التي هي أصعب الأمور،

27 راجع مسند أحمد بن حنبل 2:438، 2:495، 4:506، 2:370، 2:407، 2:416، 2:462.
28 سورة النساء (4):56.
29 لكن: ممن، ب.
30 عليه السلام: (صلعم)، ب.
31 وظهرت: فظهرت، ب.
32 فنتج: نتج، الأصل؛ منح (؟)، ب.
33 بترك: على ترك، الأصل ب.
34 قل لئن: فلو، الأصل ب.
35 الإنس والجن: الجن والإنس، الأصل ب.
36 بمثل هذا القرءان لا يأتون، -، الأصل ب.
37 سورة الإسراء (17):88.
38 تعالى: -، ب.
39 أم: -، ب.
40 قُل: فقل، الأصل؛ قل، ب.
41 مثله: من مثله، ب.
42 سورة هود (11):13.
43 سورة البقرة (2):23.
44 إذ: اذا، الأصل ب.
45 بالكل: الكل، ب.
46 المقاتلة: المقابلة، الأصل؛ المقاتلة، ب.

فالعقل[47] السليم يحكم على عجزهم عن المعارضة، فثبت أن القرآن معجزة[48] من محمد على وفق دعواه. وأما[49] الكبرى فظاهرة لا تحتاج إلى بينة. فثبت بالقياس البرهاني أن محمداً رسول الله حقاً.

## الفصل الثالث فيما يرد على بيان الصغرى وجوابه

لا يقال: إن القرآن يحتمل أنه[50] أنزل إلى نبي آخر ولم يتمكن من الدعوة ومات أو قُتِل ووقع في يد محمد صلى الله عليه وسلم[51]، أو أنه ترصد[52] حكايات القرون الماضية من الوحي السابق وغيره وما وجد من كلمات رائقة[53] ونكات فائقة[54] وجمعه ورتّبه قرآناً. وإن[55] سلمنا صحة تواتر أصل القرآن بأنه أنزل إلى محمد عليه السلام[56] لكن لا نسلم [أن] ترتّب ما فيه من السور والآيات متواتر[57] لجواز اتفاق بعض الصحابة على التحريف في بعض الآيات، فإن ابن مسعود اختلف في كون الفاتحة والمعوّذتين من القرآن، وكذا الكلام في بسم الله الرحمن الرحيم في أوائل السور. واختلاف نسخ القرآن في زمان عثمان وإحراقها إلا واحدة[58] يدل على التحريف[59]. وأيضاً لما أرسل مروان مصحف حفصة[60] إلى عبد الله بن عمر أمر بإحراقه، ونقل عن عمر أنه[61] يؤتى بالآية أو الآيتين[62]، وعن عثمان بعد فراغ المصحف إذا نظر فيه قال: أحسنتم وفيه شيء من اللحن سنقوّمه[63] بألسنتنا.

لأنا نقول: إن القرآن يشتمل على حكايات مختصة بأحوال محمد عليه السلام[64] وأزواجه وأصحابه وأوليائه وأعدائه، فكيف ينسب إلى نبي آخر؟ وأيضاً إن تغيير حرف من القرآن في هذا العصر

---

[47] فالعقل: والعقل، ب.
[48] معجزة: معجز، ب.
[49] وأما: فأما، ب.
[50] أنه: ان، الأصل ب.
[51] صلى الله عليه وسلم: (صلعم)، ب.
[52] ترصد: يترصد، الأصل ب.
[53] رائقة: رايقة، الأصل؛ رائعة، ب.
[54] فائقة: فايقة، الأصل؛ فائقة، ب.
[55] وإن: ولئن، الأصل ب.
[56] عليه السلام: (صلعم)، ب.
[57] متواتر: بتواتر، ب.
[58] واحدة: واحد، الأصل؛ واحدا، ب.
[59] التحريف: التحريق، ب.
[60] حفصة: الحفص، الأصل؛ الحفصة، ب.
[61] أنه: بأنه، الأصل ب.
[62] أو الآيتين: والآيتين، الأصل ب.
[63] سنقوّمه: سنقريه، الأصل ب.
[64] عليه السلام: (صلعم)، ب.

ممتنع، مع أن مراعاة القرآن في عهد[65] الصحابة وإن لم يكن أقوى من عهدنا[66] لم يكن أضعف، وهذا دليل قطعي على عدم التغيير والتحريف في جميع الأزمنة. واختلاف ابن مسعود في الفاتحة والمعوّذتين مروي بالآحاد لا يعارض [به] التواتر اليقيني. ولئن سلمنا لما ضرّ غرضنا لو كانت آيات التحدي منزلة من عند الله ولم تكن[67] من القرآن، وكذا الكلام في بسم الله، واختلاف المصاحف في زمان عثمان واختيار واحد منها وهو[68] مصحف زيد بن ثابت [فلأنه كان] عرضه على النبي بعد عرض عبد الله بن مسعود وأبيّ بن كعب، وبه كان يقرأ النبي ويصلي إلى أن انتقل[69] إلى جناب العزّة. فلهذا اختار المسلمون مصحف زيد وإحراقهم الباقي مخافة الاختلاف مع أن الثلاثة حقة صحيحة منزلة من عند الله تعالى. فإن الاختلاف في اللغات والترتيب لا في الأحكام والمعاني، وأشار إليه بقوله عليه السلام[70] «أنزل القرآن على سبعة أحرف» أي سبع لغات. والرواية عن عمر بأنه يؤتى بآية أو آيتين فغير مقبولة [لكونها] من باب الآحاد. فإن جامع القرآن هو النبي بنفسه عليه السلام[71] جمعاً على ترتيب ما أنزل عليه أوله ﴿إِقْرَأْ﴾[72] وآخره[73] سورة البراءة وجمعاً آخر رتّبه[74] في المدينة أوله سورة البقرة وآخره سورة الناس. وما حكوه عن عثمان أنه قال: وفيه لحن، وإن سلمنا صحة روايته، فمراده اللحن في الكتابة.

## الفصل الرابع فيما يرد على القرآن وجوابه

لا يقال: إن[75] القرآن وصفه مشتمل على البيان والإخبار[76] عما لا يفهم، وليس كذلك لأن الحروف التي في أوائل السور غير معلومة، ولأصحاب[77] التفسير ملتبسة مضطربة في شرحها مع أنه مشتمل على توضيح الواضحات والتكرار الكثير كقوله تعالى ﴿ثَلَاثَةِ [82ب] أَيَامٍ[78] فِي

---

65 عهد: عصر، ب.
66 عهدنا: هذه، الأصل ب.
67 تكن: يكن، الأصل؛ تكن، ب.
68 وهو: لأن، الأصل ب.
69 انتقل: ينتقل، الأصل ب.
70 عليه السلام: عم، ب.
71 عليه السلام: (عم)، ب.
72 سورة العلق (96):1.
73 وآخره: وآخر، الأصل؛ آخره، ب.
74 رتّبه: ثبته، ب.
75 إن: أن، ب.
76 الإخبار: الاختبار، ب.
77 ولأصحاب: وأصحاب، الأصل ب.
78 أيامٍ: -، الأصل ب.

الْحَجِّ وَسَبْعَةٍ إِذَا رَجَعْتُمْ تِلْكَ عَشَرَةٌ كَامِلَةٌ﴾[79]. وأيضاً وصفه مشتمل على أنه جامع لكل العلوم كقوله تعالى ﴿مَا فَرَّطْنَا فِي الْكِتَابِ مِن شَيْءٍ﴾[80]، وقوله ﴿وَلَا رَطْبٍ وَلَا يَابِسٍ إِلَّا فِي كِتَابٍ مُبِينٍ﴾[81]، وليس كذلك لأنا وجدناه[82] خالياً عن[83] أكثر المسائل المهمة، بل وجدناه مشتملاً على أحكام متناقضة كآيات الجبر المقابلة لآيات القدر، وآيات التشبيه المقابلة لآيات التنزيه وعلى سوء الترتيب كقوله تعالى[84] ﴿إِيَّاكَ نَعْبُدُ وَإِيَّاكَ نَسْتَعِينُ﴾[85]، فإن الاستعانة متقدمة على الفعل الذي هو التعبد. وقريش[86] لا يهمزون الكلام وجميع القراء يهمزون خلافاً لقوله تعالى ﴿وَمَا أَرْسَلْنَا مِن رَّسُولٍ إِلَّا بِلِسَانِ قَوْمِهِ﴾[87]. فكل هذه أمارات تطرّق[88] التحريف إليه، وإلا لكان مضرّاً في إثبات كونه معجزاً.

لأنا نقول: إن الحروف التي في أوائل السور معلومة لمن كان له أهلية معرفتها، كما إذ[89] سئل[90] علي كرم الله وجهه عن معاني هذه الحروف قال: «هذه أسرار تحت خاتمي». واختلاف المفسرين في تفسيرها يدل على صعوبتها لا على عدم فهمها، وكما[91] أنه مشتمل على مرموزات روحانية لأهل العرفان فهو مشتمل على توضيح الواضحات، والتكرار المناسب لأهل الظاهر والعوام الذين هم أكثر الناس.

وأيضاً إن القرآن جامع لكل العلوم مع زيادات[92] أخر من المعاني[93] القدسية والإدراكات اللاهوتية التي لا توجد في الكتب العلمية، ويشهد عليه التفسير الكبير المنسوب إلى الإمام العلامة فخر الدين الرازي نوّر الله ضريحه، فإن جميع العلوم مندرج فيه ولا يكون وافياً بتفسيره أو تفسير بعضه. واختلاف الآيات بالجبر والقدر والتنزيه والتشبيه مرتفع بتأويلات مذكورة في تفسيره

---

79 سورة البقرة (2):196.
80 سورة الأنعام (6):38.
81 سورة الأنعام (6):59.
82 وجدناه: وجدنا، الأصل؛ وجدناه، ب.
83 عن: من، ب.
84 تعالى: تع، ب.
85 سورة الفاتحة (1):5.
86 وقريش: والقريش، الأصل ب.
87 سورة ابراهيم (14):4.
88 تطرّق: تطارق، الأصل ب.
89 إذ: اذا، الأصل ب.
90 سئل: + عن، الأصل ب.
91 وكما: فكما، ب.
92 زيادات: فائدات، ب.
93 المعاني: المعين، الأصل؛ المعاني، ب.

وتفاسير غيره، وترتيب الفاتحة صحيح[94]، فإن الاستعانة التقرب إلى جنابه لا التعبد، وكما أن قريشاً[95] قوم النبي فالعرب كلهم قومه. فثبت صحة القرآن وكمال بيانه وتمام حكمته[96].

الفصل الخامس في إثبات نبوة محمد عليه السلام[97] من التوراة

المشهور أن ختم التوراة بآية تعريبها: إن قدرة الله تعالى من سينا أقبلت وأشرقت من سعير[98] ثم أطلعت من جبال فاران ومعها ربوات القديسين[99].[100] فهذه الآية تدل على طلوع نبوة من جبال فاران. والدليل الدال على أن هذا الجبل هو[101] جبل مكة هو أن إسماعيل لما فارق أباه الخليل عليهما السلام سكن إسماعيل في برية فاران ونطقت التوراة بذلك في قوله: *يشو بمدبار فاران*[102]، فقد ثبت في التوراة[103] أن جبل فاران مسكن لآل[104] إسماعيل عليه السلام[105]، وقد علم الناس جميعاً أن المشار إليه بالنبوة من ولد إسماعيل محمد عليهما السلام، وأنه بعث من مكة التى كان فيها مقام إسماعيل. فقد عرفت[106] من التوراة نبوة محمد والبشارة بمقدمه[107]، واليهود لفرط جهلهم وضلالهم[108] يسلمون المقدمتين ويجحدون في النتيجة، وقد علم من هذه الآية أن شريعة محمد أفضل الشرائع، لأن شريعة موسى منسوبة إلى الإقبال، وشريعة عيسى إلى الإشراق، وشريعة محمد[109] عليهم السلام[110] إلى الطلوع، والإقبال والإشراق من مبادئ الطلوع دون العكس، بل كمالهما في الطلوع، فقد ثبت بنص التوراة[111] أن شريعة محمد أكمل الشرائع وأتمها وهو أشرف الرسل وأفضلهم عليهم السلام.

---

[94] صحيح: صميم، ب.
[95] قريشا: القريش، الأصل ب.
[96] حكمته: حكمه، الأصل ب.
[97] عليه السلام: -، ب.
[98] سعير: سيعر، الأصل؛ سعير، ب.
[99] القديسين: القدسيين، الأصل ب.
[100] سفر التثنيه 33:2
[101] هو: -، ب.
[102] *يشو بمدبار فاران*: **וישב במדבר פראן** (سفر التكوين 21:21).
[103] التوراة: التورية، الأصل؛ التوراة، ب.
[104] لآل: آل، ب.
[105] عليه السلام: (عم)، ب.
[106] عرفت: وفت، ب.
[107] بمقدمه: المتقدمة، الأصل ب.
[108] وضلالهم: وصلالهم، ب.
[109] محمد: + الى، الأصل.
[110] عليهم السلام: -، ب.
[111] التوراة: التورية، الأصل؛ التوراة، ب.

## الفصل السادس في كمال نبوة محمد عليه السلام[112]

إذا عرفتَ بالتواتر الصحيح، [أي] تواتر إجماع الأمم والطوائف، وجود محمد عليه السلام[113] وأن القرآن ظهر عليه وهو معجز بإظهاره يمكن دعوة الناس إلى [دين] الله تعالى، والمعجز على قسمين، قولي مخصوص بالخواص، وفعلي مخصوص بالعوام. [83أ] والمعجز الفعلي لا يعمل[114] عمل القولي وبالعكس، فلكل نبي كتاب ومعجز إلا محمد صلى الله عليه وسلم[115]، فإن كتابه معجز عظيم يكفي الخواص والعوام والمتوسط، وكمال النبي وشرفه بحسب كمال معجزاته، وكمال المعجز بحسب كمال ما يظهر فيه المعجز. وليس للإنسان خاصية أشرف من النطق. وكل معجز يظهر فيه فهو أشرف المعجزات، فقد عرفت أن معجز محمد عليه السلام[116] أكمل المعجزات وهو أفضل الأنبياء. ولو فرضنا شخصاً عاقلاً خالياً عن[117] تعصب ملة ودين[118] ومذهب[119] واستمع جميع الكتب السماوية كالصحف لإبراهيم والتوراة[120] لموسى والإنجيل لعيسى والزبور لداود وغيرها من الوحي السابق واستمع الفرقان مع كمال بلاغته المشتمل[121] على المعاني الحقة الروحانية والمعقولات الصافية الحقيقية والأحكام السياسية العادلة ومعرفة مبدأ الخلق ومعادهم وكثرة حكمه وإرشاده، لا بد أن يحكم على أن هذا الكتاب أفضل الكتب الإلهية.

## الفصل السابع في بيان[122] [أن] شريعة محمد عليه السلام[123] لا تكون[124] قابلة للنسخ، وأنه[125] خاتم النبيين

112 عليه السلام: (عم)، ب.
113 عليه السلام: (عم)، ب.
114 يعمل: تعمل، الأصل؛ يعمل، ب.
115 صلى الله عليه وسلم: صلعم، ب.
116 عليه السلام: صلعم، ب.
117 عن: من، ب.
118 ودين: + في (مشطوباً)، الأصل.
119 ومذهب: فهو يذهب، ب.
120 والتوراة: والتورية، الأصل؛ التوراة، ب.
121 المشتمل: مشتمل، الأصل؛ مشتملا، ب.
122 بيان: + ان، ب.
123 عليه السلام: (صلعم)، ب.
124 تكون: يكون، الأصل؛ تكون، ب.
125 وأنه: وهو، الأصل ب.

لما[126] كان محققاً عند أرباب التحقيق أن الإنسان المؤيد بالتأييدات الإلهية الربانية والفائز بالكمالات والسعادات[127] الإنسانية إذا بلغ منتهى مراتبه يصير روحاً للعالم ومتصرفاً فيه كتصرف نفسك في بدنك، وكلما[128] كان التصرف والتعقل له أتم وأكمل كان ذاته أفضل[129] وأشرف، فنزول ثلاثة آلاف من الملائكة مسومين[130] في غزوة تارةً ونزول خمسة آلاف من الملائكة في غزوة أخرى[131] يدل على كمال تصرفه في العلويات والروحانيات. ومن كان متصرفاً في العلويات[132] لا بد من[133] أن يكون متصرفاً في السفليات دون العكس. فمن قدر على تنزيل الملائكة قدر على شق البحر والأرض[134] وتحريك الرياح المؤذية[135] والأمن من الحريق[136] وغير ذلك، وهذه المرتبة غاية كمال الإنسان، لا يمكن أن يتجاوز عن هذا الحد شخص إنساني. وشريعة هذا الشارع أفضل الشرائع وأتمها، فعلى هذا لا يمكن [أن تكون] شريعة ناسخة لشريعته، فإن شريعة الناسخ يجب أن تكون أفضل من شريعة المنسوخ، ولا أفضل من شريعته فلا ناسخ لها. فقد عرفت أن محمداً عليه أفضل الصلوات هو أكمل الخلائق ولا مرتبة أعلى من مراتبه في مدارج الكمال الإنسانية، وهو خاتم النبيين لعدم ناسخه، وهو[137] المطلوب.

## الباب الثاني في بيان شريعة اليهود وخساسة طبيعتهم وفيه عشرة فصول

### الفصل الأول في بيان نسخ شريعتهم

لا شك أن قبل موسى أنبياء عليهم السلام[138] وشرائع أخرى فشريعة موسى إما عين تلك الشرائع أو مع زيادة عليها. لا سبيل إلى الأول وإلا لكان شريعته عبثاً، ولا سبيل إلى الثاني، لأن لا معنى للزيادة في الشرع إلا تحريم ما تقدمت إباحته أو إباحة ما تقدم تحريمه كتحريم العمل في

---

126 لما: كما، ب.
127 والسعادات: والاستعلاءات، ب.
128 وكلما، وكما، ب.
129 أفضل: اكمل، ب.
130 مسومين: منزلين، ب.
131 إشارة إلى سورة آل عمران (3):124–125.
132 في العلويات: بالعلويات، ب.
133 من: -، ب.
134 إشارة إلى سورة الشعراء (26):63.
135 إشارة إلى سورة الأعراف (7):133.
136 إشارة إلى سورة الأنبياء (21):68–70، سورة العنكبوت (29):24، سورة الصفات (37):97–98.
137 وهو: فهو، ب.
138 عليهم السلام: (عم)، ب.

السبت. وليس تحريم العمل فيه لأن الله تعالى يكرهه في جميع الأزمنة، وإلا لكان مكروهاً في زمان آدم ونوح وإبراهيم عليهم السلام[139] فكانوا مرتكبين للأعمال[140] المكروهة عند الله تعالى. فإن السبت موجود في زمانهم مع العمل فيه، فتحريم العمل في السبت ليس لكراهة الله تعالى بل لخصوصية شريعة موسى عليه السلام[141] بحسب مصلحة بعض الأوقات. فجاء عيسى بشريعة أخرى ناسخة لتحريم عمل السبت[142] في زمان آخر بحسب مصلحة. ولا يرد هذا بمثل تحريم الخمر فإن إباحته[143] مكروهة في نفس الأمر عند الله تعالى لا بحسب وقت معين، لأن الخمر مزيل[144] للعقل الذي هو أشرف خواص الإنسان ومضرّ[145] في مصلحة المعاش والمعاد بالنسبة إلى كل شخص بخلاف عمل السبت. فثبت أن شريعة عيسى ناسخة لشريعة موسى وشريعة محمد ناسخة للشريعتين[146]. فشريعة موسى منسوخة[147] مرتين. وهو المطلوب.

الفصل الثاني [83ب] في بيان أن شريعة عيسى ناسخة لشريعة موسى من التوراة[148]

إذا عرفت أن ختم التوراة[149] بآية تعريبها[150] هذا: إن قدرة الله تعالى أقبلت من سينا[151] ثم أشرقت من سعير[152] ثم أطلعت من جبال[153] فاران.[154] واليهود يعلمون كلهم أن جبل سعير[155] هو الجبل الذي فيه بنو العيص الذين آمنوا بعيسى، وكان في هذا الجبل مقام المسيح. وأيضاً مسطور في التوراة[156] وهذا تعريبه: ينتظرون قائماً يأتيهم من ولد داود عليه

139 عليهم السلام: (عم)، ب.
140 للأعمال: الأعمال، ب.
141 عليه السلام: (عم)، ب.
142 لتحريم عمل السبت: لعمل تحريم السبت، الأصل ب.
143 إباحته: اباحتها، ب.
144 لأن الخمر مزيل: لأنها مزيلة، ب.
145 ومضرّ: ومضرّة، ب.
146 للشريعتين: لشريعتين، الأصل ب.
147 منسوخة: ناسخة ومنسوخة، الأصل ب.
148 التوراة: التورية، الأصل؛ التوراة، ب.
149 التوراة: التورية، الأصل؛ التوراة، ب.
150 تعريبها: تعريبه، الأصل؛ تعريبها، ب.
151 سينا: سنا، الأصل ب.
152 سعير: سيعر، الأصل؛ سعير، ب.
153 جبال: جبل، ب.
154 سفر التثنية 33:2.
155 سعير: سيعر، الأصل؛ سعير، ب.
156 التوراة: التورية، الأصل، التوراة، ب.
157 عليه السلام: (عم)، ب.
158 راجع اشعياء 11:1.

السلام[157]، [158] وهذا المنتظر هو المسيح. وقد كان للأنبياء أمثال أشاروا بها إلى جلالة دين المسيح وخضوع الجبّارين لأهل ملته وإتيانه بالنسخ العظيم. فمن ذلك أن الأسد يأكل في ذلك الزمان التبن كالبقر. [159] [160] واليهود لفرط جهلهم لم يفهموا من تلك الأمثال إلا الصور الحسّية لقصور عقلهم عن إدراك المعاني العقلية ومرموزات الأنبياء، فتولوا عن الإيمان بالمسيح مع معجزاته وكمالاته وينتظرون الأسد حتى يأكل التبن كالبقر[161]. فقد[162] جعل الله في تضليلهم[163] طريقاً مما يتناولون[164] في أيديهم من نص تنزيلهم[165]، وأعماهم عنه ليكون حجة عليهم، فثبت بما في أيديهم أن عيسى عليه السلام[166] ناسخ لشريعتهم.

الفصل الثالث في بيان أن شريعة موسى مع كونها[167] منسوخة فائتة[168] عنهم

اعلم أن شريعة موسى لا تثبت إلا بشهادة الأمم الذين شهادتهم إن كانت مقبولة عند اليهود فهم يتفقون على تكفيرهم وتضليلهم وتحريفهم في كتاب الله تعالى. فإن قالوا: لا نقبل شهادة أحد غير اليهود، فلم يبق لهم تواتر إلا من طائفتهم وهم أقل الطوائف عدداً وأكثرهم جهلاً وكذباً مع تواتر الغزاة عليهم. فيصير تواترهم وشرعهم كذلك أضعف التواتر والشرائع. وإن[169] سلمنا صحة أصل تواترهم لكن لا نسلم صحة تواتر هذه التوراة التي في أيديهم، فإن حفظها لم يكن عندهم فرضاً ولا[170] سنة، فلو كان فرضاً لا يحفظ إلا واحد بعد واحد من الهارونيين فصلاً من التوراة[171] أو فصولاً، فإذ[172] انقرضت دولتهم انطمست[173] آثارهم بسبب المصافات والغارات وقتل أكابرهم وعلمائهم وحرق كتبهم مكرراً، فلما جاء عزير رأى فناء هيكلهم وزوال دولتهم ورفع كتابهم جمع

159 كالبكر: كالبكرة، ب.
160 اشعياء 11:7.
161 كالبكر: كالبكرة، ب.
162 فقد: قد، الأصل؛ -، ب.
163 تضليلهم: تصليلهم، ب.
164 يتناولون: يتداولون، ب.
165 تنزيلهم: يزيلهم، الأصل.
166 عليه السلام: (عم)، ب.
167 كونها: كونه، الأصل؛ كونها، ب.
168 فائتة: فاتت، الأصل ب.
169 وإن: ولئن، الأصل؛ وان، ب.
170 ولا: ولكن، الأصل؛ و[لم] يكن، ب.
171 التوراة: التورية، الأصل؛ التوراة، ب.
172 فإذ: فإذا، الأصل ب.
173 انطمست: وانطمست، الأصل ب.

من محفوظاته[174] ومن الفصول التي يحفظها الكهنة مع زيادة ونقصان بحسب أغراضه ومصلحة الوقت. فهي[175] بالحقيقة كتاب عزير [و]ليس كتاب الله تعالى، وتواترهم يشهد على صحة هذه القضية. فقد عرفت أن شريعة موسى مع كونها منسوخة مفقودة بينهم.

الفصل الرابع في بيان تحريف آخر بعد تحريف عزير

اعلم أن التوراة[176] كما هي[177] مشهورة أربع نسخ مختلفة، اثنتان منها عند النصارى، وبينهما اختلاف لفظي ومعنوي لا تكادان[178] تتطابقان[179]. والنسخة الثالثة يسمونها توراة السبعين منسوبة إلى سبعين شخصاً من أشياخهم، وتخالف الأولين بكثير من الأحكام وتخالف لما في أيديهم من تواريخ الأعمار[180] المذكورة في أوائل التوراة[181] وفي غيرها. وامتنع أن[182] يحكم العقل بصحة كلها[183]، والتواتر الصحيح لا يشهد بصحة أحد على التعيين دون الباقي، فلا اعتماد على كتابهم وشريعتهم. وهذا الاختلاف يدل على أن التوراة[184] المنزلة[185] على موسى مفقودة بينهم لكثرة التحريف، والاختلاف بين الأربع أكثر من أن يستند إلى سهو الكتّاب[186] والناسخين. ولا يصح أن يقال أن هذا الاختلاف كاختلاف قراءات القرآن، قراءة ابن مسعود وأبيّ وغيرهما، فإن اختلاف القراءات واللغات لا يقتضي تعدد الكتاب وتغيير الأحكام والمعنى.

الفصل الخامس فيما وجدنا في التوراة مما يدل على التحريف

وهو أن هذه التوراة[187] التي في أيديهم مشتملة على اتصاف الله تعالى بصفات لا يليق وصفه بها كندم الله تعالى على خلق البشر وشقّ عليه،[188] وكالتجسيم والتشبيه والكفريات التي

---

174 محفوظاته: محفوظطاته، الأصل؛ محفوظاته، ب.
175 فهي: فهو، ب.
176 التوراة: التورية، الأصل.
177 كما هي: ما هل (؟)، الأصل؛ ...، ب.
178 لا تكادان: لا يكاد ان، ب.
179 تتطابقان: يتطابقا، الأصل.
180 الأعمار: الأعمال، الأصل ب.
181 التوراة: التورية، الأصل؛ التوراة، ب.
182 أن: ام، ب.
183 كلها: كلمتها، ب.
184 التوراة: التورية، الأصل؛ التوراة، ب.
185 المنزلة: المنزل، الأصل؛ المنزلة، ب.
186 الكتّاب: الكاتب، الأصل ب.
187 التوراة: التورية، الأصل؛ التوراة، ب.
188 سفر التكوين 6:6.

تستبعدها[189] العقول أن تضاف إلى ذاته، وعلى قصص وحكايات لا فائدة في ذكرها [84أ] لا في الدنيا ولا في[190] الآخرة، مثل قصة[191] القبائل التي[192] تتفرع من نوح وأسمائهم ومواضعهم، وكذلك[193] حكاية أولاد سعير[194] ووصف الملوك الذين ملكوا في ادوم وعدد منازل سلوك[195] بني إسرائيل في أرض الشام، فالعقل المستقيم يأبى أن تصدر[196] أمثال هذه الكلمات عن حضرة الربوبية بلا فائدة عامة ولا خاصة، مع خلوها عن[197] ذكر حال الثواب والعقاب الأخرويين، وهو من أهم مهمات التنزيل والأصل الأعظم في التشريع. فلو كانت هذه التوراة[198] من عند الله تعالى لما جاز خلوها من التصريح[199] بذلك والتأكيد فيه والعدول عنه إلى الدنيويين[200] اللذين[201] قد أكثر ذكرهما[202] في التوراة. فإن الدنيا زائلة سعادتها، وشقاوتها غير معتبرة عند أرباب البصيرة. والتجربة اقتضت أن النعيم في الدنيا غير مختص بالصالحين، والشقاء[203] لا يختص[204] بالضالين[205]. وفي[206] صلواتهم [أدعية] لفقوها بعد زوال الدولة عنهم، ولهم أصوام غير الصوم المفروض[207] في التوراة[208] واستكثار[209] [من] النسوان والذهب والفضّة، وداود استكثر من النسوان ومن الذهب والفضّة استكثاراً عظيماً، مع أن الزيادة والنقصان على ما فيها من الفرائض منهي [عنه] في التوراة[210]. فكل ذلك يدل على تحريفات ونسخ بعض أحكام التوراة[211]، فلا يكون كتابهم وشريعتهم معتبرة عند أصحاب التحقيق.

---

189 تستبعدها: يستبعدها، ب.
190 في: -، ب.
191 قصة: صفة، الأصل ب.
192 التي: -، ب.
193 وكذلك: ولذلك، الأصل؛ ولا كل، ب.
194 سعير: سيعر، الأصل؛ سعير، ب.
195 سلوك: ملوك، الأصل ب.
196 تصدر: يصدر، الأصل ب.
197 عن: من، ب.
198 التوراة: التورية، الأصل؛ التوراة، ب.
199 التصريح: التشريع، ب.
200 الدنيويين: + من، ب.
201 اللذين: الذين، الأصل؛ اللذين، ب.
202 ذكرهما: ذكرهم، ب.
203 والشقاء: والشقاوة، ب.
204 يختص: تختص، ب.
205 بالضالين: بالصالحين، الأصل؛ بالضالين، ب.
206 وفي: ومن، ب.
207 الصوم المفروض: صوم المفروض، الأصل؛ الصوم المفروض، ب.
208 التوراة: + ولا يجوز للملك، ب.
209 واستكثار: استكثار، ب.
210 التوراة: التورية، الأصل؛ التوراة، ب.
211 التوراة: التورية، الأصل؛ التوراة، ب.

## الفصل السادس في بيان جهلهم وحمقهم

لما كان شرف[212] الإنسان بالعلوم والمعارف الحقيقية والملكات المرضية، فكل طائفة أعلم وأعدل فهم أشرف الطوائف والأمم. ولا شك أن للمسلمين حكماً وعلوماً[213] وتصانيف لم تكن[214] لطائفة[215] منها أخرى. وأما جميع ما نسب إلى اليهود مع أن استفادوه[216] من علوم غيرهم لا[217] يكون عشراً من الفنون الحكمية التي استخرجها حكماء اليونان والنبط والهند. وأما المسلمون فقد[218] دوّنوا جميع علوم الطوائف واستخرجوا[219] من علوم أنفسهم علوماً دقيقة ومعارف غامضة يستحيل أن يقف أحد من الناس[220] على جميع ما صنفوه[221] لكثرة علومهم وسعة معلوماتهم. فقد بطل قول اليهود بأنهم وآباءهم[222] أعقل الناس وأفضلهم وأشرفهم، فإن مَن هذه صفته يجب أن يستدل بآثاره على فضائله. وقد علم من ذلك إبطال ما يدعونه من محبّة الله تعالى لهم، ويزعمون أن الله تعالى يحبّهم دون جميع الناس ويحب طائفتهم وسلالتهم ولا يختار الصالحين إلا منهم، لأن الله تعالى لا يحب الجاهلين المفسدين[223]، بل هو علاّم يحب كل عليم.

## الفصل السابع في بيان بلادتهم وعدم استعدادهم من التوراة

اعلم أن هذه التوراة[224] التي في أيديهم قد شهدت على جهلهم وإفلاسهم في الفطانة، وهذا تعريب ما هو مسطور في التوراة[225]: إن اليهود لشعب عادم الرأي وليس فيهم فطانة وحكمة[226]، وهذه الطائفة مع فنون[227] جهلهم وشقاوتهم وذهاب الدولة ونحوستهم وتفرّق شملهم وعلمهم

---

212 شرف: -، ب.
213 حكماً وعلوماً: حكم وعلوم، الأصل؛ حكما وعلوما، ب.
214 تكن: يكن، الأصل ب.
215 لطائفة: (لطائفة)، ب.
216 استفادوه: استفادوهم، الأصل، استفادوا، ب.
217 لا: فلا، ب.
218 فقد: قد، الأصل؛ فقد، ب.
219 واستخرجوا: واستخرجهم، الأصل؛ واستخرجوا، ب.
220 من الناس: -، ب.
221 صنفوه: صنفوهم، الأصل؛ صنفوه، ب.
222 وآباءهم: وآبائهم، الأصل ب.
223 لأن الله لا يحب الجاهلين المفسدين: إشارة إلى قوله تعالى ﴿إنَّ اللهَ لاَ يُحِبُّ المُفْسِدِينَ﴾ (سورة القصص (28):77) وقوله تعالى ﴿وَاللهُ لا يُحِبُّ المُفْسِدِينَ﴾ (سورة المائدة (5):64).
224 التوراة: التورية، الأصل؛ التوراة، ب.
225 التوراة: التورية، الأصل؛ التوراة، ب.
226 سفر التثنية 32:28.
227 فنون: فضول، ب.

بالغضب الممدود عليهم يقولون كل يوم في صلاتهم[228] أنهم[229] أبناء الله وأحبّاؤه، ويمثّلون أنفسهم بعناقيد العنب وسائر الأمم بالشوك المحيط بأعالي حيطان الكرم، ويعتقدون أنهم أشرف الطوائف وأعلمهم مع أن التوراة[230] ناطقة بجهالتهم[231] وبلادتهم. وليس للإنسان مرض نفساني أصعب من الجهل المركب وعدم الفطنة ﴿وَمَن[232] لَّمْ يَجْعَلِ اللهُ لَهُ نُوراً فَمَا لَهُ مِن نُّورٍ﴾.[233]

## الفصل الثامن في بيان مذلّتهم بين الطوائف

اعلم أن هذه الطائفة لكثرة نحوستهم وشقاوتهم قد استولى الله[234] عليهم جميع الطوائف مثل الكسدانيين والبابليين والفرس واليونان والنصارى وأهل الإسلام، وكل واحد من هذه الطوائف قصدهم أشد القصد وطلب استئصالهم[235] وبالغ في إحراق بلادهم وإحراق كتبهم، لا يبقى لهم مدينة ولا جيش ولا شريعة ولا كتب، وأشد على اليهود من جميع هذه [الممالك] ما نالهم من ملوكهم العصاة مثل احاب[236] وأحزيا وامصيا[237] ويهورام[238] ويربعام[239] وغيرهم [84ب] من ملوك الإسرائيليين الذين قتلوا الأنبياء وأحرقوا كتبهم وعبدوا الأصنام وبنوا لها البيع العظيمة والهياكل وعكفوا على عبادتها وتركوا أحكام التوراة وشرعها مدداً طويلة وأعصاراً متصلة، وليس تطرّق[240] هذه الآفات إلا لخبث[241] سيرتهم وكثرة ذلّتهم ورذيلتهم كما جاء في حقهم في تنزيل العزيز ﴿وَضُرِبَتْ عَلَيْهِمُ الذِّلَّةُ وَالْمَسْكَنَةُ وَبَآءُو بِغَضَبٍ مِنَ اللهِ﴾[242]، والله أعلم.

---

228 صلاتهم: صلوتهم، الأصل؛ صلاتهم، ب.
229 أنهم: -، ب.
230 التوراة: التورية، الأصل؛ التورية، ب.
231 بجهالتهم: بجهالاتهم، الأصل؛ بجهالتهم، ب.
232 ومن: من، ب.
233 سورة النور (24):40.
234 الله: -، ب.
235 استئصالهم: استيصالهم، الأصل ب.
236 احاب: اخاراث، الأصل؛ اخاب، ب.
237 وامصيا: وامضيا، الأصل؛ وامصيا، ب.
238 ويهورام: وبهورام، الأصل؛ ويهورام، ب.
239 ويربعام: ويرثعام، الأصل؛ ويربعام، ب.
240 تطرّق: يطاق، الأصل ب.
241 لخبث: لخبت، ب.
242 سورة البقرة (2):61.

## الفصل التاسع في شرارة أنفسهم

اعلم أن الفرس واليونان طال [ما] يمنعون اليهود عن الصلوات والعبادات لمعرفتهم بأن معظم صلوات هذه الطائفة دعاء على جميع الأمم بالبوار والهلاك، وعلى العالم بالخراب سوى بلادهم التي هي أرض كنعان، ﴿وَمَا دُعَاءُ الكَافِرِينَ إلاَّ في ضَلاَلٍ﴾[243]. ولا ينبغي لأحد من العلماء أن يستبعد اتفاق كافة هذه الأمة[244] على الباطل والعناد، وتعاونهم على الشر والفساد، ولا يحكم العقل على خيرية[245] أن أفضل أعمالهم الذي هو الصلاة متضمن[246] لشر[247] جميع الخلائق وخراب العالم. ومن خاصية الشرير أن يرجع الشر إلى نفسه، فلهذا لا تكون بلادهم معمورة ولا حياتهم آمنة من البليات ولا أنفسهم سالمة من الآفات، بل منغصة بالإذلال والإيذاء، لا تزال مخوفة من شر الأعداء مع فوت آخرتهم[248] عنهم، كما جاء في القرآن المجيد ﴿وَمَنْ أَعْرَضَ عَن ذِكْرِي فَإِنَّ لَهُ مَعِيشَةً ضَنكا وَنَحْشُرُهُ يَوْمَ القِيَامَةِ أَعْمَى﴾[249].

## الفصل العاشر في أن أحق[250] الناس بأن ينسب إلى السفاهة والسقوط طائفة اليهود

فإنهم مع عدم استعدادهم لتلك[251] المعقولات والمعاني الحقة متسرعون[252] إلى قبول الباطل، فإن آباءهم كانوا يشاهدون الآيات الحسّية والنار السماوية والمعجزات الظاهرة ما لم يره غيرهم من الأمم، وهم مع ذلك يهمّون برجم موسى وهارون[253] عليهما السلام في كثير من الأوقات. والدليل المستقل على حماريتهم اتخاذهم العجل في أيام موسى عليه السلام[254] وتركهم[255] شريعته

243 سورة الرعد (13):14.
244 الأمة: الأمم، الأصل؛ الأمة، ب.
245 خيرية: خيريتهم، الأصل؛ خيرية، ب.
246 متضمن: متضمنا، الأصل؛ متضمن، ب.
247 لشر: لسرّ، ب.
248 آخرتهم: اخراهم، ب.
249 سورة طه (20):124.
250 أحق: حق، الأصل ب.
251 لتلك: لادراك، ب.
252 متسرعون: متسرعا، الأصل؛ متسرعون، ب.
253 وهارون: وهرون، الأصل؛ وهارون، ب.
254 عليه السلام: (عم)، ب.

والإغماض عن[256] كمالاته ومعجزاته. والدليل القوي على خسّة طبيعتهم اختيارهم العود[257] إلى مصر والرجوع من حضرة[258] موسى عليه السلام[259] إلى العبودية الباطلة لفرعون ليشبعوا من أكل البصل والقثاء والثوم والعدس، لأنهم كانوا غير مستحقين للمنّ والسلوى ثم عبادتهم الأصنام بعد عصر يوشع بن نون، ثم عبادتهم الكبشين من الذهب الذين عملهما[260] يربعام[261] في نابلس ونقلهم الحج إلى الكبشين وتركهم الزيارة[262] إلى بيت المقدس مع ابشالوم الولد العاصي لداود عليه السلام[263] ويعاونونه في قتل داود على الباطل، فهذه الرذائل والأفعال المذمومة تدل[264] على رداءة جوهرهم وكدورة روحانيتهم وخساسة طبيعتهم ومذلّة طينتهم، ﴿أُولَئِكَ يَلْعَنُهُمُ[265] اللهُ وَيَلْعَنُهُمُ اللاَعِنُونَ﴾[266].

هذا آخر[267] ما أردنا إيراده في هذه الرسالة والله أعلم.

[255] وتركهم: وتركوا، الأصل؛ وترك، ب.
[256] عن: من، الأصل؛ عن، ب.
[257] العود: العودة، ب.
[258] حضرة: نفرة، ب.
[259] عليه السلام: (عم)، ب.
[260] عملهما: + في (مشطوبا)، الأصل.
[261] يربعام: بارثغام، الأصل؛ يربعام، ب.
[262] الزيارة: لزيارة، الأصل؛ الزيارة، ب.
[263] عليه السلام: (عم)، ب.
[264] تدل: يدل، الأصل؛ تدل، ب.
[265] يلعنهم: لعنهم، الأصل؛ يلعنهم، ب.
[266] سورة البقرة (2):159.
[267] آخر: كفى، ب.

# ABBREVIATIONS AND BIBLIOGRAPHY


Abu l-Barakāt Hibat Allāh al-Baghdādī: *Al-Kitāb al-Muʿtabar fī l-ḥikma*. Isfahan [2]1373/1994.

Abu l-Maʿālī Muḥammad b. Rāfiʿ Salāmī: *Taʾrīkh ʿulamāʾ Baghdād al-musammā bi-Muntakhab al-mukhtār*. Ed. ʿAbbās al-ʿIzāwī. Baghdad 1357/1938.

Adana = Adana İl Halk Kütüphanesi, Adana
Catalogue:
1985–86. *Türkiye Yazmalari Toplu Kataloğu. The Union Catalogue of Manuscripts in Turkey*. 01. Adana İl Halk Kütüphanesi ve Müzesi 1–3. Ankara.

Adang, Camilla: *Muslim Writers on Judaism and the Hebrew Bible. From Ibn Rabban to Ibn Hazm*. Leiden 1996.

-Afandī, ʿAbd Allāh b. ʿĪsā: *Riyāḍ al-ʿulamāʾ* 1–6. Ed. Aḥmad al-Ḥusaynī. Qum 1401/1981.

Āghā Buzurg al-Ṭihrānī: *Ṭabaqāt aʿlām al-shīʿa* 1–2. Beirut 1390–91/1971; 3–5, ed. ʿAlī Naqī Munzawī. Beirut 1392–95/1972–75; 6, ed. ʿAlī Naqī Munzawī. Tehran 1362/1983.

Ahmed III = Sultan Ahmed III Collection (held in Topkapı Saray Müzesi Kütüphanesi), Istanbul
Catalogue:
→ Topkapı Saray Müzesi Kütüphanesi

Ākhūnd = Madrasa-yi Ākhūnd-i Hamadān (Kitābkhāna-yi Gharb), Hamadān
Catalogue:
1974. Jawād Maqṣūd: "Fihrist-i nuskhahā-yi khaṭṭī-yi kitābkhāna-yi Gharb – Madrasa-yi Ākhūnd-i Hamadān." In Muḥammad Rawshan, Jawād Maqṣūd and Parvīz Adhkāʾī: *Fihrist-i nuskhahā-yi khaṭṭī-yi kitābkhānahā-yi Rasht u Hamadān*. Tehran 1353, pp. 1239–1618.

-ʿAlawī, Muḥammad Mahdī: "Shubhat Hibat Allāh Ibn Kammūna al-Yahūdī." *Lughat al-ʿarab* 7 (1929), p. 368.

ʿAlawī, Sayyid Aḥmad: *Sharḥ al-Qabasāt*. Ed. Ḥāmid Nājī Iṣfahānī. Tehran 1376/1997.

Alper, Ömer Mahir: *Alkın Hazzı. İbn Kemmûne'de Bilgi Teorisi*. Istanbul 2004.

—— "Özü dolayisiyla varliği zorunlu'nun kanitlanmasi hakkinda." *İstanbul Üniversitesi İlahiyat Fakültesi Dergisi Sayı* 9 (2004), pp. 139–49.

-Ālūsī, Shihāb al-Dīn Maḥmūd: *Gharāʾib al-ightirāb wa-nuzhat al-albāb fī l-dhahāb wa-l-iqāma wa-l-iyāb*. Ed. Khayr al-Dīn al-Ālūsī. Baghdad 1317/1899–1900.

-Amīn, Muḥsin: *Aʿyān al-shīʿa* 1–11. Beirut 1403/1983.

Amīr al-Muʾminīn = Maktabat Amīr al-Muʾminīn, Najaf
Catalogue:
2002. al-Sayyid ʿAbd al-ʿAzīz al-Ṭabāṭabāʾī: "Fihrist makhṭūṭāt maktabat Amīr al-muʾminīn al-ʿāmma al-Najaf al-ashraf (13)." *Turāthunā* 71–72 (1423), pp. 245–85.
2003–4. al-Sayyid ʿAbd al-ʿAzīz al-Ṭabāṭabāʾī: "Fihris makhṭūṭāt Amīr al-Muʾminīn al-ʿāmma, al-Najaf al-ashraf (14)." In: *Turāthunā* 73–74 (1424), pp. 216–72.

Amitai, Reuven: "The Conversion of Tegüder Aḥmad to Islam." *Jerusalem Studies in Arabic and Islam* 25 (2001), pp. 15–43.
[Anonymous]: *Al-Ḥawādith al-jāmiʿa wa-l-tajārib al-nāfiʿa*. Ed. Muṣṭafā Jawād [as a work by Ibn al-Fuwaṭī]. Baghdad 1351/1932.
Āqā Muḥammad ʿAlī Bihbahānī: *Khayratiyya dar ibṭāl-i ṭarīqa-yi ṣūfiyya*. Ed. Sayyid Mahdī Rajāʾī. Qum 1412/1991.
—— *Rādd-i shubahāt al-kuffār*. Ed. Sayyid Mahdī Rajāʾī. Qum 1413/1992.
Āstān-i Quds = Kitābkhāna-yi Markazī-yi Āstān-i Quds-i Raḍawī, Mashhad
Catalogues:
1926–85. *Fihrist-i Kitābkhāna-yi Āstān-i Quds-i Raḍawī* 1–21. Mashhad 1305–64.
1990. Muḥammad Āṣif Fikrat: *Fihrist-i alifbāʾī-yi kitābkhāna-yi markazī-yi Āstān-i Quds-i Raḍawī*. Mashhad 1369.
Atıf Efendi = Atıf Efendi Kütüphanesi, Istanbul
Catalogue:
1892. *Defter-i Kütüphane-i Atıf Efendi*. Istanbul 1310.
1987. *Atıf Efendi Kütüphanesi. I. Kısım Eserler Kataloğu. II. Kısım Eserler Kataloğu*. Istanbul.
AUB = American University of Beirut
Catalogue:
1985. Yūsuf Q. Khūrī: *Al-Makhṭūṭāt al-ʿarabiyya al-mawjūda fī maktabat al-Jāmiʿa al-Amīrikīya fī Bayrūt. Dalīl murattab ḥasab niẓām Diyuwī [Dewey] al-ʿushrī li-taṣnīf al-kutub wa-tansīqihā, maʿa malāḥiq wa-kashshāfāt li-l-ʿanāwīn wa-l-muʾallifīn*. Beirut.
Awqāf ʿāmma = Maktabat al-Awqāf al-markaziyya, Baghdad
Catalogue:
1973–74. ʿAbd Allāh al-Jubūrī: *Fahras al-makhṭūṭāt al-ʿarabiyya fī maktabat al-awqāf al-ʿāmma fī Baghdād* 1–4. Baghdad 1393–94.
Ayasofya = Ayasofya Collection (held in the Süleymanie Kütüphanesi), Istanbul
Catalogue:
1887. *Defter-i Kütüphane-i Ayasofya*. Istanbul 1304.
Bacha [Bāshā], Habib [Ḥabīb]: *Ḥawāshi (Notes) d'Ibn al-Maḥrūma sur le "Tanqīḥ" d'Ibn Kammūna* [Ḥawāshī Ibn al-Maḥrūma ʿalā Kitāb Tanqīḥ al-abḥāth li-l-milal al-thalāth li-Ibn Kammūna]. Etude et édition critique. Rome 1984.
—— "Tanqīḥ al-abḥāth li-l-milal al-thalāth." *Parole de l'Orient* 11 (1971), pp. 151–62.
Baneth, Dawid Hartwig: "Ibn Kammuna." *Monatsschrift für Geschichte und Wissenschaft des Judentums* 69 (1925), pp. 295–311.
Bankipore = Khuda Bakhsh Oriental Public Library, Bankipore
Catalogues:
1908–. *Catalogue of the Arabic and Persian Manuscripts in the Oriental Public Library at Bankipore*. Calcutta/Patna.
Barkhʷāh, Ansīyah: *Taṣḥīḥ-i Risāla-yi Azaliyyat al-nafs wa-baqāʾihā taʾlīf-i Ibn Kammūna*. MA thesis: Madrasa-yi Murtaḍā Muṭahharī (Sipahsālār), 1379/2000.
Baron, Salo Willmayer: *Religious controls and dissensions* (= A Social and religious history of the Jews, vol. 5). New York ²1957.
Berkeley Magnes Museum = The Judah L. Magnes Museum, Berkeley, California
Berlin = Staatsbibliothek zu Berlin
Catalogues:
1878–97. Moritz Steinschneider: *Verzeichniss der Hebräischen Handschriften* 1–2 Berlin.

Berman, Lawrence V.: [Review of] "Ibn Kammūna's *Examination of the Three Faiths.*" *Journal of the American Oriental Society* 91 (1971), pp. 333–34.
Biblioteca Angelica = Biblioteca Angelica, Rome
Catalogue:
1878. Ignazio Guidi: *Catalogo dei codici siriaci, arabi, etiopici, turchi e copti della Biblioteca Angelica*. Florence.
Blumenthal, David: "An Arabic Translation to 'Hilkhot Yesode Ha-Torah'." [Hebrew] *Da'at* 14 (1985), pp. 113–47.
Bodleian = Bodleian Library, Oxford
Catalogues:
1787. Joannes Uri: *Bibliothecae Bodleianae codicum manuscriptorum orientalium videlicet hebraicorum, chaldaicorum, syriacorum, aethiopicorum, arabicorum, persicorum, turcicorum, copticorumque*. Pars prima. Oxford.
1821–35. Alexander Nicoll and Edward Bouverie Pusey: *Bibliothecae Bodleianae codicum manuscriptorum orientalium catalogi partis secundae* 1–2. Oxford.
1886. Adolf Neubauer: *Catalogue of the Hebrew Manuscripts in the Bodleian Library and in the College Libraries of Oxford. Including mss. in other languages, which are written with Hebrew characters, or relating to the Hebrew language or literature; and a few Samaritan mss.* Oxford [reprinted 1994].
British Library = The British Library: Oriental and India Office Collection, London
Catalogues:
1849–71. William Cureton and Charles Rieu: *Catalogus codicum manuscriptorum orientalium qui in Museo Britannico asservantur. Pars secunda. Codices Arabicos amplectans* 1–2. London [reprinted Hildesheim 1979].
1877. Otto Loth: *A Catalogue of the Arabic Manuscripts in the Library of the India Office.* London.
1894. Charles Rieu: *Supplement to the catalogue of the Arabic Manuscripts in the British Museum*. London [reprinted Hildesheim 1979].
1912. Alexander George Ellis and Edward Edwards: *A Descriptive List of the Arabic Manuscripts Acquired by the Trustees of the British Museum since 1894.* London 1912.
Carullah = Carullah Collection (held in the Süleymanie Kütüphanesi), Istanbul
Chester Beatty = The Chester Beatty Library and Gallery of Oriental Art, Dublin
Catalogue:
1955–66. Arthur Arberry: *The Chester Beatty Library: A Handlist of the Arabic Manucripts* 1–8. Dublin.
*Collected Papers on Islamic Philosophy and Mysticism*. Eds. Mehdi Mohaghegh and Hermann Landolt. Tehran 1971.
*Collected Texts and Papers on Logic and Language*. Eds. Mehdi Mohaghegh and Toshihiko Izutsu. Tehran 1974.
Corbin, Henry: *En islam iranien. Aspects spirituels et philosophiques. II Sohrawardī et les platoniciens de Perse*. Paris 1971.
—— *La philosophie iranienne islamique aux xvii^e^ et xviii^e^ siècles*. Paris 1981.
Cottrell, Emily: "Šams al-Dīn al-Šahrazūrī et les manuscrits de *'La Promenade des âmes et le jardin des réjouissances: Histoire des philosophes.' (Nuzhat al-arwāḥ wa-rawḍat al-afrāḥ fī ta'rīḫ al-ḥukamā')*." *Bulletin d'Etudes Orientales* 56 (2006), pp. 1–35.
Crone, Patricia: *Medieval Islamic Political Thought*. Edinburgh 2004.

Daiber, Hans: *Bibliography of Islamic Philosophy* 1–2. Leiden 1999.
*Dāʾirat al-maʿārif-i buzurg-i islāmī* 1–. Ed. Kāẓim Mūsawī Bujnūrdī. Tehran 1374/1996–.
Damat İbrahim Paşa = Damat İbrahim Paşa Collection (held in the Süleymanie Kütüphanesi), Istanbul.
Dānishgāh = Kitābkhāna-yi Markazī u Markaz-i Asnād-i Dānishgāh-i Tihrān, Tehran
Catalogues:
1951–85. Muḥammad Taqī Dānishpazhūh and ʿAlī Naqī Munzawī: *Fihrist-i nuskhahā-yi khaṭṭī-yi kitābkhāna-yi markazī u Markaz-i Asnād-i Dānishgāh-i Tihrān* 1–18. Tehran 1330–64.
1969–84. Muḥammad Taqī Dānishpazhūh: *Fihrist-i mīkrūfīlmhā-yi khaṭṭī-yi kitābkhāna-yi markazī-yi dānishgāh-i Tihrān* 1–3. Tehran 1348–63.
Dānishkadah = Kitābkhāna-yi Dānishkada-yi Ilāhiyyāt u Maʿārif-i Islāmī-yi Dānishgāh-i, Tehran
Catalogue:
1966–69. Muḥammad Bāqir Ḥujjatī and Muḥammad Taqī Dānishpazhūh: *Fihrist-i nuskhahā-yi khaṭṭī-yi Dānishkada-yi Ilāhiyyāt u Maʿārif-i Islāmī-yi Dānishgāh-i Tihrān* 1–2. Tehran 1345–48.
*Dānishnāma-yi Īrān u Islam. Encyclopaedia of Iran and Islam* 1–3. Ed. Ehsan Yarshater. Tehran 1354/1975–1370/1991.
*Dānishnāma-yi jahān-i Islām* [*Encyclopaedia of the World of Islam*] 1–. Tehran 1375/1996–.
Dānishpazhūh, Muḥammad Taqī: *Fihrist-i mīkrūfīlmhā-yi khattī-yi kitābkhāna-yi markazī dānishgāh-i Tihrān* 1–3. Tehran 1348–63/1969–84.
Danner, Victor: [Review of] "Ibn Kammūna's *Examination of the Three Faiths.*" *Journal of the American Academy of Religion* 42 (1974), pp. 565–66.
-Dashtakī al-Shīrāzī, Ghiyāth al-Dīn Manṣūr: *Ishrāq hayākil al-nūr li-kashf ẓulumāt shawākil al-ghurūr*. Ed. ʿAlī Awjabī. Tehran 1382/2003.
-Dashtakī, Ṣadr al-Dīn: *Ḥāshiya ʿalā Sharḥ al-Tajrīd al-jadīd*. Ms. Majlis 1998.
—— "Risāla fī shubhat jadhr al-aṣamm." Ed. Aḥad Farāmarz Qarāmalikī. *Kheradnāme-ye Ṣadrā* 2 v/vi (1375/1997), pp. 74–82.
David ben Joshua: [Commentary on Maimonides' *Mishneh Tōrāh*]. Ms. RNL II Firk. Yevr.-Arab. I 1497.
—— *Al-Murshid ilā l-tafarrud wa-l-murfid ilā l-tajarrud*. Ed. and transl. into Hebrew Paul B. Fenton. Jerusalem 1985.
-Dawānī, Jalāl al-Dīn Muḥammad: *Ḥāshiya jadīda ʿalā l-Tajrīd*. Ms. Āstān-i Quds 19827.
—— "Nihāyat al-kalām fī ḥall 'kull kalāmī kādhib'." Ed. Aḥad Farāmarz Qarāmalikī. *Nāme-ye Mufīd* 2 i (1375/1996), pp. 97–134.
—— "Risālat ithbāt al-wājib al-jadīda." In *Sabʿ Rasāʾil li-l-ʿAllāma Jalāl al-Dīn Muḥammad al-Dawānī (al-mutawaffā 908 h.) wa-l-Mullā Ismāʿīl al-Khʷājūʾī al-Iṣfahānī (al-mutawaffā 1173 h.)*. Ed. Sayyid Aḥmad Tūysirkānī. Tehran 1381/2002, pp. 115–70.
—— "Risālat ithbāt al-wājib al-qadīma." In *Sabʿ Rasāʾil li-l-ʿAllāma Jalāl al-Dīn Muḥammad al-Dawānī (al-mutawaffā 908 h.) wa-l-Mullā Ismāʿīl al-Khʷājūʾī al-Iṣfahānī (al-mutawaffā 1173 h.)*. Ed. Sayyid Aḥmad Tūysirkānī. Tehran 1381/2002, pp. 67–114.
—— *Thalāth rasāʾil* [*Tafsīr Sūrat al-kāfirūn*; *Shawākil al-ḥūr*; *Unmūdhaj al-ʿulūm*; *wa-bi-dhaylihī Risālat Hayākil al-nūr*]. Ed. Sayyid Aḥmad Tūysirkānī. Mashhad 1411/1991.
*Dharīʿa* = Āghā Buzurg al-Ṭihrānī: *Al-Dharīʿa ilā taṣānīf al-shīʿa* 1–25. Beirut 1403–6/1983–86.

Dībājī, Ibrāhīm: "Aḥwāl u āthār-i Mullā Shamsāyi-yi Gīlānī." In *Collected Papers on Islamic Philosophy and Mysticism.* Eds. Mehdi Mohaghegh and Hermann Landolt. Tehran 1971, pp. 53–96.

Dietrich, Ernst Ludwig: [Review of] "Saʿd b. Manṣūr Ibn Kammūna, *Examination of the Inquiries into the three faiths*... Edited by Moshe Perlmann." *Oriens* 21–22 (1968–69), pp. 440–41.

*The Encyclopaedia of Islam.* New Edition 1–11. Leiden/London 1960–2002.

*Encyclopaedia Iranica* 1–. London [a.o.] 1985–.

*Encyclopaedia Judaica* 1–16. Jerusalem 1982.

*Encyclopaedia Judaica. Das Judentum in Geschichte und Gegenwart* 1–10. Berlin 1928–34.

*Encyclopaedia of Religion* 1–16. New York/London 1987.

Endress, Gerhard: "Philosophische Ein-Band-Bibliotheken aus Isfahan." *Oriens* 36 (2001), pp. 10–58.

Esad Efendi = Esad Efendi Collection (held in the Süleymanie Kütüphanesi), Istanbul
Catalogue:
1885. *Defter-i Kütübhane-i Esad Efendi.* Istanbul 1262.

Ess, Josef van: *Der Wesir und seine Gelehrten.* Wiesbaden 1981.

Falk, Zeev: "Said ben Mansur Ibn Kammuna." *Religionen in Israel* 1 ii (1995), pp. 22–24.

Farāmarz Qarāmalikī, Aḥad: "Kitābshināsī-yi taḥlīlī-yi pārādox-i durūghgū nazd-i dānishmandān-i musalmān." *Majalla-yi ʿIlmī-yi Ilāhiyyāt u Maʿārif-i Islāmī* 1 ii (1379/2000), pp. 25–47.

—— "Āghā Ḥusayn Khʷānsārī u muʿammā-yi jadhr-i aṣamm." *Kheradnāme-ye Ṣadrā* 3 x (1376/1998), pp. 77–89.

—— "Kamāl al-Dīn Maḥmūd Bukhārī u muʿammā-yi jadhr-i aṣamm." *Kheradnāme-ye Ṣadrā* 4 xiv (1377/1999), pp. 53–57.

—— "Muʿammā-yi jadhr-i aṣamm dar ḥawza-yi Shīrāz." *Kheradnāme-ye Ṣadrā* 4 iv (1377/1999), pp. 80–85.

—— "Muʿammā-yi jadhr-i aṣamm nazd-i ḥukamāʾ-yi dawra-yi Iṣfahān." *Kheradnāme-ye Ṣadrā* 3 xi (1377/1998), pp. 40–45.

—— "Muʿammā-yi jadhr-i aṣamm nazd-i mutakallimūn." *Kheradnāme-ye Ṣadrā* 2 v/vi (1375/1997), pp. 67–73.

—— "Mukātabahā-yi Dawānī u Dashtakī dar ḥall-i muʿammā-yi jadhr-i aṣamm." *Kheradnāme-ye Ṣadrā* 2 viii & ix (1376/1997), pp. 95–101.

Fâtih = Fâtih Collection (held in the Süleymanie Kütüphanesi), Istanbul.

Feyzullah Efendi = Feyzullah Efendi Collection (held in Bayezit Devlet Kütüphanesi), Istanbul
Catalogues:
1892. *Defter-i Kütüphane-i Feyzullah Efendi ve şeyh Murat ve Kalkandelenli Ismail Ağa.* Istanbul 1310.
1978–79. Ḥamīd Majīd Haddū: "Al-Makhṭūṭāt al-ʿArabiyya fī l-Maktaba al-Waṭaniyya bi-Istānbūl: Khizānat Fayḍ Allāh Efendi." *Al-Mawrid* 7 ii (1978), pp. 311–64; 8 i (1979), pp. 305–48.
1983. M.S. Tayşi: "Şeyhülislâm Feyzullah Efendi ve Feyziye Medresesi." *Türk Dünyasi Araştirmalari* 23 (1983), pp. 9–100.

Fenton, Paul B.: "A Judaeo-Arabic Commentary on Maimonides' *Mišnē Tōrā* by Rabbi David Ben Joshua Maimonides (circa 1335–1414)." [Hebrew] In *Heritage and*

*Innovation in Medieval Judaeo-Arabic Culture. Proceedings of the Sixth Conference of the Society for Judaeo-Arabic Studies.* Eds. Joshua Blau and David Doron. Ramat Gan 2000, pp. 145–60.

—— "The Literary Legacy of David Ben Joshua, Last of the Maimonidean Negīdim." *The Jewish Quarterly Review* 25 i (1984), pp. 1–56.

Fink, Christine: *Ibn Kammuna über die Grundbegriffe der Philosophie. Analyse eines Kapitels aus seinem Werk 'Das Neue in der Philosophie'.* MA thesis Frankfurt/Main 1997.

Fischel, Walter: "Arabische Quellen zur Geschichte der babylonischen Judenheit im 13. Jahrhundert." *Monatsschrift für Geschichte und Wissenschaft des Judentums* 79 [N.F. 43] (1935), pp. 302–22.

Flügel = Österreichische Nationalbibliothek, Vienna
Catalogue:
1865–67. Gustav Flügel: *Die arabischen, persischen und türkischen Handschriften der Kaiserlich-Königlichen Hofbibliothek zu Wien* 1–3. Wien.

Fujii, Hideo and Kazumi Oguchi: *Lost Heritage. Antiquities Stolen from Iraq's Regional Museums.* Fascicle 3. Tokyo 1996.

GAL = Carl Brockelmann: *Geschichte der arabischen Litteratur* 1–2. Zweite den Supplementbänden angepasste Auflage 1–2. Leiden 1943–49.

GALS = Carl Brockelmann: *Geschichte der Arabischen Litteratur. Supplementbände* 1–3. Leiden 1937–42.

GCAL = Georg Graf: *Geschichte der christlichen arabischen Literatur* 1–5. Vatican 1944–53.

Gharawiyya = Al-Rawḍa al-Ḥaydariyya Collection (Khizānat al-Rawḍa al-Ḥaydariyya), also known as al-Khizāna al-Gharawiyya or Khizānat al-Mashhad al-ʿAlawī, Najaf
Catalogues:
1959. Ḥusayn ʿAlī Maḥfūẓ: "Fihris al-khizāna al-Gharawiyya bi-l-Najaf fī mashhad Amīr al-Muʾminīn al-Imām ʿAlī b. Abī Ṭālib ʿalayhi al-salām." *Revue de l'Institut des Manuscrits Arabes* 5 (1378), pp. 23–30.
1971. Aḥmad al-Ḥusaynī: *Fihrist makhṭūṭāt al-Rawḍa al-Ḥaydariyya fī l-Najaf al-Ashraf.* Najaf 1391.

-Ghazzālī, Abū Ḥāmid Muḥammad: *Iḥyāʾ ʿulūm al-dīn* 1–5. Beirut n.d.

Gilli-Elewy, Hend: *Bagdad nach dem Sturz des Kalifats. Die Geschichte einer Provinz unter ilḫânischer Herrschaft (656–735/1258–1335).* Berlin 2000.

Gobillot, Geneviève: "Ibn Kammûna. Les trois anneaux de la prophétie?" *Tsafon. Revue d'études juives du Nord* 50 (automne 2005–hiver 2006) [Hommage à Jean-Marie Delmaire], pp. 23–48.

Goitein, Shlomo Dov: *A Mediterranean Society. The Jewish Community of the World as Portrayed in the Documents of the Cairo Geniza* 1–6. Berkeley 1999.

—— "A Report on Messianic Troubles in Baghdad in 1120–21." *The Jewish Quarterly Review* 43 (1952–53), pp. 57–76.

Goldziher, Ignaz: "Saʿd b. Manṣūr ibn Kammūna's Abhandlung über die Unvergänglichkeit der Seele." In *Festschrift zum achtzigsten Geburtstage Moritz Steinschneider's.* Leipzig 1896, pp. 110–14 [reprinted in Ignaz Goldziher: *Gesammelte Schriften.* Hildesheim 1970, vol. 4, pp. 32–36].

Goodman, Lenn Evan: [Review of] "Ibn Kammūna's *Examination of the Three Faiths.*" *Philosophy East West* 22 (1972), pp. 487–88.

-Ḥāʾirī, ʿAbd al-Ḥusayn [et al.]: *Fihrist-i Kitābkhāna-yi Majlis-i Shūrā-yi Millī* 1–22. Tehran 1305–57/1926–78.

Ḥamīd, Ḥamīd: "Naẓarī bih mīrāth-i maʿnawī-yi Ibn Kammūna. Zindigī, andīsha u shubahāt." *Īrānshināsī* 11 iii (1999), pp. 632–43.

Hamidiye = Hamidiye Collection (held in the Süleymanie Kütüphanesi), Istanbul.

Hekimoğlu Ali Paşa = Hekimoğlu Ali Paşa Collection (held in the Süleymanie Kütüphanesi), Istanbul
Catalogue:
1894. *Defter-i Hekimoğlu Ali Paşa Kütüphanesi*. Istanbul 1311.

d'Herbelot, Barthélemy: *Bibliothèque orientale ou dictionnaire universel: contenant generalement tout ce qui regarde la connaissance des peuples de l'Orient*. Paris 1697.

Hindūshāh b. Sanjar al-Ṣāḥibī, Abu l-Faḍl: *Ṣaḥāḥ al-ʿajam*. Ed. Ghulām Ḥusayn Baygdilī. Tehran 1361/1982.

—— *Tajārib al-salaf dar tārīkh*. Taṣwīr-i nuskha-yi muʾarrakh-i 846 h.q. Ed. Amīr Sayyid Ḥasan Rawḍātī. Isfahan 1361/1982.

—— *Tajārib al-salaf dar tawārīkh-i khulafāʾ u wuzarā-yi īshān*. Ed. ʿAbbās Iqbāl. Tehran 1313/1934 [reprinted Tehran 1344/1965].

Hirschfeld, Hartwig (ed.): *Arabic Chrestomathy in Hebrew Characters. With a Glossary*. London 1892.

—— *Das Buch al-Chazarî des Abû-l-Ḥasan Jehuda Hallewi im arabischen Urtext sowie in der hebräischen Übersetzung des Jeduda ibn Tibbon*. Leipzig 1887 [reprinted Jerusalem 1970].

Hirschfeld, Leo: *Saʿd b. Manṣûr ibn Kammûnah und seine polemische Schrift* تنقيح الأبحاث للملل الثلاث. Inaugural-Dissertation zur Erlangung der Doctorwürde vorgelegt der Hohen Philosophischen Fakultät der Universität Heidelberg. Berlin 1893.

Horten, Max: *Das philosophische System von Schirázi*. Strassburg 1913.

—— "Die philosophischen und theologischen Ansichten von Lahígi (ca. 1670) nach seinem Werke 'Die aufgehenden Sterne der Offenbarung', einem Kommentar zur Dogmatik des Ṭusi (1273 *)." *Der Islam* 3 (1912), pp. 91–131.

Ḥujjatī, Muḥammad Bāqir and Muḥammad Taqī Dānishpazhūh: *Fihrist-i nuskhahā-yi khaṭṭī-yi kitābkhāna-yi Dānishkada-yi Ilāhiyāt u Maʿārif-i Islāmī Dānishgāh-i Tihrān* 1–2. Tehran 1345–48/1966–69.

-Ḥurr al-ʿĀmilī, Muḥammad b. al-Ḥasan: *Amal al-āmil* 1–2. Ed. Aḥmad al-Ḥusaynī. Baghdad 1385/1965.

-Ḥusaynī, Aḥmad: *Fihrist makhṭūṭāt al-Rawḍa al-Ḥaydariyya fī l-Najaf al-Ashraf*. Najaf 1391/1971.

Ibn al-Fuwaṭī: *Majmaʿ al-ādāb fī muʿjam al-alqāb* 1–6. Ed. Muḥammad al-Kāẓim. Tehran 1416/1995–96.

—— *Talkhīṣ majmaʿ al-ādāb fī muʿjam al-alqāb* 1–4. Ed. Muṣṭafā Jawād. Damascus 1963–65.

Ibn Kammūna, Saʿd b. Manṣūr: *Al-Jadīd fī l-ḥikma*. Ed. Ḥamīd Marʿīd al-Kabīsī. Baghdad 1403/1982.

—— "Maqāla fī anna l-nafs laysat bi-mizāj al-badan wa-lā kāʾina ʿan mizāj al-badan." In Insīyah Barkh$^{w}$āh: *Taṣḥīḥ-i Risāla-yi Azaliyyat al-nafs wa-baqāʾihā taʾlīf-i Ibn Kammūna*. MA thesis: Madrasa-yi Murtaḍā Muṭahharī (Sipahsālār), 1379/2000, pp. 125–65.

—— "Maqāla fī anna wujūd al-nafs abadī wa-baqāʾahā sarmadī." In Insīyah Barkh$^{w}$āh: *Taṣḥīḥ-i Risāla-yi Azaliyyat al-nafs wa-baqāʾihā taʾlīf-i Ibn Kammūna*. MA thesis: Madrasa-yi Murtaḍā Muṭahharī (Sipahsālār), 1379/2000, pp. 166–78.

—— "Maqāla fī anna wujūd al-nafs abadī wa-baqāʾahā sarmadī." In Leon Nemoy

(ed.): "Ibn Kammunah's Treatise on the Immortality of the Soul." *Ha-Rofe ha-ʿIvri. The Hebrew Medical Journal* 1 (1962), pp. 239–213.
—— "Maqāla fī l-taṣdīq bi-anna nafs al-insān bāqiya abadan." In Insīyah Barkhʷāh: *Taṣḥīḥ-i Risāla-yi Azaliyyat al-nafs wa-baqāʾihā taʾlīf-i Ibn Kammūna.* MA thesis: Madrasa-yi Murtaḍā Muṭahharī (Sipahsālār), 1379/2000, pp. 72–124.
—— "Al-Maṭālib al-muhimma min ʿilm al-ḥikma." Ed. Sayyid Ḥusayn Sayyid Mūsawī. *Kheradnāma-yi Ṣadrā* 8 xxxii (1382/2003), pp. 64–86.
—— *Tanqīḥ al-abḥāth li-l-milal al-thalāth* → Perlmann, Moshe (ed.): *Saʿd b. Manṣūr Ibn Kammūna's Examination of the Inquiries into the Three Faiths.*
—— *Al-Tanqīḥāt fī sharḥ al-Talwīḥāt.* Ed. Sayyid Ḥusayn Sayyid Mūsawī. PhD thesis: Dānishgāh-i Tihrān, 1375–76/1996–97.
—— *Al-Tanqīḥāt fī sharḥ al-talwīḥāt. Refinement and commentary on Suhrawardī's* Intimations. *A Thirteenth Century Text on Natural Philosophy and Psychology.* Eds. Hossein Ziai and Ahmed Alwishah. Costa Mesa, California 2003.
Ibn Mubārakshāh al-Bukhārī, Shams al-Dīn Muḥammad: *Sharḥ Hikmat al-ʿayn.* Kazan 1319/1901.
Ibn al-Nafīs: *Al-Muhadhdhab fī l-kuḥl al-mujarrab.* Eds. Muḥammad Ẓāfir al-Wafāʾī and Muḥammad Rawwās Qalʿajī. Rabat 1988.
Ibn Quṭlūbughā, Zayn al-Dīn Qāsim: *Tāj al-tarājim fī ṭabaqāt al-Ḥanafiyya.* Ed. Gustav Fluegel. Leipzig 1862.
—— *Tāj al-tarājim fī man ṣannafa min al-Ḥanafiyya.* Ed. Ibrāhīm Ṣāliḥ. Beirut 1412/1992.
Ibn al-Sāʿī: *Al-Jāmiʿ al-mukhtaṣar.* Baghdad 1934.
Ibn Sīnā: *Al-Ishārāt wa-l-tanbīhāt.* Ed. Mujtabā al-Zāraʿī. Qum 1381/2002.
—— *Al-Mabdaʾ wa-l-maʿād.* Ed. ʿAbd Allāh Nūrānī. Tehran 1363/1984.
Ibn Taghrībirdī, Abu l-Maḥāsin Yūsuf: *Al-Manhal al-ṣāfī wa-l-mustawfā baʿda l-wāfī* 1–8. Ed. Muḥammad Muḥammad Amīn. Cairo 1984–90.
IMHM = The Institute of Microfilmed Hebrew Manuscripts at the Jewish National and University Library, Jerusalem.
*Istanbul Umumî kütüphaneleri Yazmalari Sergisi. 15.-22.IX.1951. Bu sergi XXII. Milletlerarasi Müsteşrikler Kongresi için tertip edilmiştir.* Istanbul 1951.
Istanbul University = Istanbul Üniversitesi Kütüphanesi, Istanbul
Catalogue:
1951–53. Fehmi Edhem Karatay: *Istanbul Üniversitesi Kütüphanesi Arapça bazmalar alfabe kataloğu.* Cilt I. 2 Fascs. Istanbul.
ʿIzzat, ʿAbd al-ʿAzīz: *Ibn Miskawayh. Falsafatuhū al-akhlāqiyya wa-maṣādiruhā.* Cairo 1946.
Jewish Theological Seminary = The Jewish Theological Seminary, New York.
John Rylands = John Rylands University Library, Manchester
Catalogue:
1934. Alphonse Mingana: *Catalogue of the Arabic manuscripts in the John Rylands Library.* Manchester.
*Jung-i Mahdawī. Taʾrīkh-i kitābat az 753 qamarī be baʿd. Chāp-i ʿaksī az rū-yi nuskha-yi khaṭṭī-yi Kitābkhāna-yi Duktur Aṣghar Mahdawī.* Tehran 1380/2002.
Kadīwar, Muḥsin and Muḥammad Nūrī: *Maʾkhadh shināsī-yi ʿulūm-i ʿaqlī. Manābiʿ chāpī-yi ʿulūm-i ʿaqlī az ibtidāʾ tā 1375* 1–3. Tehran 1378/1999.
Kaḥḥāla, ʿUmar Riḍā: *Muʿjam al-muʾallifīn* 1–14. Beirut n.d.
*Kashf al-ẓunūn* = Ḥājjī Khalīfa: *Kashf al-ẓunūn ʿan asāmī al-kutub wa-l-funūn. Lexicon bibliographicum et encyclopaedicum a Mustafa Ben Abdallah Katib Jelebi dicto et nomine Haji Khalfa celebrato* 1–7. Ed. Gustav Flügel. Leipzig 1835–58.

-Kātibī, Najm al-Dīn: *Ḥikmat al-ʿayn*. Transl. [into Persian] ʿAbbās Ṣadrī. Tehran 1375/1991.

-Khafrī, Shams al-Dīn Muḥammad b. Aḥmad: *Taʿlīqa bar Ilāhiyyāt Sharḥ al-tajrīd-i Mullā ʿAlī Qūchshī*. Ed. Fīrūzah Sāʿatchiyān. Tehran 1382/2004.

Kholeif, Fathalla: *A Study on Fakhr al-Dīn al-Rāzī and his Controversies in Transoxania*. Beirut [2]1984.

Khwānsārī, Aqā Ḥusayn b. Jamāl al-Dīn: "Al-Ḥāshiya ʿalā l-Shifāʾ (Ilāhiyyāt)." Ed. Ḥāmid Nāji Iṣfahānī. In *Dabīrkhāna-yi kongiri-yi Āqā Ḥusayn Khwānsārī*. Qum 1378/1999.

Kochius, Carolus Fridericus: "Observatio nova de vero scriptore libri in 3 impostoribus mohammedano." In *Annales Academiae Juliae ex editis et manuscriptis monumentis compositi*. Julio, Augusto, September 1721, § viii.

Köprülü = Köprülü Kütüphanesi, Istanbul

Catalogues:

1886. *Köprülü-zâdeh Mehmed Paşa'nın Kütüphanesinde mahfuz kütüb-i mevcûdenin defteri*. Istanbul 1303.

1986. *Fihris makhṭūṭāt Maktabat Kūprīlī. Catalogue of manuscripts in the Köprülü Library* 1–3. Prepared by Ramazan Şeşen, Cevat İzgi and Cemil Akpınar. Presented by Ekmeleddin İhsanoğlu. Istanbul 1406.

Kritzeck, James: [Review of] "Ibn Kammūna's *Examination of the Three Faiths*." *Journal of Near Eastern Studies* 32 (1973), pp. 506–7.

Lāhījī, ʿAbd al-Razzāq: *Gawhar-i murād*. Ed. Muʾassassa-yi taḥqīqāti-yi Imām Ṣādiq. Qum 1383/2004.

Lâleli = Lâleli collection (held at the Süleymanie Kütüphanesi), Istanbul

Catalogue:

1894. *Defter-i Kütüphane-i Lâleli*. Istanbul 1311.

Landberg = Carlo Landberg collection of Arabic manuscripts at the Beinicke Library at Yale University

Catalogue:

1956. Leon Nemoy: *Arabic Manuscripts in the Yale University Library*. New Haven.

Lane, George: *Early Mongol rule in thirteenth-century Iran. A Persian renaissance*. London 2003.

Langermann, Y. Tzvi: "Ibn Kammūna and the 'New Wisdom' of the Thirteenth Century." *Arabic Sciences and Philosophy* 15 (2005), pp. 277–327.

—— "Arabic Writings in Hebrew manuscripts. A preliminary relisting." *Arabic Sciences and Philosophy* 6 (1996), pp. 137–60.

Leiden = Rijksuniversiteit te Leiden, Bibliotheek (University Library)

Catalogues:

1851–77. *Catalogus Codicum Orientalium Bibliothecae Academiae Lugduno-Batavae*. Eds. R.P.A. Dozy (vols. 1, 2), P. de Jong and M.J. de Goeje (vols. 3, 4), M.J. de Goeje (vol. 5), M.Th. Houtsma (vol. 6).

1957. Petrus Voorhoeve: *Handlist of Arabic manuscripts in the Library of the University of Leiden and other collections in the Netherlands*. Leiden.

Los Angeles = University Research Library (University of California), Los Angeles

Catalogue:

1983. Muḥammad Taqī Dānishpazhūh: "Fihrist-i nuskhahā-yi khaṭṭī-yi Dānishgāh-i Lus Āngilis." *Nuskhahā-yi khaṭṭī* (= *Nashriyya-yi Kitābkhāna-yi Markazī u Markaz-i Asnād-i Dānishgāh-i Tihrān*) 11–12 (1362), pp. 1–112.

Maʿhad = Institute of Arabic Manuscripts (Maʿhad al-Makhṭūṭāt al-ʿArabiyya), Cairo
Catalogue:
1954–80. Fuʾād Sayyid [et al.]: *Fihris al-makhṭūṭāt al-muṣawwara fī Maʿhad Iḥyāʾ al-Makhṭūṭāt al-ʿArabiyya* 1–14. Cairo.

Mahdawī, Yaḥyā: *Fihrist-i nuskhahā-yi muṣannafāt-i Ibn Sīnā.* Tehran 1333/1954.

Mahdī Bayānī = Mahdī Bayānī Collection, Tehran (the collection was dispersed after the owner's death in 1346/1968. Parts of the collection are apparently now in the Āstān-i Quds Library in Mashhad and in the Majlis-i Sinā Library in Tehran)
Catalogues:
1961. Ḥasan ʿAlī Maḥfūẓ: *"Makhṭūṭāt al-Duktūr Mahdī Bayānī fī Tihrān." Revue de l'Institut des Manuscrits Arabes* 7 (1961), pp. 3–6.
1961a. Mahdī Bayānī: "Majmūʿa-yi munshaʾāt." *Rāhnimā-yi Kitāb* 4 (1340), pp. 239–44.
1961b. Mahdī Bayānī: "Fihrist-i mujmalī az manẓūmāt-i Fārsī." *Nuskhahā-yi khaṭṭī* (= *Nashriyya-yi Kitābkhāna-yi Markazī u Markaz-i Asnād-i Dānishgāh-i Tihrān*) 1 (1340), pp. 8–17.
1969: Muḥammad Taqī Dānishpazhūh: "Chand nuskhah az marḥūm-i Bayānī." *Nuskhahā-yi khaṭṭī* (= *Nashriyya-yi Kitābkhāna-yi Markazī u Markaz-i Asnād-i Dānishgāh-i Tihrān*) 6 (1348), pp. 691–95.

Maḥfūẓ = Private collection of Dr. Ḥusayn Alī Maḥfūẓ (now held at the Saddam Manuscripts Library [Dār Ṣaddām li-l-Makhṭūṭāt]), Baghdad
Catalogue:
1960. ʿAlī Ḥusayn Maḥfūẓ: "Khizānat al-Duktūr Ḥusayn ʿAlī Maḥfūẓ bi-l-Kāẓimiyya." *Revue de l'Institut des Manuscrits Arabes* 6 (1960), pp. 15–58.

Maḥfūẓ, Ḥusayn ʿAlī: "Fihris al-khizāna al-Gharawiyya bi-l-Najaf fī mashhad Amīr al-Muʾminīn wa-l-Imām ʿAlī b. Abī Ṭālib ʿalayhi al-salām." *Revue de l'Institut des Manuscrits Arabes* 5 (1959), pp. 23–30.

—— "Nafāʾis al-makhṭūṭāt al-ʿarabiyya fī Īrān." *Revue de l'Institut des Manuscrits Arabes* 3 (1957), pp. 31–78.

Maimonides: *Mishneh Torah. The Book of Knowledge* [*Sefer ha-Madaʿ*]. Edited according to the Bodleian (Oxford) Codex with Introduction, Biblical and Talmudical References, Notes and English Translation by Moses Hyamson. Jerusalem 1962 [New corrected edition Jerusalem/New York 1981].

Majlis = Majlis-i Shūrā-yi Islāmī Library (Shumāra-yi yik), Tehran
Catalogue:
1926–78. ʿAbd al-Ḥusayn al-Ḥāʾirī [et al.]: *Fihrist-i Kitābkhāna-yi Majlis-i Shūrā-yi Millī* 1–22. Tehran 1305–57.

Majlis-i Sinā = Majlis-i Shūrā-yi Islāmī Library (Shumāra-yi du), Tehran (formerly called Kitābkhāna-yi Majlis-i Sinā)
Catalogues:
1976. Muḥammad Taqī Dānishpazhūh and Bahāʾ al-Dīn ʿIlmī Anwārī: *Fihrist-i kitābhā-yi khaṭṭī-yi Majlis-i Sinā.* Vol. 1. Tehran 1355.
1980. Muḥammad Taqī Dānishpazhūh and Bahāʾ al-Dīn ʿIlmī Anwārī: *Fihrist-i kutub-i khaṭṭī-yi kitābkhāna-yi Majlis-i Shūrā-yi Islāmī (Shumāra-yi du).* Vol. 2. Tehran 1359.

Malik = Milli Malik Library, Tehran
Catalogues:
1973–90. Īraj Afshār and Muḥammad Taqī Dānishpazhūh: *Fihrist-i kitābhā-yi khaṭṭī-yi Kitābkhāna-yi Malik* 1–8. Tehran 1352–69.

Marʿashī = Āyat Allāh Marʿashī Najafī Library, Qum

Catalogue:
1975–. Aḥmad al-Ḥusaynī, Sayyid Maḥmūd Marʿashī Najafī [et al.]: *Fihrist-i Kitābkhāna-yi ʿUmūmī-yi Haḍrat-i Āyat Allāh al-ʿuẓmā Najafī Marʿashī* 1–. Qum.

Margoliouth, George: *Catalogue of the Hebrew and Samaritan Manuscripts in the British Museum*. London 1905.

—— "A Muhammadan Commentary on Maimonides' Mishneh Torah." *The Jewish Quarterly Review* (1901), pp. 488–502.

Martin, Christopher J.: "Obligations and Liars." In *Sophisms in Medieval Logic and Grammar. Acts of the Ninth European Symposium for Medieval Logic and Semantics, held at St Andrews, June 1990*. Ed. Stephen Read. Dordrecht/Boston/London 1993, pp. 357–81.

Maʿrūf, Nājī: *Taʾrīkh ʿulamāʾ al-Mustanṣiriyya* 1–2. Baghdad 1379/1959.

*Metzler Lexikon jüdischer Philosophen. Philosophisches Denken des Judentums von der Antike bis zur Gegenwart*. Eds. Andreas B. Kilcher, Otfried Fraisse and Yossef Schwartz. Stuttgart/Weimar 2003.

Miller, Larry B.: "A Brief History of the Liar Paradox." In *Of Scholars, Savants, and their Texts. Studies in Philosophy and Religious Thought. Essays in Honor of Arthur Hyman*. Ed. Ruth Link-Salinger. Frankfurt/Main 1989, pp. 173–82.

Mīnuwī = Kitābkhāna-yi Mīnuwī, Tehran
Catalogue:
1996. Muḥammad Taqī Dānishpazhūh and Īraj Afshār: *Fihristwāra-yi Kitābkhāna-yi Mīnuwī u Kitābkhāna-yi Markazī-yi Pazhūhishgāh*. Tehran 1374.

Mīnuwī, Mujtabā: "Az khazāʾin-i Turkiyah." *Majalla-yi Dānishkada-yi Adabiyyāt* 4 ii (1335/1956), pp. 42–75; 4 iii (1335/1956), pp. 53–89 [Reprinted in *Mīnuwī bar gustara-yi adabiyyāt-i fārsī. Majmūʿa-yi maqālāt*. Ed. Māhmunīr Mīnuwī. Tehran 1381/200, pp. 199–261].

Mīr, Muḥammad Taqī: *Buzurgān-i nami-yi pārs*. Shiraz 1368/1989.

Mīr Dāmād: *Muṣannafāt-i Mīr Dāmād mushtamil bar dah ʿunwān az kitābhā u risālahā u ijāzahā u nāmahā*. Volume 1. Ed. ʿAbd Allāh Nūrānī. Tehran 1381/2003.

—— "Taʿlīqāt manṭiqiyya." Ed. ʿAbd Allāh Nūrānī. In *Collected Texts and Papers on Logic and Language*. Eds. Mehdi Mohaghegh and T. Izutsu. Tehran 1974, pp. 287–307.

Mishkāt = Collection of manuscripts donated by Sayyid Muḥammad Mishkāt to the Central Library of Tehran University
Catalogue:
1951–85. Muḥammad Taqī Dānishpazhūh and ʿAlī Naqī Munzawī: *Fihrist-i nuskhahā-yi khaṭṭī-yi kitābkhāna-yi markazī u Markaz-i Asnād-i Dānishgāh-i Tihrān* 1–18. Tehran 1330–64. [Vol. 3 i: *Fihrist-i kitābkhāna-yi ihdāʾi-yi Āqā-yi Muḥammad Mishkāt bih Kitābkhāna-yi Dānishgāh-i* Tihrān. Tehran 1332/1953–54].
1976. Muḥammad Shīrwānī: *Fihristwāra-yi nuskhahā-yi khaṭṭī-yi majmūʿa-yi Mishkāt*. Tehran.

Miskawayh, Abū ʿAlī Aḥmad b. Muḥammad b. Yaʿqūb: *Al-Fawz al-asghar*. Ed. Ṣāliḥ ʿUḍayma. Traduction française et notes par Roger Arnaldez. Tunis 1987.

MS = Aḥmad b. Muṣṭafā Ṭāshkuprī Zadeh: *Miftāḥ al-saʿāda wa-miṣbāḥ al-siyāda fī mawḍūʿāt al-ʿulūm* 1–3. Cairo 1968.

Mudarris Raḍawī, Muḥammad Taqī: *Aḥwāl u āthār-i Khʷāja Naṣīr al-Dīn Ṭūsī*. Tehran 1334/1955–56.

Mudarrisī Zanjānī, Muḥammad: *Sarguzasht u ʿaqāʾid-i falsafī-yi Khʷāja Naṣīr al-Dīn Ṭūsī*. Tehran 1379/2000.

Muḥammad al-Samāwī = Collection of manuscripts donated by Shaykh Muḥammad al-Samāwī to the Imām al-Ḥakīm Public Library (Maktabat al-Imām al-Ḥakīm al-ʿāmma), Najaf.

Muḥammadī, Afsānah: *Risāla fī shubhat jadhr al-aṣamm 'kull kalāmī kādhib'*. MA Thesis, University of Tehran, 1381/2002.

*Muʿjam al-turāth al-kalāmī* = al-Lajna al-ʿilmiyya fī muʾassassat al-Imām al-Sādiq: *Muʿjam al-turāth al-kalāmī. Muʿjam yatanāwalu dhikr asmāʾ al-muʾallafāt al-kalāmiyya (al-makhṭūṭāt wa-l-maṭbūʿāt) ʿabra l-qurūn wa-l-maktabāt allatī tatawaffiru fīhā nusakhihā* 1–5. Qum 1423/1381/[2002].

*Mukhtārāt* = Şeşen, Ramazan [Ramaḍān Shīshīn]: *Mukhtārāt min al-makhṭūṭāt al-ʿarabiyya al-nādira fī maktabāt Turkiyā*. Istanbul 1997.

Munzawī, Aḥmad: *Fihrist-i nuskhahā-yi khaṭṭī-yi fārsī* 1–6. Tehran 1969–74.

Munzawī, ʿAlī Naqī: *Ibn Kammūna wa-falsafatuhū fī l-nubuwwa*. PhD thesis, Université Saint Joseph, Beirut 1972. *[Al-Qism al-awwal: al-Madkhal wa-l-naṣṣ al-kāmil li-Tanqīḥ al-abḥāth; al-Qism al-thānī: al-Taʿlīqāt ʿalā Tanqīḥ al-abḥāth li-l-milal al-thalāth]*.

—— "Tanqīḥ al-abḥāth-i Ibn Kammūna." *Waḥīd* 6 v (1348/1969), pp. 369–71; 6 vi (1348/1969), pp. 503–14; 6 xii (1348/1969), pp. 1079–83.

-Mūsawī, Mūsā: *Min al-Suhrawardī ilā l-Shīrāzī*. Beirut 1979.

Mūsawī, Sayyid Ḥusayn Sayyid: "Mullā Ṣadrā u Ibn Kammūna." In *Mullā Ṣadrā u muṭālaʿāt-i taṭbīqī. Majmūʿa-yi maqālāt-i hamāyish-i jahānī-yi ḥakīm Mullā Ṣadrā*. Vol. 5. Tehran 1381/2002, pp. 46–53.

-Muwaqqit, ʿAlāʾ al-Dīn Abu l-Ḥasan al-Dīn Abu l-Ḥasan ʿAlī b. Ṭaybugha: *Sharḥ Sefer ha-maddaʿ*. Ms. British Library Add 27,294 [IMHM microfilm no. 6086].

—— *Sharḥ Sefer ha-maddaʿ*. Ms. RNL Firk. Yevr.-Arab. I 1426 [IMHM microfilm no. 55193].

-Nakhjawānī, Najm al-Dīn Abū Bakr Muḥammad: *Ajwiba ʿan iʿtirāḍāt Fakhr al-Dīn al-Rāzī li-l-qānūn*. Ms. Ayasofya 3721, ff. 1a–66b.

Naṣīr al-Dīn al-Ṭūsī: *Ajwibat al-masāʾil al-naṣīriyya mushtamil bar 20 risāla*. Ed. ʿAbd Allāh Nūrānī. Tehran 1383/2005.

—— *Akhlāq-i muḥtashamī*. Ed. Muḥammad Taqī Dānishpazhūh. Tehran 1377/1998.

—— *Talkhīṣ al-muḥaṣṣal al-maʿrūf bi-Naqd al-Muḥaṣṣal*. Beirut 1405/1985.

*Nawādir* = Şeşen, Ramazan [Ramaḍān Shīshīn]: *Nawādir al-makhṭūṭāt al-ʿarabiyya fī maktabāt Turkiyā* 1–3. Beirut 1975.

-Nayrīzī, Najm al-Dīn Maḥmūd: *Miṣbāḥ al-arwāḥ fī kashf ḥaqāʾiq al-alwāḥ*. Ms. Ragıp Paşa 853.

Nemoy, Leon: *The Arabic Treatise on the Immortality of the Soul by Saʿd ibn Manṣūr ibn Kammūna (XIII century). Facsimile Reproduction of the Only Known Manuscript (Cod. Landberg 510, fol. 58–70) in the Yale University Library. With a Bibliographical Note*. New Haven 1944.

—— "A Forgotten 13th Century Jewish Physician – Ibn Kammuna." *Ha-Rofe ha-ʿIvri. The Hebrew Medical Journal* 2 (1947), pp. 159–57.

—— "Contributions to the Textual Criticism of Judah ha-Levi's Kitāb al-Khazarī." *The Jewish Quarterly Review* 26 (1936), pp. 221–26.

—— "Heʿarot le-ḥibur shel Ibn Kammunah ʿal ha-hevdelim ben ha-Rabanim ve-ha-Karaʾim." *Tarbiz* 24 (1953–54), pp. 343–53.

—— "Ibn Kammunah rofe yehudi." *Ha-Rofe ha-ʿIvri. The Hebrew Medical Journal* 2 (1947), pp. 123–25.

—— "Ibn Kammunah u-masato ʿal hishaʾarut ha-nefesh." *Ha-Rofe ha-ʿIvri. The Hebrew Medical Journal* 1 (1962), pp. 131–36.

—— "Ibn Kammunah's Treatise on the Differences between the Rabbanites and the Karaites." *Jewish Quarterly Review* 13 (1972–73), pp. 97–135, 222–46.
—— (ed.) "Ibn Kammunah's Treatise on the Differences between the Rabbanites and the Karaites." *Proceedings of the American Academy for Jewish Research* 36 (1968), pp. 107–65.
—— "Ibn Kammūna's Treatise on the Immortality of the Soul." In *Ignace Goldziher Memorial Volume*. Eds. Samuel Löwinger, Alexander Scheiber and Joseph Somogyi. Jerusalem 1958, vol. 2, pp. 83–99.
—— (ed.) "Ibn Kammunah's Treatise on the Immortality of the Soul." *Ha-Rofe ha-ʿIvri. The Hebrew Medical Journal* 1 (1962), pp. 239–213.
—— "New Data for the Biography of Saʿd Ibn Kammūnah." *Revue des études juives* 123 (1964), pp. 507–10.
—— "Remarks on Ibn Kammūna's Treatise on the Differences between the Rabbanites and the Karaites." [Hebrew] *Tarbiz* 24 (1955), pp. 343–53, vi (English summary) [reprinted in *Sefer yovel mugash le-khevod Yisrael Elfenbeyn*. Ed. Yehudah L. Maimon. Jerusalem 1962, pp. 201–8].
—— [Review of] "Perlmann, Moshe, Ed. *Saʿd Ibn Manṣūr Ibn Kammūnah. Examination of the Inquiries into the Three Faiths*." *Jewish Social Studies* 31 i (1969), pp. 333–35.
—— [Review of] "Ibn al-Maḥrūma's Ḥawāshī sur Ibn Kammūna." *The Jewish Quarterly Review* 77 i (1986), pp. 82–83.
Netzer, Amnon: "Rashīd al-Dīn and his Jewish Background." *Irano-Judaica* III. Jerusalem 1994, pp. 118–26.
Niewöhner, Friedrich: "*Die Wahrheit ist eine Tochter der Zeit*. Ibn Kammūna's historisch-kritischer Religionsvergleich aus dem Jahre 1280." In *Religionsgespräche im Mittelalter*. Eds. Bernard Lewis and Friedrich Niewöhner. Wiesbaden 1992, pp. 357–69.
-Niʿma, ʿAbd Allāh: *Falāsifat al-shīʿa. Ḥayātuhum wa-ārāʾuhum*. Beirut 1962.
Nūrānī, ʿAbd Allāh: "Muṭāraḥāt manṭiqiyya bayn al-Kātibī wa-l-Ṭūsī." In *Collected Texts and Papers on Logic and Language*. Eds. Mehdi Mohaghegh and Toshihiko Izutsu. Tehran 1974, pp. 279–86.
Nuruosmaniye = Nuruosmaniye Kütüphanesi, Istanbul
Catalogue:
1886. *Nuruosmaniye Kütüphanesi'nde mahfuz kütüb-i mevcûdenin defteridir*. Istanbul 1303.
-Oraibi, Ali Ahmed: *Shiʾi Renaissance: A Case Study of the Theosophical School of Bahrain in the 7th/13th Century*. Diss. McGill University 1992.
Paret, Rudi: [Review of] "Saʿd b. Manṣūr Ibn Kammūna's *Examination of the Inquiries into the Three Faiths*... Edited by Moshe Perlmann." *Der Islam* 46 (1970), pp. 102–4.
Perlmann, Moshe: "Ibn Maḥrūma, a Christian opponent of Ibn Kammūna." *American Academy for Jewish Research* 2 (1965) [= Harry A. Wolfson Jubilee Volume], pp. 641–55.
—— (ed.) *Saʿd b. Manṣūr Ibn Kammūna's Examination of the Inquiries into the Three Faiths. A Thirteenth-Century Essay in Comparative Religion*. Berkeley/Los Angeles 1967.
—— *Ibn Kammuna's Examination of the Three Faiths. A Thirteenth-Century Essay in Comparative Study of Religion*. Translated from the Arabic, with an Introduction and Notes. Berkeley/Los Angeles 1971.
—— "'Proving Muhammad's Prophethood', A Muslim critique of Ibn Kammuna." [Hebrew] In *Hagut Ivrit Ba'Amerika. Studies on Jewish Themes by Contemporary American Scholars*. Eds. Menahem Zohori, Arie Tartakover and Haim Ormian. Vol. III. Tel Aviv 1974, pp. 75–97.

—— "The Medieval Polemics Between Islam and Judaism." In *Religion in a Religious Age*. Ed. Shlomo Dov Goitein. New York 1974, pp. 103–38.

Pertev Paşa = Pertev Paşa Collection (held in the Süleymanie Kütüphanesi), Istanbul.

Pines, Salomon: "La 'philosophie orientale' d'Avicenne et sa philosophie contre les Bagdadiens." *Archives d'Histoire Doctrinale et Littérature du Moyen Age* 27 (1952), pp. 5–37.

Pohl, Stephan: *Zur Theosophie im nachmongolischen Iran. Leben und Werk des Ǧalāladdīn ad-Dawwānī (gest. 902/1502)*. Bochum 1997 [unpublished manuscript].

Pourjavady, Nasrollah: "Opposition to Sufism in Twelvershiism." In *Islamic Mysticism Contested. Thirteen Centuries of Controversies and Polemics*. Eds. Frederick de Jong and Bernd Radtke. Leiden 1999, pp. 614–23.

Pourjavady, Nasrollah and Peter Lamborn Wilson: *Kings of Love. The Poetry and History of the Niʿmatullāhī Sufi Order*. Tehran 1978.

Pourjavady, Reza: "Kitābshināsī-yi āthār-i Jalāl al-Dīn Dawānī." *Maʿārif* 15 i/ii (1377/1998), pp. 81–138.

Pourjavady, Reza and Sabine Schmidtke: "Muslim Polemics against Judaism and Christianity in 18th Century Iran. The Literary Sources of Āqā Muḥammad ʿAlī Bihbahānī's (1144/1732–1216/1801) *Rādd-i shubahāt al-kuffār*." *Studia Iranica* 35 (2006), pp. 69–94.

—— "Quṭb al-Dīn al-Shīrāzī's (d. 710/1311) *Durrat al-Tāj* and Its Sources. (Studies on Quṭb al-Dīn al-Shīrāzī I)." *Journal Asiatique* 292 i–ii (2004), pp. 309–28.

—— "Some notes on a new edition of a medieval philosophical text in Turkey. Shams al-Dīn al-Shahrazūrī's *Rasāʾil al-Shajara al-ilāhiyya*." *Die Welt des Islams* 46 (2006), pp. 76–85.

-Qalqashandī, Shihāb al-Dīn: *Ṣubḥ al-aʿshā fī ṣināʿat al-inshāʾ* 1–14. Cairo 1913–16.

-Qummī, ʿAbbās: *Al-Kunā wa-l-alqāb* 1–3. Tehran 1368/1989.

Qūshchī, ʿAlī: *Sharḥ Tajrīd al-kalām*. Lithograph edition. Tehran 1301/[1883–84].

Quṭb al-Dīn al-Rāzī: "al-Risāla al-maʿmūla fī l-taṣawwur wa-l-taṣdīq." In *Risālatān fī l-taṣawwur wa-l-taṣdīq*. Ed. Mahdī Sharīʿatī. Qum 1416/[1995–96].

Quṭb al-Dīn al-Shīrāzī: *Durrat al-tāj*. Ed. Sayyid Muḥammad Mishkāt. Tehran 1317–20/1938–41 [³1369/1990] [containing *fātiḥa*, *jumla* 1, 2, 3, and 5].

—— *Durrat al-tāj*. Ed. Sayyid Ḥasan Mishkān Ṭabasī. Tehran 1324/1945 [containing *jumla* 4].

—— *Durrat al-tāj*. Ed. Māhdukhtbānū Humāʾī. Tehran 1369/1991 [containing sections three and four of the *khātima*].

—— *Sharḥ Ḥikmat al-ishrāq*. Lithograph edition by Asad Allāh Harātī. Tehran 1313–15/1895–97 [reprinted without place and date].

—— *Sharḥ Ḥikmat al-ishrāq*. Eds. ʿAbd Allāh Nūrānī and Mahdī Muḥaqqiq. Tehran 1380/2001.

—— *Sharḥ Ḥikmat al-ishrāq*. Ms. Turhan Valide 207.

Ragıp Paşa = Ragıb Paşa Kütüphanesi (held in the Süleymanie Kütüphanesi), Istanbul
Catalogue:
1892. *Defter-i Kütüphane-i Râgıp Paşa*. Istanbul 1310.

Rahman, Fazlur: *The Philosophy of Mullā Ṣadra (Ṣadr al-Dīn al-Shīrāzī)*. Albany 1975.

Rawḍātī = Private collection of manuscripts of Muḥammad ʿAlī Rawḍātī, Isfahan.
Catalogue:
1958. Muḥammad ʿAlī Rawḍātī: *Fihrist-i kutub-i khaṭṭī-yi kitābkhānahā-yi Iṣfahān*. Isfahan 1377.

Reisman, David C.: *The Making of the Avicennan Tradition. The Transmission, Contents, and Structure of Ibn Sīnā's* al-Mubāḥaṯāt *(The Discussions)*. Leiden 2000.

Rescher, Nicholas: *The Development of Arabic Logic*. Pittsburgh 1964.

Ritter I = Ritter, Hellmut: "Philologika IX. Die vier Suhrawardī. Ihre Werke in Stambuler Handschriften." *Der Islam* 24 (1937), pp. 270–86 [reprinted in *Beiträge zur Erschließung arabischer Handschriften in Istanbul und Anatolien*. Ed. F. Sezgin in collaboration with M. Amawi, D. Bischoff, C. Ehrig-Eggert, A. Jokosha, E. Neubauer. Frankfurt/Main 1986, vol. 2, pp. 94–110].

Ritter II = Ritter, Hellmut: "Philologika IX. Die vier Suhrawardī. (Fortsetzung und Schluß.)" *Der Islam* 25 (1939), pp. 35–86 [reprinted in *Beiträge zur Erschließung arabischer Handschriften in Istanbul und Anatolien*. Ed. F. Sezgin in collaboration with M. Amawi, D. Bischoff, C. Ehrig-Eggert, A. Jokosha, E. Neubauer. Frankfurt/Main 1986, vol. 2, pp. 111–62].

Ritter/Uppsala = Hellmut Ritter Microfilm Collection of Uppsala University Library Catalogue:
1992. *Catalogue of the Arabic Manuscripts in the Hellmut Ritter Microfilm Collection of Uppsala University Library, including later accessions*. By Berhard Lewin and Oscar Löfgren. Ed. Mikael Persenius. Uppsala.

Roggema, Barbara: "Epistemology as Polemics. Ibn Kammuna's Examination of the Apologetics of the Three Faiths." In *The Three Rings. Textual Studies in the Historical Trialogue of Judaism, Christianity, and Islam*. Eds. Barbara Roggema, Marcel Poorthuis and Pim Valkenberg. Leuven 2005, pp. 47–70.

—— "Jewish-Christian debate in a Muslim context. Ibn al-Maḥrūma's notes on Ibn Kammūna's *Examination of the Inquiries into the Three Faiths*." In *All Those Nations... Cultural Encounters Within and With the Near East. Studies presented to Han Drijvers at the occasion of his sixty-fifth birthday by colleagues and students*. Collected and edited by Herman J.L. Vanstiphout with the assistance of Wout J. van Bekkum, Geerd Jan van Gelder and Gerrit J. Reinink. Groningen 1999, pp. 131–39.

Roper, Geoffrey (ed.): *World Survey of Islamic Manuscripts* 1–4. London 1992–94.

Rosen, Victor: *Les manuscrits arabes de l'Institut des Langues Orientales*. St. Petersburg 1877 [reprinted Amsterdam 1971].

Rosenkranz, Simone: "Judentum, Christentum und Islam in der Sicht des *Ibn Kammûna*." *Judaica. Beiträge zum Verstehen des Judentums* 52 (1996), pp. 4–22.

*Routledge Encyclopaedia of Philosophy* 1–10. London 1998.

Rudolph, Kurt: [Review of] "Ibn Kammūna, Sa'd B. Manṣūr: Examination of the Inquiries into the three Faiths...Edited by Moshe Perlmann." *Orientalistische Literaturzeitung* 67 ix/x (1972), pp. 462–65.

*Sab' Rasā'il li-l-'Allāma Jalāl al-Dīn Muḥammad al-Dawānī (al-mutawaffā 908 h.) wa-l-Mullā Ismā'īl al-Khwājū'ī al-Iṣfahānī (al-mutawaffā 1173 h.)*. Ed. Sayyid Aḥmad Tūysirkānī. Tehran 1381/2002.

Ṣadr al-Dīn Maḥallāṭī = Kitābkhāna-yi Āqā-yi Ṣadr al-Dīn Maḥallāṭī, Shiraz (the collection is now part of the Library of the Khānaqāh-i Aḥmadiyya, Shiraz)
Catalogue:
1967. Muḥammad Taqī Dānishpazhūh: "Fihrist-i kitābkhānahā-yi 'umūmī u khuṣūṣī-yi Shīrāz." *Nuskhahā-yi khaṭṭī (= Nashriyya-yi Kitābkhāna-yi Markazī u Markaz-i Asnād-i Dānishgāh-i Tihrān)* 5 (1346), pp. 271–74.

Saliba, George: *The Astronomical Work of Mu'ayyad al-Dīn al-'Urḍī. A Thirteenth Century Reform of Ptolemaic Astronomy. Kitāb al-Hay'ah*. Edition and Introduction. Beirut 1990.

—— "The original source of Quṭb al-Dīn al-Shīrāzī's Planetary Model." In idem: *A History of Arabic Astronomy. Planetary Theories during the Golden Age of Islam*. New York/London 1994, pp. 119–34.

Samaw'al al-Maghrībī: *Ifḥām al-Yahūd. Silencing The Jews*. Edited and Translated by Moshe Perlmann. New York 1964.

*Samaw'al al-Maghribī's (d. 570/1175) Ifḥām al-yahūd. The Early Recension*. Introduced and edited by Ibrahim Marazka, Reza Pourjavady and Sabine Schmidtke. Wiesbaden 2006.

Schmidtke, Sabine: "The Doctrine of the Transmigration of Soul according to Shihāb al-Dīn al-Suhrawardī (killed 587/1191) and his Followers." *Studia Iranica* 28 (1999), pp. 237–54.

—— "Studies on Sa'd b. Manṣūr Ibn Kammūna (d. 683/1284): Beginnings, Achievements, and Perspectives." *Persica. Annual of the Dutch-Iranian Society* 29 (2003), pp. 105–21.

—— *Theologie, Philosophie und Mystik im zwölferschiitischen Islam des 9./15. Jahrhunderts. Die Gedankenwelten des Ibn Abī Ǧumhūr al-Aḥsā'ī (um 838/1434–35 – nach 906/1501)*. Leiden 2000.

Schwarb, Gregor: "The commentary of 'Alā' al-Dīn Abū l-Ḥasan 'Alī b. Ṭaybugha (d. 793/1391) on Hilkhot Yesodey ha-Torah, I–IV: a scientific commentary." (Forthcoming).

—— "Découverte d'un nouveau fragment du *Kitāb al-muġnī fī 'abwāb al-tawḥīd wa-l-'adl* du Qāḍī 'Abd al-Ǧabbār al-Hamaḏānī dans une collection karaïte de la British Library." *Mélanges de l'Institut Dominicain d'Etudes Orientales du Caire* 27 (2006) (in press).

Şehid Ali Paşa = Şehid Ali Paşa Collection (held in the Süleymanie Kütüphanesi), Istanbul
Catalogue:
1894. *Kiliç Ali Paşa Kütüphanesi defteri*. Istanbul 1311.

-Shabībī, Muḥammad Riḍā: *Mu'arrikh al-'Irāq Ibn al-Fuwaṭī* 1–2. Baghdad 1378/1958.

—— *Turāthunā al-falsafī. Ḥājatuhū ilā l-naqd wa-l-tamḥīṣ*. Baghdad 1385/1965.

-Shādhilī, Ṣadaqa b. Ibrāhīm: *Al-'Umda al-kuḥliyya fī l-amrāḍ al-baṣariyya*. Ms. Munich 834.

-Shahrazūrī, Shams al-Dīn Muḥammad b. Maḥmūd: *Kanz al-ḥikma* [being a Persian translation of *Nuzhat al-arwāḥ wa-rawḍat al-afrāḥ*]. Transl. Ḍiyā' al-Dīn Durrī. Tehran 1937.

—— *Rasā'il al-Shajara al-ilāhiyya fī 'ulūm al-ḥaqā'iq al-rabbāniyya*. Al-Mujallad al-thānī fī l-'ulūm al-ṭabī'iyya. Ed. Najafqulī Ḥabībī. Tehran 1384/2006.

—— *Sharḥ Ḥikmat al-ishrāq. Commentary on the Philosophy of Illumination*. Ed. Hossein Ziai. Tehran 1372/1993.

-Shīrāzī, Ṣadr al-Dīn: [Glosses on Quṭb al-Dīn al-Shīrāzī's *Sharḥ Ḥikmat al-ishrāq*]. Lithograph edition by Asad Allāh Harātī. Tehran 1313–15/1895–97 [reprinted without place and date].

—— *al-Ḥikma al-muta'āliya fī l-asfār al-'aqliyya al-arba'a* 1–9. Beirut [5]1419/1999.

—— *Majmū'a-yi rasā'il-i falsafī-yi Ṣadr al-muta'allihīn*. Ed. Ḥāmid Nājī Iṣfahānī. Tehran 1375/1996.

—— *Sharḥ al-Hidāya al-athīriyya*. Lithograph edition. Tehran 1313/[1895–96].

—— *Sih risāla falsafī*. Ed. Sayyid Jalāl Āshtiyānī. Qum 1378/1999.

—— *Ta'līqāt 'alā l-Shifā'*. Lithograph edition. Qum 1399/[1978–79].

—— "Risālat al-taṣawwur wa-l-taṣdīq." In *Risālatān fī l-taṣawwur wa-l-taṣdīq*. Ed. Mahdī Sharīʿatī. Qum 1416/[1995–96].

Shīrāz = Kitābkhāna-yi Millī-yi Fārs, Shiraz
Catalogue:
1972. ʿAlī Naqī Bahrūzī and Muḥammad Ṣādiq Faqīrī: *Fihrist-i kutub-i khaṭṭī-yi Kitābkhāna-yi Millī-yi Fārs*. Volume 2, az Shumāra-yi 401 tā 780. Shiraz 1351.

Shushtarī, Nūr Allāh: *Majālis al-muʾminīn* 1–2. Ed. Aḥmad ʿAbd al-Munāfī. Tehran 1375–76/1955–56.

Sirat, Colette: *A History of Jewish Philosophy in the Middle Ages*. Cambridge/Paris 1990.

Sklare, David: "A Guide to Collections of Karaite Manuscripts." In *Karaite Judaism. A Guide to its History and Literary Sources*. Ed. Meira Polliack. Leiden 2003, pp. 893–924.

Spade, Paul Vincent: *The Medieval Liar. A Catalogue of the* Insolubilia-*Literature*. Toronto 1975.

—— "The Origins of the Medieval Insolubilia-Literature." *Franciscan Studies* 33 (1973), pp. 292–309 [reprinted in Paul Vincent Spade: *Lies, Language and Logic in the Late Middle Ages*. London 1988, part II].

Spuler, Bertold: *Die Mongolen in Iran*. Berlin ²1955.

St. Petersburg RNL Firk. = Abraham Firkovitch Collection (held in the Russian National Library), St. Petersburg
Catalogues:
1987. Viktor V. Lebedev: *Arabskie Socienija v evreiskoj grafike. Katalog rukopeseij*. Leningrad.
1991. Paul B. Fenton: *A Handlist of Judeo-Arabic Manuscripts in Leningrad. A Tentative Handlist of Judeo-Arabic Manuscripts in the Firkovic Collection*. Based on the card catalogue of the State Public Library prepared and updated by A.A. Harkavy, F. Kokovtsov, R. Ravrebe, A. Vasilev, V. Lebedev. Jerusalem.

Steinschneider, Moritz: *Die Arabische Literatur der Juden. Ein Beitrag zur Literaturgeschichte der Araber, großenteils aus handschriftlichen Quellen*. Frankfurt/Main 1902 [reprinted Hildesheim 1986].

—— *Polemische und apologetische Literatur in arabischer Sprache zwischen Muslimen, Christen und Juden, nebst Anhängen verwandten Inhalts*. Leipzig 1877 [reprinted Hildesheim 1966].

—— "Schriften der Araber in hebräischen Handschriften, ein Beitrag zur arabischen Bibliographie." *Zeitschrift der Deutschen Morgenländischen Gesellschaft* 47 (1893), pp. 335–84.

—— *Verzeichniss der Hebraeischen Handschriften* 1–2. Berlin 1878–79.

Stroumsa, Sarah: *Freethinkers of Medieval Islam. Ibn al-Rāwandī, Abū Bakr al-Rāzī, and Their Impact on Islamic Thought*. Leiden 1999.

—— "On Jewish Intellectuals Who Converted in the Early Middle Ages." In *The Jews of Medieval Islam. Community, Society, and Identity*. Ed. Daniel Frank. Leiden 1995, pp. 179–97.

-Suhrawardī, Shihāb al-Dīn: "Kalimat al-taṣawwuf." *In Œuvres philosophiques et mystiques. Tome IV*. Textes édités avec prolégomènes en persan par Dr. Najaqoli Habibi. Prolégomènes en anglais par Seyyed Hossein Nasr. Tehran 1380/2001, pp. 99–139.

—— *Kitāb al-Lamaḥāt*. Ed. Amīl al-Maʿlūf. Beirut ²1991.

—— *Le Livre de la Sagesse Orientale. Kitâb Ḥikmat al-Ishrâq. Commentaires de Qoṭbaddîn Shîrâzî et Mollâ Ṣadrâ Shîrâzî*. Traduction et notes par Henry Corbin. Etablies et introduites par Christian Jambet. Paris 1986.

—— *Manṭiq al-talwīḥāt*. Ed. ʿAlī Akbar Fayyāḍ. Tehran 1334/1955.

—— *Opera Metaphysica et Mystica I.* Ed. Henry Corbin. Istanbul 1945.
—— *Opera Metaphysica et Mystica II* (= *Œuvres philosophiques et mystiques*). Ed. Henry Corbin. Tehran 1952.
—— *Opera Metaphysica et Mystica III* (= *Œuvres philosophiques et mystiques II. Œuvres en persan*). Ed. Henry Corbin. Tehran 1970.
—— *Œuvres philosophiques et mystiques. Tome IV.* Textes édités avec prolégomènes en persan par Dr. Najaqoli Habibi. Prolégomènes en anglais par Seyyed Hossein Nasr. Tehran 1380/2001.
—— *The Philosophy of Illumination.* A New Critical Edition of the Text of Ḥikmat al-ishrāq with English Translation, Notes, Commentary, and Introduction by John Walbridge and Hossein Ziai. Provo, Utah 1999.
—— *Risālat Maqāmāt al-Ṣūfiya (The Ṣūfī Stations).* Edited with Introduction and Notes by Emile Maalouf [Amīl al-Maʿlūf]. Beirut 1993.
*Süleymaniye Kütüphanesi Eser Adı Kataloğu* 1–28. Istanbul 1996.
*Tanqīḥ* → Perlmann, Moshe (ed.): *Saʿd b. Manṣūr Ibn Kammūna's Examination of the Inquiries into the Three Faiths.*
Taymūriyya = al-Khizāna al-Taymūriyya (held in the Dār al-Kutub al-Qawmiyya al-Miṣriyya, the Egyptian National Library), Cairo
Catalogue:
1947–50. *Fihris al-Khizāna al-Taymūriyya* 1–4. Cairo 1366–69.
Topkapı Saray = Topkapı Saray Müzesi Kütüphanesi, Istanbul
Catalogue:
1962–66. Fehmi Edhem Karatay: *Topkapı Sarayı Müzesi Kütüphanesi. Arapça yazmalar kataloğu* 1–4. Istanbul.
Tunkābunī, Muḥammad b. ʿAbd al-Fattāḥ Sarāb: *Risāla fī taḥrīr shubhat al-tawḥīd.* Ms. Mahdawī 592/2, ff. 3a–6a.
—— *Risālat shubhat jadhr al-aṣamm.* Ms. Mahdawī 592/1, ff. 8b-11a.
Turhan Valide = Turhan Valide Collection (held in the Süleymanie Kütüphanesi), Istanbul
*Türkiye Diyanet Vakfı İslâm Ansiklopedisi* 1–. Istanbul 1988–.
Ullmann, Manfred: *Die Medizin im Islam.* Leiden/Köln 1970.
Urvoy, Dominique: *Les penseurs libres dans l'Islam classique. L'interrogation sur la religion chez les penseurs arabes indépendants.* Paris 1996.
Utas, Bo (ed.): *Tarīq ut-taḥqīq. A Sufi Mathnavi ascribed to Ḥakīm Sanāʾī of Ghazna and probably composed by Aḥmad b. al-Ḥasan b. Muḥammad an-Naxčavānī.* A critical edition, with a history of the text and a commentary. Lund 1973.
Vajda, Georges: "Le milieu juif à Bagdad." *Arabica* 9 (1962), pp. 389–93.
Vazīrī = Sayyid ʿAlī Muḥammad Vazīrī Library, Yazd
Catalogue:
1971–79. Muḥammad Shīrwānī: *Fihrist-i nuskhahā-yi khaṭṭī-yi kitābkhāna-yi Vazīrī-yi Yazd* 1–5. 1350–58.
Walbridge, John: *The Philosophy of Qutb al-Dīn Shīrāzī. A Study in the Integration of Islamic Philosophy.* PhD thesis: Harvard University Cambridge, Mass., 1983.
—— [Review of] "Ibn Kammūna: *Al-Tanqīḥāt fī Sharḥ al-Talwīḥāt. Refinement and Commentary on Suhrawardī's Intimations*..." *Iranian Studies* 38 iii (2005), pp. 520–1.
—— *The Science of Mystic Lights. Qutb al-Dīn Shīrāzī and the Illuminationist Tradition in Islamic Philosophy.* Cambridge, Mass. 1992.

Waṣṣāf [ʿAbd Allāh b. Faḍl Allāh al-Shīrāzī]: *Tārīkh-i Waṣṣāf*. Lithograph edition. Bombay 1269/1853.

Wisnovsky, Robert: "The Nature and Scope of Arabic Philosophical Commentary in Post-Classical (ca. 110–1900 AD) Islamic Intellectual History. Some Preliminary Observations." In *Philosophy, Science and Exegesis in Greek, Arabic and Latin Commentaries*. Volume Two. Eds. Peter Adamson, Han Baltussen and M.W.F. Stone. London 2004, pp. 149–91.

Wolff, Johann Christoph: *Bibliotheca Hebraea*. Hamburg 1715–33.

Wüstenfeld, Ferdinand: *Geschichte der arabischen Aerzte und Naturforscher. Nach den Quellen bearbeitet*. Göttingen 1840.

Yahuda = Abraham Shalom Yahuda Collection, Jewish National and University Library (JNUL), Jerusalem
Catalogue:
1997. Efraim Wust: *A Catalogue of the Arabic Manuscripts in the A.S. Yahuda Collection. Jewish National and University Library, Jerusalem*. Limited Preliminary Edition. Jerusalem.

Yeni Câmi = Yeni Câmi collection (held in the Süleymanie Kütüphanesi), Istanbul
Catalogue:
1883. *Yeni Câmi Kütüphanesi ʾnde mahfuz kütüb-i mevcûdenin defterdir*. Istanbul 1300.

Yūsuf Ağa = Yūsuf Ağa collection (held in the Süleymanie Kütüphanesi), Istanbul.

Ẓāhiriyya = Dār al-kutub al-Ẓāhiriyya collection (since 1985 being part of the Asad National Library [Maktabat al-Asad al-Waṭaniyya]), Damascus.
Catalogues:
1970. ʿAbd al-Ḥamīd al-Ḥasan: *Fihris makhṭūṭāt Dār al-kutub al-Ẓāhiriyya. Al-Falsafa wa-l-manṭiq wa-ādāb al-baḥth*. Damascus 1390.
1987. Ṣalāḥ Muḥammad al-Khiyamī and Muḥammad Muṭīʿ al-Ḥāfiẓ: *Al-Fihris al-ʿāmm li-makhṭūṭāt Dār al-kutub al-Ẓāhiriyya*. Damascus 1407.

Zaryāb-Khūʾi, ʿAbbās: "Sih nukta dar bara-yi Rashīd al-Dīn Faḍl Allāh." In *Proceedings of the Colloquium on Rashīd al-Dīn Faḍl Allāh. Tehran-Tabriz, 11–16 Aban 1348 (2–7 November 1969)*. Tehran 1971, pp. 123–35.

Ziai, Hossein: *Knowledge and Illumination. A Study of Suhrawardī's Ḥikmat al-Ishrāq*. Altanta 1990.

—— "The Illuminationist Tradition." In *History of Islamic Philosophy*. Eds. Seyyed Hossein Nasr and Oliver Leaman. London/New York 1996, vol. 1, pp. 465–96.

-Ziriklī, Khayr al-Dīn: *Al-Aʿlām* 1–8. Beirut 1980.

# INDEX OF NAMES AND PLACES

This index covers names of persons, locations, groups and institutions. The Arabic definite article (al-/-), the transliteration symbols for the Arabic letters *hamza* and *ʿayn*, and the distinction between different letters transliterated by the same Latin character are ignored for purpose of alphabetization.

# INDEX OF MANUSCRIPTS

# INDEX OF IBN KAMMŪNA'S WORKS